Water Quality Management for Coastal Aquaculture

Water Quality Management for Coastal Aquaculture

By

Sukumar Bandyopadhyay

Former Professor and Emeritus Fellow

Indian institute of Technology

Kharagpur

2008

DAYA PUBLISHING HOUSE

Delhi - 110 035

ISBN10 81-7035-538-9
ISBN13 978-81-7035-538-0

Published by : **Daya Publishing House**
1123/74, Deva Ram Park
Tri Nagar, Delhi - 110 035
Phone: 27383999
Fax: (011) 23260116
e-mail : dayabooks@vsnl.com
website : www.dayabooks.com

Showroom : 4760-61/23, Ansari Road, Darya Ganj,
New Delhi - 110 002
Phone: 23245578, 23244987

Laser Typesetting : **Classic Computer Services**
Delhi - 110 035

Printed at : **Chawla Offset Printers**
Delhi - 110 052

PRINTED IN INDIA

Om Tat Sat

Dedicated to

My Parents

***Pratima* (my wife)**

and

Children

Acknowledgement

The writing of the text book has been supported by All India Council for Technical Education (AICTE), a statutory body of the Government of India, under its Emeritus Fellowship Program, which utilizes services of highly qualified and experienced superannuated professors of Technical Institutions/Universities in stimulating and achieving excellence in technical education. The author gratefully acknowledges the grant.

In the preparation of the manuscript, various source materials have been used and duly acknowledged at the proper places. Of special mention, the author would like to acknowledge Dr. C. E. Boyd, Department of Fisheries and Allied Aquacultures, Auburn University, U. S. A. The author first conceived of formulating a course on water quality management and then writing a comprehensive text on the subject after coming across the vast literature and contribution of Dr. Boyd in the field. The author borrowed heavily from Boyd with due citation and acknowledgement.

A number of people rendered help and suggestion in the preparation of this book, and the author is grateful to them. They are his colleagues, students, research scholars and staffs in the Department and in the Institute.

Grateful thanks are due to wife Pratima and children for their constant support, encouragement and cooperation in this endeavor.

S. Bandyopadhyay

Preface

Of late, modern aquaculture has developed as a new science, but its technological and engineering application is still confronting the challenges of commercialization. Although growth of aquaculture has taken place tremendously as a food producing activity, it has brought about some associated engineering and environment problems. Aquaculture particularly in the coastal belt has come under increasing criticism due to adverse social and environmental impacts and questionable sustainability because of disease outbreaks and its related problems. Many poorly planned and operated farms and use of inappropriate technology in both small-scale and large-scale sectors have led to localized environmental impacts and social conflicts with other coastal and inland resource users.

There exists considerable knowledge on biology of many aquatic animals and their breeding and culture techniques. While this knowledge helped the industry in the initial stages with the necessary engineering support from the traditional disciplines like civil and agricultural engineering it did not meet much success. On the other hand need for an integrated development of aquaculture technology and engineering for growth of the industry was very much felt. Although few texts and reference books appeared lately in the market, none has addressed the basic management issues such as the fundamental aspects of water resource and environment problems, dynamics of the aquatic systems in a water body, and impact and conflict in the surrounding environment. As a teacher in Aquacultural Engineering in a premier Technological Institute in India, the author tried to formulate a course on 'Water Quality Management' in the recent past and found that there is a lack of text book intended to give clear understanding of the basic problems and the practical aspects for the students undergoing manpower training in the relevant disciplines as well as for the research workers, managers and operators in the industry.

The present treatise is basically a text book on water quality management with a comprehensive view on water resources that would be used for aquaculture in the coastal belt. It deals with various water quality aspects ranging from water sources and their characteristics through water quality and culture dynamics to treatment systems for prevention of environment impact in the surrounding water bodies. The topics on water resources have been discussed as a perspective of coastal aquaculture with a view to presenting the scenario of water availability and water quality in the coastal areas vis-à-vis the environmental impact and conflict with other users. The various water bodies in the coast such as estuaries, lakes, wetland and marine areas, and their general characteristics and usefulness in coastal aquaculture have been highlighted. Although the book is primarily meant for coastal aquaculture management, the author did not intend to distinguish between coastal and inland aquaculture so far pond dynamics and pond treatment are concerned.

For comprehensive treatment, the contents of the book have been divided into three parts: water resources for coastal aquaculture, water quality and water quality management. Part I of the book addresses the different sources of water such as surface and ground waters, and coastal environment with ecological divisions and coastal regulation zones. Characteristics of water resources, in general and inherent in the coastal aquatic ecosystem, have been described. Water resource use in coastal areas encompasses mainly coastal fisheries and aquaculture, but it highlights multiple uses like agriculture, forestry and waste disposal. Impacts of resource use on the coastal environment with potential and specific cases have been discussed.

Part II on water quality deals with physico-chemical, bio-physical and biological characteristics and their interactions on the dynamics of the systems in a water body. Aquaculture pond dynamics with respect to nutrients such as fertilizers, oxygen, natural food organisms, and artificial feeds are the major topics of the treatise in this part. Fundamental principles of equilibrium relationships and process kinetics discussed under this part would be helpful in clear understanding of the dynamic processes. Several numerical problems also have been introduced.

Part III on water quality management breaks up the topics under pond treatment for control and management of aquatic environment during culture, and farm effluent treatment before discharge into the surrounding water bodies. Several treatment methods have been described with emphasis on theory, process fundamentals and equipment design. Related numerical problems have been given as examples.

The textbook has been basically planned for use by the students (both under-graduate and post-graduate) of fishery science and technology, aquaculture and aquacultural engineering, marine science, water science and technology, coastal engineering and related fields. In fact, the book is directed at multi-level readership with multi-disciplinary approach, and for academic/professional requirement. The topics best describe natural resource management in general and environmental science, in particular, and so would help in dissemination of knowledge in the broad area of coastal environment management. A set of norms and standards for the discharge of effluents as per the guideline of the Govt. of India's new regulations has been given in the book, which would help in sustainable and environment-friendly aquaculture technology and optimum utilization of the resources and return to the farmers.

S. Bandyopadhyay

Contents

Acknowledgements *vii*

Preface *ix*

Chapter 1: Introduction **1**

1.1 Water and Land Resource Use

1.2 Environmental Impact from Agriculture and Aquaculture

1.3 Food Production and Fisheries

1.4 Perspective of Water Quality Management in Aquaculture

PART I: WATER RESOURCES FOR COASTAL AQUACULTURE

Chapter 2: Water Resources **15**

2.1 Sources of Water

2.1.1 Surface Water

1. Ponds, Lakes and Reservoirs

2. Streams and Rivers

3. Sea or Saltwater

2.1.2 Ground Water

2.2 Coastal Environment

2.2.1 Coastal Areas and Zones

2.2.2 Ecological Divisions
1. Marine Environment
2. Rocky Shore
3. Sandy and Muddy Shores
4. Brackish Water or Estuarine Environment
5. Marshes and Mangroves
2.2.3 Coastal Regulation Zone
2.3 Characteristics of Water Resources
2.3.1 Environmental Characteristics of Coastal Water
1. Carrying Capacity and Standing Crop
2. Primary Productivity and Food Chain
3. Principles Governing the Coastal Water Ecosystem
4. Aquatic Biodiversity
5. Ecological Factors
2.3.2 General Characteristics of Source Water
1. Water Temperature and Circulation
2. Dissolved Oxygen Content
3. pH and Carbon Dioxide
4. Nutrients and Organic Substances
5. Plant and Animal Community
6. Ground Water Characteristics
7. Summary

Chapter 3: Water Resource Use in Coastal Area **52**
3.1 Coastal Fisheries and Aquaculture
3.1.1 Coastal Fisheries
1. Types of Fisheries
2. Inland Capture Fisheries
3. Marine Fisheries
3.1.2 Coastal Aquaculture
1. Types of Aquaculture Production Systems
2. Species Cultured in Coastal Waters
3. Operation of Coastal Aquaculture Farms
3.2 Multiple Use of Coastal Resources
3.2.1 Coastal Agriculture
1. Constraints Affecting Coastal Agriculture
2. Crop Selection for Salt-affected Soils

3.2.2 Coastal Forestry
1. Types of Coastal Forests
2. Socio-economic Values of Coastal Forests
3. Special Characteristics of Coastal Forestry
3.2.3 Waste Disposal and Pollution in Coastal Areas
1. Sources of Pollution
2. Types of Contaminants and Pollutants
3. Major Examples of Coastal Pollution

Chapter 4: Impact of Coastal Resource Use on the Environment 89
4.1 Impacts on Coastal Environment
4.1.1 Alterations and Destruction of Habitats
4.1.2 Effects of marine Pollution on Human Health
4.1.3 Hypernutrification and Eutrophication
4.1.4 Decline of Fish Stocks and Other Renewable Resources
4.1.5 Changes in Sediment Flows
4.2 Potential and Specific Cases of Impacts
4.2.1 Agricultural Activities
4.2.2 Capture Fisheries and Coastal Aquaculture Activities
4.2.3 Multiple Activities
4.3 Integrated Ecosystem Approach for Resource Use
References

PART II: WATER QUALITY

Chapter 5: Water Quality Characteristics 109
5.1 Water Quality Parameters
5.1.1 Classification of Water Quality Parameters
1. Dissolved Oxygen
2. Primary Productivity and Nutrients
3. Temperature
4. Salinity
5. Suspended Solids
6. pH
7. Alkalinity and Hardness
8. Dissolved Gases
9. Biological Parameters

5.2 Fundamental Principles
5.2.1 Equilibrium Relationships
1. Some Thermodynamic Concepts of Equilibria
2. Ionic Equilibrium in Water
3. Ionization of Acids and Bases
4. Solubility Relationship
5.2.2 Process Kinetics
1. Rate of a Chemical Reaction
2. Kinetic Models of Homogeneous Reactions
3. Effect of Temperature on Reaction Rate
5.2.3 Biological Reaction Systems
1. Kinetics of Enzyme Catalyzed Reactions
2. Kinetics of Microbial Growth

Chapter 6: Aquaculture Pond Dynamics **152**
6.1 Dynamics of Nutrientsin Pond Ecosystem
6.1.1 Nitrogen Cycle
6.1.2 Phosphorus Cycle
6.1.3 Carbon Cycle
6.2 Dynamics of Dissolved Oxygen in Pond Water
6.2.1 Biological Processes
1. Photosynthetic Oxygen Production
2. Oxygen Requirements of Fish
3. Diurnal Changes of Oxygen Concentration in Ponds
6.2.2 Diffusional Oxygen Transfer by Natural Aeration
6.2.3 DO Concentration Balance in Pond Water during Culture
1. Channel Catfish Pond
2. Trout Pond
3. Warm Water Fish
6.3 Dynamics of Fertilized Pond
6.3.1 Effects of Fertilization on Pond Dynamics
1. Changes in Acidity due to Nitrogen Fertilizer
2. Effects of Fertilization on Phosphorus Cycle
3. Plants and Invertebrates
6.4 Dynamics of Limed Pond
6.4.1 Effects of Liming on Pond Dynamics
1. Increase in Total Alkalinity
2. Increase in Concentration of Total Available Carbon Dioxide

3. Increase in Total Hardness
4. Effect on Activity of Microorganisms
5. Increase in the Availability of Mud Phosphate
6.4.2 Effects of Liming on Plankton and Invertebrates
6.5 Dynamics of Fed Pond
6.5.1 Types of Feeding and Feeding Options
6.5.2 Supplementary Diet Feeding
6.5.3 Complete Diet Feeding
6.5.4 Feed Conversion, Utilization and Waste Production
1. Material Balance of Feed Utilization
2. Nutrients and Solids Budget
3. Waste Components
4. COD Balance
6.5.5 Waste Production from Fertilization
6.5.7 Residues of Chemicals
6.5.8 Effects of Wastes on Culture Environment
6.5.9 Relationship of Water Quality with Feeding Rate
References

PART III: WATER QUALITY MANAGEMENT

Chapter 7: Introduction 197
7.1 Culture Systems
7.1.1 Types of Culture Systems
1. Open System
2. Semi-closed System
3. Closed System
7.1.2 Basic Approach of Closed System
7.2 Treatment Methods
7.2.1 Pond Management Methods
7.2.2 Recirculating Methods

Chapter 8: Fertilization of Ponds 202
8.1 Fertilizers
8.1.1 Types, Properties and Sources of Fertilizers
1. Types and Sources
2. Properties
8.1.2 Requirement of Fertilizers
1. Principle
2. General Guidelines for Fertilizer Requirement

8.2 Application of Fertilizers
8.2.1 Types of Fertilizer
8.2.2 Application Rate
8.2.3 Method of Fertilizer Application
1. Platform Method
2. Nylon Cloth or Bag Method
3. Application of Liquid Fertilizers
8.3 Organic Manures
8.3.1 Types of Manures
8.3.2 Method of Straight Manual Application of Manures
1. Frequency of Application
2. Methods
3. Manure Application through Integrated Farming of Livestock

Chapter 9: Liming of Ponds 213
9.1 Lime Requirement andLiming Rate
9.1.1 Calculation of Liming Rate
1. Technique Employed on Agricultural Crop
2. Technique Based on Exchange Acidity of Soil
9.2 Liming Methods
9.2.1 Liming of Ponds
1. Liming Materials
2. Methods of Application
9.2.2 Liming of Acid-sulphate Soils

Chapter 10: Aeration 220
10.1 Aeration Fundamentals
10.1.1 Theory of Oxygen Transfer
1. Factors Affecting Volumetric Oxygen Transfer Coefficient ($k_L a$)
2. Evaluation of $k_L a$ by Aeration Experiment
3. Measurement of DO
4. Standard Oxygen Transfer Rate and Aeration Efficiency
10.1.2 Rating of Aeration Systems under Field Conditions
10.2 Aeration Systems
10.2.1 Types of Aerators
1. Classification
2. Surface Aerators
3. Diffused Air Systems
4. Gravity Aerators

10.2.2 Types of Aeration
1. Emergency Aeration
2. Supplemental or Continuous Aeration
3. Aeration to Prevent Thermal and Oxygen Stratification
4. Aeration of Source Water
10.2.3 Comparative Performance of Various Aerators
1. Aeration Rate and Efficiency
2. Oxygen Saturation and Oxygen Transfer
10.3 Fish Production
10.4 Aeration Process and Aerator Design
10.4.1 Computation of Oxygen Demand and Supplemental Aeration Requirement
1. Average Daily Oxygen Demand
2. Maximum Daily Oxygen Demand
3. Oxygen Supplied by Water Flow
4. Supplemental Oxygen Demand
10.4.2 Surface Aerator Design
1. Practical Approach
2. Simulation Approach

Chapter 11: Feed Management 246
11.1 Feeding Options
11.1.1 Pond Fertilization and Supplemental Feeding
1. Feed Ingredients
2. Supplementary Feeds
11.1.2 Complete Diet Feeding
1.Types of Feed
2. Formulation
3. Preparation
11.2 Feeding Methods
11.2.1 Feeding Rate and Frequency
1. Feeding Rate
2. Feeding Frequency
11.2.2 Feeding Tables
11.2.3 Feeding Devices
1. Hand-feeding or Manual Feeding
2. Automatic Feeders

Chapter 12: Effluent Treatment Systems 267

12.1 Types of Waste Materials in Aquaculture Effluents

12.1.1 Suspended Solids

12.1.2 Nutrient and BOD

12.1.3 Pathogens

12.2 Treatability of Aquaculture Effluents

12.2.1 Load and Concentration of Pollutants

12.2.2 Pollution Potential of Effluents

12.2.3 Comparison of Effluents from Different Culture Systems

1. Intensive Aquaculture Systems

2. Semi-intensive Aquaculture Systems

3. Extensive Aquaculture Systems

12.3 Effluent Standards and Regulations

12.3.1 Effluents Standards

12.3.2 Guidelines and Codes of Conduct

12.3.3 Codes of Practice

1. Farm Effluents

2. Site Characteristics for Discharge Regulations

12.3.4 General Regulations of Coastal Farm

12.4 Effluent Treatment Practices

12.4.1 Treatment Technologies in Use

1. Solids Removal from the Pond Bottoms

2. Solids Removal by Sedimentation Ponds

3. Solids Removal by Filtration

4. Solids Removal in Cage Farms

5. Biological Treatment

6. Sludge Treatment

12.4.2 Effluent Treatment in Shrimp Farming Systems

1. Effluent Treatment Scheme of Aquaculture Authority of India

2. Environment-Friendly Scheme for Intensive Farming

3. Closed–Recirculating Shrimp Farming

Chapter 13: Solids Removal 286

13.1 Screening

13.1.1 Types of Screens

13.1.2 Typical Design Characteristics and Data

13.2 Mechanical Filtration
13.2.1 Types of Filters
1. Gravity Filters
2. Rapid Filters
3. Diatomaceous Earth Filter
13.2.2 Filtration Process
1. Solids Removal Mechanisms
2. Mathematical Analysis
3. Computation of Head-loss
4. Filtration Process Variables
13.3 Sedimentation of Solids
13.3.1 Types of Settling
13.3.2 Types of Sedimentation Tanks or Basins
13.3.3 Mathematical Analysis of Settling
1. Settling Velocity Analysis
2. Removal Efficiency of a Basin

Chapter 14: Biological Filtration 306
14.1 Principle of Ammonia Removal by Nitrification
14.1.1 Organisms
14.1.2 Reactions
14.1.3 Environmental Factors Affecting Nitrification Rate
1. Ammonia Concentration
2. Dissolved Oxygen Concentration
3. Temperature Changes
4. pH Changes
5. Effect of Minerals and Chemicals
6. Filter Media Types
14.2 Filter Design
14.2.1 Filter Configuration
1. Submerged Filters
2. Trickling Filters
3. Rotating Media Filters
14.2.2 Operating Parameters
1. Flow Distribution
2. Hydraulic Loading
3. Duty Cycle

14.2.3 Comparison of Existing Designs of Biofilters
14.2.4 Filter Design Procedure
1. Ammonia Mass Balance
2. Nitrate-Nitrogen Mass balance
3. DO Mass Balance
4. DO Mass Balance in Biofilter

Chapter 15: Disinfection 324

15.1 Methods of Disinfection
15.1.1 Chlorination Process
1. Forms of Chlorine
2. Chemistry of Chlorination
3. Disadvantages of Chlorination
4. Chlorine Removal
5. Chlorine Compounds Used in Practice
15.1.2 Potassium Permanganate Treatment
15.2 Mechanisms and Kinetics of Disinfection
References

Appendices 341

Index 351

1

Introduction

Water is a precious and most commonly used natural resource for human survival and industrial growth. Civilizations and habitats often developed along the banks of several important rivers for easy availability of water for drinking, farming and transportation. Water played a very important role for human survival and welfare, which not only depended on its regular availability but also on control of floods and droughts resulting deaths of human life and livestock. From time immemorial proper water control and use of rainfall for maximizing agricultural production were considered most for national security and well being. Importance of water resource in the development process was clearly recognized from very ancient times in different parts of the world.

The oceans covering about three quarters of the earth's surface are the great natural water resources, and awed ancient man by their size and power. In early days geographical exploration of the oceans began in a search for reliable trade routes and culminated in discovery on new continents and passages. Since the dawn of civilization the oceans and coastal areas have played a major role in the development of human settlement. Today, as in the past, mankind depends in many ways up on the sea. For every nation located on a coast, the sea provides an inexpensive means of communication and transport, a source of food, and deposits of petroleum or valuable minerals. As human settlements developed, the world's coastal areas came under pressure with population growth, urbanization and industrial development, and trade and capital flows. Historically, the cities in the coastal area have grown from port development and expanded both along the waterfront and into the hinterland.

Of late, the world's ever growing human population has necessitated utilizing all the available natural resources for increased production of food. Traditionally man uses the basic natural resources of land and water for food production through agriculture and animal husbandry. In course of time fisheries and aquaculture have become the two other major food production systems in addition to agriculture and animal husbandry. Agriculture is the oldest and aquaculture is the newest food production systems on earth. On the global basis agriculture has been remarkably successful in providing 95 per cent of all animal and plant protein and 99 per cent of the calories consumed by humans with per capita food production being higher than 30 years ago (FAO, 2000). On the other hand, aquaculture contributed about 30 per cent of the total fish supply globally and supplied over one-third of the total animal protein intake in most leading aquaculture producing countries in Asia (FAO, 1997; 2000).

1.1 Water and Land Resource Use

Of the various water resources of the world, 97.5 per cent by volume is saline as ocean water, and 2.5 per cent is actually fresh. About one per cent of this fresh water is available in lakes, rivers, wetlands and the soil. Water provided by the fresh water systems is the most important resource used by households, agriculture and industry. Agriculture production uses more fresh water than any other activity. Worldwide about 93 per cent of the fresh water is consumed by agriculture for irrigation (WRI, 2003 a).

Ground water has always been considered to be a readily available water source for use in households, agricultural and industrial use. In general it is thought that ground water is more important in arid or semi-arid areas and surface water as more important in humid areas. However, convenient availability, excellent natural quality and relatively low capital cost of development have revealed the worldwide importance of ground water. In the United States ground water accounts for about 50 per cent of livestock and irrigation use and about 40 per cent of public water supplies. Many of the major cities of Europe and Latin America are dependent on ground water. In Asia and Africa surface water is used in most of the large cities, but rural areas are dependent on ground water. A survey of the Economic and Social Commission for Asia and the Pacific (ESCAP) in 1990 showed that many countries depend on ground water for drinking, industrial and agricultural supply (Table 1.1).

Table 1.1: Percentage of Total Water Supply from Groundwater in Different Sectors of Water Use in Selected Countries in ESCAP Region

Country	Drinking Water Supply	Municipal Water Supply	Industrial Water Supply	Agricultural Water Supply
Australia		20 (for municipal, industrial)		13
Bangladesh	2 (for drinking, municipal, industrial)			20
China	18			
India	80 (rural)	50	50–80	45
Japan	14	–	6	4
Republic of Korea	35 (rural)	8 (urban)	–	8
Malaysia	25 (for drinking, municipal)		3	2
Mongolia	82	100	99	60
Nepal	60	80	80	20
Pakistan	–	–	–	44
Philippines	60	80	50	–
Thailand	50	10	20	15

Source: Adapted from Das Gupta (1998).

Historically, agricultural production has increased mainly by bringing more land into farming. But there are limitations of cropland not already being farmed and growing competitions from land uses for industrial, commercial and residential development. In fact, densely populated parts of India, China, Indonesia, Egypt and West Europe have reached limits of expansion for agriculture many years ago. Total agricultural land has remained constant in South Asia, but has actually decreased in many industrialized countries. It increased in China, Brazil and West Asia (FAO, 2000). However, intensification of agriculture has been rapid with significant increase of irrigated area and input of fertilizers over the past three decades (Wood *et al.*, 2000). Worldwide irrigated land accounts for only 5.5 per cent of all agricultural land and 17.5 per cent of cropland. But in some areas such as in China and India agriculture is very extensive holding together about 41 per cent of the world's irrigated area (FAO, 2000).

In coastal areas many river deltas and plain land are very productive agricultural areas supporting the economic development of the people. The most developing countries are dependent on use of natural resources in coastal areas for agriculture, fisheries and forestry besides mining, oil and gas extraction, tourism and ocean transport. In particular, fisheries comprising both capture and aquaculture depend on natural resources available in a variety of environments, ranging from the high seas through natural inland water bodies to excavated ponds. In coastal area most capture fisheries and aquaculture are heavily dependent on the coastal land and water resources. Before the jurisdiction at the international level came into enforcement almost all of the world's deep-sea fisheries were operated under conditions of free and open access, without any control, and thus resulted in over-exploitation of resources. The coastal fisheries operate usually under conditions of common property regimes, which also failed in many cases due to legal, social and technological problems. In both cases excessive exploitation beyond the optimum level led to decline of stocks and long-term productivity (Scialabba, 1998).

Two types of coastal aquaculture farms, land and water-based, use land and water in different ways, as shown in Table 1.2. Not all types of land are suitable for aquaculture. Productive agricultural land is not easily available for aquaculture. Wastelands or wetlands are not truly waste lands, as they play important role in maintenance of ecosystem of the area and contribute to aquatic biodiversity and preservation of wildlife. Water-based systems do not, of course, utilize land. However, they occupy areas of water bodies that can result in competition for resources.

Table 1.2: Land and Water Use Variation in Coastal Aquaculture

Farms	*Type*	*Land*	*Water*	*Other Users*
Land based	Inland pond	Agricultural Ground	Surface	Agriculture
	Coastal pond	Wetland or wasteland	Tide or floodwater High water table	Forestry Wildlife
Water based	Intertidal		Estuaries, lagoons, lakes	Forestry Wildlife
	Marine		Open sea water	Marine life

Surface fresh waters are most commonly used for freshwater aquaculture. Ground water resources are considerably more abundant but are usually appear to be less used than surface waters in the world. Growing demand for water from aquaculture industry results in increased competition with other users such as irrigation and drinking water supply. Coastal waters, particularly near fresh water sources, are used for aquaculture, but with risk of instability due to pollution as well as fluctuations in salinity and temperature. There are some non-competitive uses of water resources for aquaculture such as water unsuitable for drinking and irrigation, water in homestead type fish farming used for irrigation, and integrated rice-fish culture.

1.2 Environmental Impact from Agriculture and Aquaculture

Coastal areas constitute aquatic ecosystems ranging from rivers through estuaries to the sea. These ecosystems have either been damaged or come under potential threats of damage from pollution and over-utilization by several human activities. Activities using the coastal resources are shown in Table 1.3, and types of pollution created in coastal areas are given in Table 1.4. Agriculture, forestry, fisheries and aquaculture are the traditional resource-based activities in the coastal areas. The potential environmental changes of the agriculture, fisheries and aquaculture activities on coastal ecosystems are given in Table 1.5.

Table 1.3: Activities Using Coastal Resources

Activity	*Major Use of Resources*
Agriculture	Land
Forestry	Land
Fisheries	Water
Aquaculture	Land and water
Waste disposal	Water
Urban development, tourism and industry	Land and water
Port and harbor	Land and water

Table 1.4: Types of Pollution in Coastal Areas

Type of Pollution	*Pollutant*
Air pollution, outside catchment sources	Nitrogen, phosphorus and ammonia going into the water body from atmospheric deposition
Water pollution, land-based within catchment sources	Nitrogen and phosphorus from sewage and agricultural sources, industrial pollutants
Water pollution, coastal and marine sources	Nitrogen and phosphorus from coastal sewage outfalls, oils and contaminated water from ships, outside catchment pollutants carried via long shore currents

Reports on environmental impact assessment shows that presently more than half of the coastal areas in most global regions are at moderate to high risk (FAO, 2000). The coastal ecosystems are highly productive and provide a wide range of 'goods and services' linked to human well-being. The Pilot Analysis of Global Ecosystems (PAGE) of the World Resources Institute (2000) has taken stock of the coastal ecosystems that can be outlined in the following points (Burk *et al.*, 2000):

1. About 40 percent of the global population lives within 100 km of coastline comprising only 22 percent of the total land area.
2. Population pressure, development activities, agriculture and aquaculture expansion are decreasing all habitats ranging from mangroves to coral reefs at an alarming rate. About half of the world's mangrove forests have already been lost.
3. Fish production capacity of the coastal ecosystems providing as much as 95 percent of world's marine fish harvest has decreased due to over-fishing, destructive trawling techniques and damage of nursery grounds.
4. Rising pollution levels have combined with increasing use of synthetic chemicals and fertilizers.

Table 1.5: Potential Environmental Changes and Impacts of Agriculture, Fisheries and Aquaculture on Ecosystem

Activity	*Environmental Change*	*Impact on Ecosystem*
	1. Estuary and inshore water quality impacts	
Agriculture development by flood control and river flow alterations	Increased salinity, decreased circulation, sediment trapping, erosion	Decreased fish yield, increased fish diseases
Agricultural pesticides, chemical use in aquaculture	Water pollution, coral bleaching	Fish kill, toxic fish consumption
Fertilizer use	Hypernutrification, eutrophication	Reduced fish yield
Agro-industries	Organic and toxic pollution	Reduced fish yield, reduction in tourism
	2. Ground water quality and quantity	
Agriculture, aquaculture	Depletion of coastal aquifers, salt water intrusion	Ground water contamination
	3. Coastal wetland (including mangrove) impacts	
Agriculture, aquaculture	Draining or dyking of mangroves, loss of coastal protection, toxic drainage	Loss of wetland, forest and wildlife production, fish kill, increased storm damage and erosion
Over-cropping, grazing in coastal watersheds	Erosion, increased sedimentation, turbidity in floodplain	Decreased fish yield, obstruction of navigation channels, damage of corals and beaches
	4. Fishing	
Fishing with explosives, with trawling and other moving gears	Perturbation of benthos, coral reef and other habitats degraded and destroyed	Decreased fish yields, decreased recreation values

Source: Extracted from GESAMP. 1991; Barg, 1997 and Scialabba, 1998.

Assessment of conditions and changing capacity of food production, water quality and quantity, and biodiversity in global ecosystems shows that:

1. Annual marine catch declined to about 0.6 percent in 1995-96, and about 75 percent of the major fisheries are fully or over-exploited.
2. Declining coastal habitats such as mangroves, wetlands and sea grasses are losing their pollutant-filtering capacity, thereby increasing harmful algal blooms and hypoxia in water bodies. More than 60 kinds of algal toxins are known by 1999.
3. Surface water quality conditions appear to have degraded in almost all regions of the world.
4. People now withdraw about half of the readily available surface water and that use is growing twice as fast as population. Ground water, which accounts for about 20 percent of global water withdrawals, is getting depleted and polluted.
5. Habitat loss, disease, invasive species and coral bleaching all characterize decline in biodiversity with resultant threats to many marine animals.
6. Modification of shorelines by development activities has become detrimental to some beaches by altering currents and sediment supply and increasing impact from storm surges.

In brief, the environmental impacts with respect to actual damage done or potential threats created by all forms of activities are as follows:

1. Decreased crop and fish yields, contamination of fish, fish disease and fish kill.
2. Loss of wetland, forest and wildlife production.
3. Ground water contamination, and depletion of coastal aquifers.
4. Increased coastal erosion, flooding and storm surges.
5. Multi-user conflict of coastal resources, and
6. Decreased tourism and recreation attraction.

1.3 Food Production and Fisheries

Globally agriculture produces sufficient food to provide every person with minimum calorie and nutrition. On average food supplies are 24 percent higher per person than in 1961 and real prices are 40 percent lower (Wood *et al.*, 2000). However, about 790 million people in the developing world are still suffering from chronic malnutrition, which is 33 percent in Sub-Saharan Africa, 31 percent in the Caribbean and 23 percent in South Asia (FAO, 1999).

With the expected increase of population by one-third to about 7500 million in 2020 over the 1995 population the global demands of cereals and meat have been projected to increase by 40 and 58 percent, respectively. This increase would be 85 percent from the developing countries, with an additional increase in demand for significant progress in poverty alleviation during this period. However, agriculture faces an enormous challenge to meet the food needs of the projected population increase over the next 20 years. Limitations of cropland and intensification of agriculture by several means to increase output has raised concerns such as stresses of increased erosion, soil nutrient depletion, salinization, and water logging of soils (Wood *et al.*, 2000). There have been several negative impacts of agriculture on fisheries and human health, as has been discussed earlier.

Eventually water crisis reaches critical proportions in large parts of the world. About 460 million people in the world are currently suffering from serious shortages and an additional 25 percent of the population may come under water scarcity problems, if the current consumption levels continue. Competition of agriculture with other water uses will be most critical in developing countries. Experts predict that water availability will be one of the major constraints for development of human society in the 21st century (WRI, 2003 a).

Water bodies are the most productive areas on earth. The oceans, in particular, constitute over 90 percent of the world's living biomass, holding 4/5 of the total protein food in the world. The coastal zones are the areas of highest biological productivity on earth. A comparison on aquatic production of various water bodies and production potentials of aquaculture systems (Table 1.6) shows that animal production in water bodies can be as high as or even higher than meat production in terrestrial system (Wheaton, 1977; Srivastava, 1991). As aquaculture depends on self-renewable natural living resources, production becomes much higher in culture systems than in natural waters.

Global Fish Production Scenario

Global production from capture fisheries and aquaculture supplied about 103 million tons of food fish, with per caput availability of 16.3 kg, in 2003 (Figure 1.1 and Table 1.7). The top ten countries China, Peru, United States, Indonesia, Japan, Chile, India, Russian Federation, Thailand and Norway contribute about 60 percent of the world catch. While total capture fisheries production remained

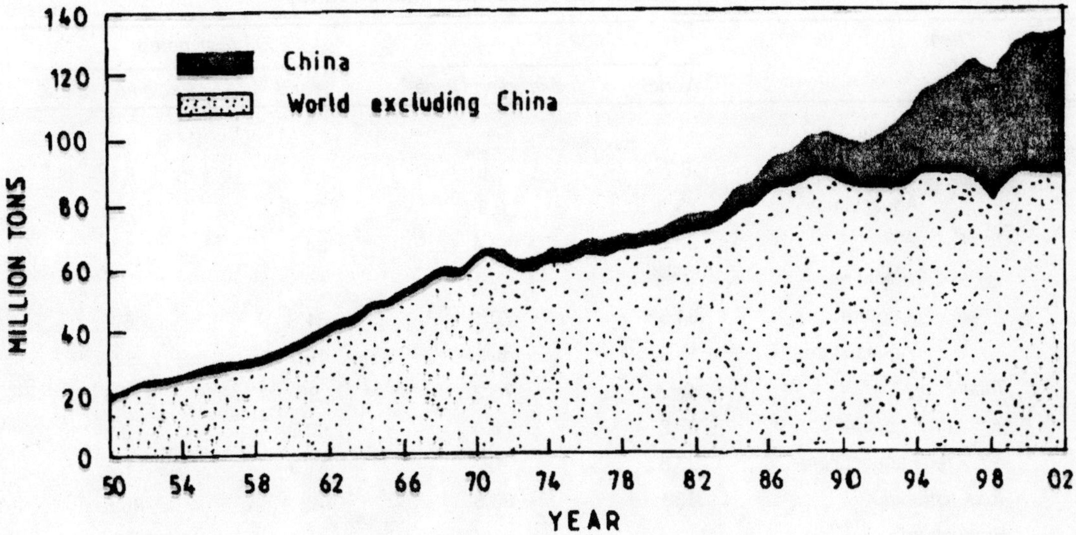

Figure 1.1: Global Fish Production from Capture Fisheries and Aquaculture (From FAO, 2004)

stable during 1999–2002, marine catch declined by 2.6 percent in 2002, and fluctuated between 80 and 86 million tons during 1993–2003. There are considerable variations in marine landings among regions, with decreases in the northwest and southeast Pacific, and eastern central and northwest Atlantic regions but increases in the tropical regions of the Indian and Pacific Oceans. Catches of oceanic species mainly from high seas continued to increase during 2000–2002.

Total inland fisheries catch remained stable in 2000-2002, with Asia and Africa contributing about 90 percent of the total production. The trends in the production shows growth over 2000 production in Asia, Africa and South America, but fall in Europe, North and Central America, and Oceania (Table 1.8). Further, the bulk production in the world (68.1 percent) was supplied by developing countries other than China, and only 6.1 percent from developed countries.

Table 1.6: Aquatic Productions of Water Bodies

Water Body	*Production (kg/ha yr)*
Natural waters	
Oceans (through fishing)	Around 2
Lakes	200–250
Brackish water pond, Philippines	500–1000
Culture systems	
Fertilized and waste fed pond	2450–5600
Fertilized pond, China	3140–6725
Flowing water, carp-catfish	up to 1,125,000
Animals on land	
Beef or mutton	225
Pork on cultivated land	560

Source: Extracted from Wheaton, 1977 and Srivastava, 1991.

Table 1.7: Global Fish Production (Million Tons)

Item	*2002*		*2003 (Estimated)*	
	World	*Excluding China*	*World*	*Excluding China*
Production: Inland				
Capture	8.7	6.5	9	6.5
Aquaculture	23.9	6.9	25.2	7.5
Total:	32.6	13.4	34.2	14
Marine				
Capture	84.5	70.1	81.3	67
Aquaculture	15.9	5.1	16.7	5.5
Total:	100.4	75.2	98	72.5
Total capture	93.2	76.6	90.3	73.5
Total aquaculture	39.8	12	41.9	13
Total fisheries	133	88.6	132.2	86.5
Production:				
Human consumption:	100.7	65.5	103	66.8
Per caput:	16.2	13.2	16.3	13.3

Note 1: Excluding aquatic plant production

Note 2: World production without China's statistics made separately because of too high and uncertainty of its production

Source: FAO, 2004.

Table 1.8: Continent-Wise Contribution of Inland Capture Fisheries Production

Continent	*Contribution in 2002 (%)*	*Growth/Fall Over 2000 (%)*
Asia	65.5	0.6
Africa	24.0	2.0
South America	4.3	9.0
Europe	4.0	-18.0
North and Central America	2.0	–9.8
Oceania	0.2	–0.5

Source: FAO, 2004.

Aquaculture continued to grow at an average rate of 8.9 percent per year since 1970 and supplied 29.9 percent of total production in 2002. In 2002 total world production was 51.4 million tons by quantity with annual increase of 6.1 percent over the 2000 figure. The world production trend as well as of major species groups are shown in Figure 1.2.

The extraordinary growth rates in 1980s and 1990s were slowed down, but overall growth rate during 1970–2002 remained substantial at 8.9 percent, and growth in crustacean production increased

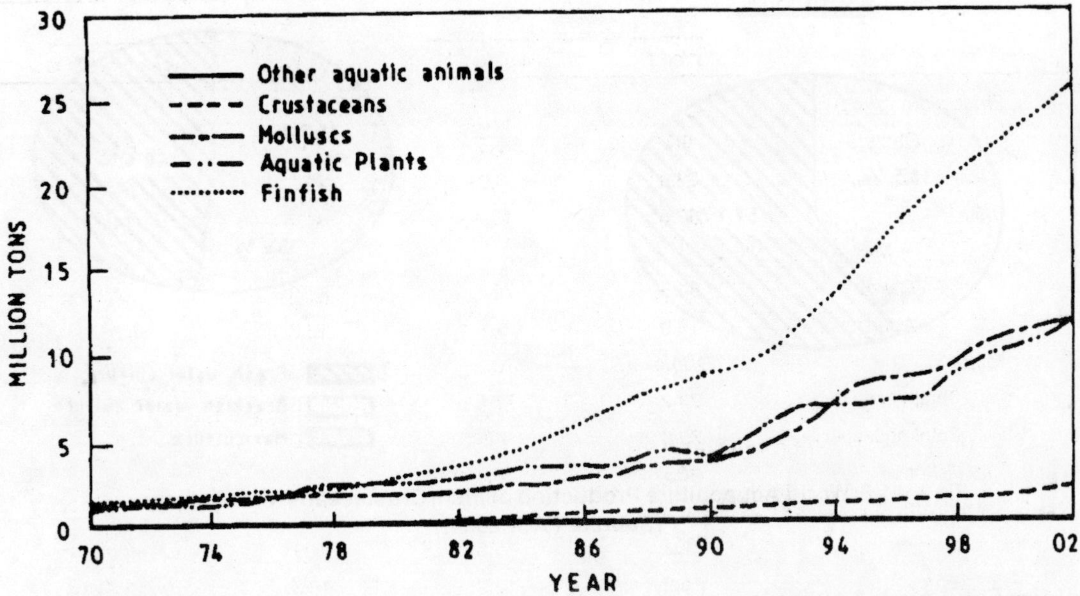

Figure 1.2: Trends in World Aquaculture Production in Major Species Groups (From FAO, 2004)

in 2000-2002 to 11.0 percent. Among the ten major species groups, the carps and other cyprinids far exceeded the production of other species groups, contributing nearly 42 percent of total production of all groups. Breakdown by environment shows that freshwater environment contributes most (57.7 percent by quantity and 48.4 percent by value), followed by mariculture (36.5 percent by quantity and 35.7 percent by value). Brackish water contribution is only 5.8 percent by quantity, but 15.9 percent by value due to high value crustaceans and fish (Figure 1.3).

The top ten countries in aquaculture production (APR values within bracket) are China (6.3), India (6.2), Indonesia (7.7), Japan (4.2), Bangladesh (9.4), Thailand (6.5), Norway (6.2), Chile (18.0), Viet Nam (0.8) and the U.S.A.(4.4) in that order. Like capture fisheries production, contribution of aquaculture production from developing countries increased from 58.8 percent in 1970 to 90.7 percent in 2002, with growth rate at 10.4, which has exceeded the growth in developed countries at 4.0. By continent, the Asian countries contributed to 91.2 percent of production quantity and 82.1 percent of the value in 2002.

Over the period 1961–2001 the world fish consumption has increased more than triple. Forecast studies indicate that there would not be any shortage of fish supplies and per capita supplies are likely to grow in the coming decades. In the context of stagnation in slow growth from capture fisheries only aquaculture expansion can meet this growing global demand. By 2015 aquaculture production will contribute 74 million tons out of total production of 179 million tons. Asia will continue to produce the bulk of aquaculture output by 2020.

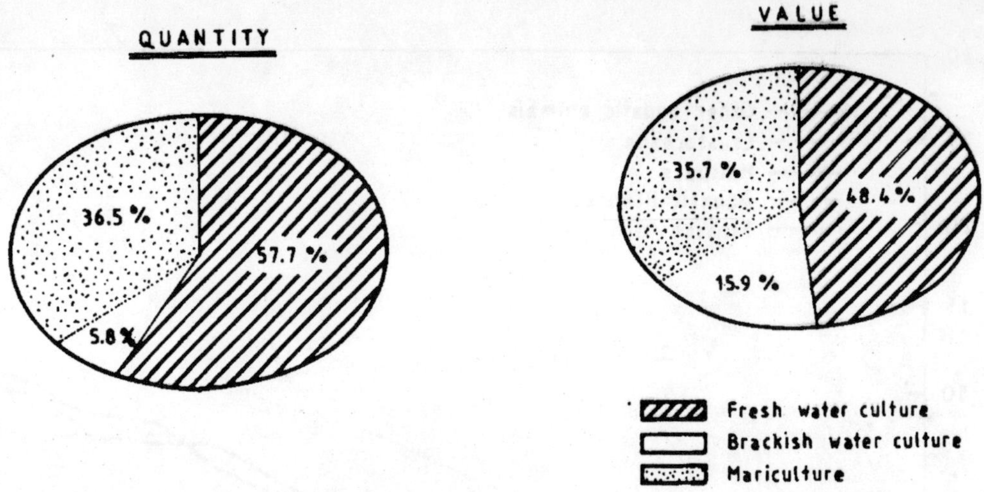

Figure 1.3: World Aquaculture Production of Fish, Crustaceans and Molluscs (*Source*: FAO, 2004)

Protein Food and Aquaculture

Aquaculture has been recognized as the fastest growing food production system. The aquaculture food fish sector has grown at an APR of 8.9 since 1970 compared with only 1.2 for capture fisheries and 2.8 for terrestrial livestock meat production over the same period. Comparison between different meat production systems is shown in Table 1.9. In terms of global meat production aquaculture production ranks fourth.

Table 1.9: Annual Per cent Rate (APR) of Farmed Terrestrial and Aquatic Meat Production

Type	*APR*	*Per cent of Global Meat Production*
Aquatic meat/1	8.9	6.2
Chicken meat	5.3	20.9
Pig meat	3.4	37.6
Mutton and lamb	1.4	3.2
Beef and veal	0.9	24.0
Others	–	8.1

Source: FAO, 1997, 2003 /1 Equivalent aquatic meat after gutting/shelling.

Overall fish supplies at least 20 percent of the average per capita animal protein intake to more than 2600 million people. In most Asian countries (with the exception in China), over one-third of the total animal protein intake is supplied by food fish. Food fish is an important dietary component because of their easy availability and lower cost than the most other animal protein sources. The important fact is that over 85 percent of total aquaculture food fish production was contributed by developing countries, and particularly by low-income-food-deficit countries, which supplied over 76 percent of the total food fish output from aquaculture (FAO, 1997). Aquaculture production greatly surpassed population growth. World average per capita food fish supply (based on China-reported growth) from aquaculture increased from 0.7 kg in 1970 to 6.4 kg in 2002, the average annual growth rate being 7.2 percent.

Aquaculture supplies over 220 different farmed aquatic animal and plant species in 2002, unlike terrestrial farming system producing a limited number of animal and plant species. This represents an opportunity to an entrepreneur to develop a new "aquaculture" product, as so many markets exist for different fish product. In developed countries new perception of the value of fish has developed as "health" food, "environmentally safe" food, and as luxury dish for occasional eating, in small quantities and at high prices (FAO, 2004).

In many countries, the role of aquaculture rather than of capture fisheries in contributing to food security has been recognized as the national priority of development and realized as an "innovative motor of growth" with several additional benefits. Global forecasts conclude "aquaculture sector could replicate the expansion of agriculture". However, many factors affect this sector, which is susceptible to meteorological, pathological and economic shocks. Therefore much will depend on the development planning and progress monitoring in marketing a commodity and expanding its production simultaneously (FAO, 2004).

1.4 Perspective of Water Quality Management in Aquaculture

We have shown in section 1.1 that potential environmental changes and impacts of fisheries and aquaculture besides that of agriculture are related to water quality problems of the surface and ground water bodies in the coastal areas and damage of wetlands. In many nations engaged in coastal aquaculture water pollution by aquaculture farm effluents has become the most common complaint and concerns of the environment activists (Boyd and Gautier, 2000 and Boyd, 2003).

Aquaculture production for fish and shrimp conducted mostly in ponds discharge effluents into the surrounding water bodies in response to water exchange during culture and water draining during harvest. Use of fertilizers and feeds for promotion of pond culture production increases inputs of nutrients, suspended solids and BOD in the culture environment as well as in the surrounding water bodies after pond effluent discharge. As culture is intensified both in the scale and capacity of farming, the amount of effluent generated also grows and reaches the limit of ecosystem's natural capacity of water purification. In fact, 'self pollution' results due to pumping of water into the ponds, previously discharged to the water bodies such as estuaries, which contains increased nutrients and contaminants and lower dissolved oxygen concentration (Csavas, 1994). This creates negative impacts on overall aquaculture yields by promoting infectious diseases, noxious algal blooms, toxic substances and sediment loads. Shrimp aquaculture industry suffered a serious set back due to disease outbreak in Taiwan (Lin, 1989 and Liao, 1992), Thailand (Briggs and Funge-Smith, 1994), China (Wang *et al.*, 1995), India (FAO/NACA, 1995) and several other countries. It now appears that these diseases were a direct consequence of ecosystem mismanagement resulting from poor farm design, inadequate water quality management in the farm and effluent discharge without treatment in the adjacent water bodies.

In view of this, water quality management in aquaculture in general and in coastal aquaculture in particular becomes critical for commercially viable and ecologically sustainable industry. Water quality management raises three general questions on: environment requirements of the culture species, maintenance of these requirements, and the environmental impacts of large scale aquaculture operations. "Water quality", as defined as a relative term, depends on the use of water for the intended purpose. For aquaculture of fish and crustaceans the water should possess certain characteristics, which are capable of supporting the growth of the culture species. These characteristics are certain parameters that provide basic information required for control and sound management of the culture system with an objective to maximize production.

The culture water should also maintain the sanitary standards necessary to allow the harvesting species to be used for animal and human consumption. In this respect the water should not be polluted by dumping raw sewage water or using the pond for washing, bathing or any other domestic cleaning purpose. For defining water quality pollution of water may mean as any addition to the water, which is detrimental to the culture species to be grown in the water or to the consumer of that species. For example, the raw sewage may not be a pollutant for fish life, but the pathogens are a pollutant.

The nature of aquatic environment shows that water is one of the best natural solvents on earth. It can dissolve many substances it comes in contact with. Hence aquatic animals are exposed to more dissolved substances (both beneficial and harmful) than terrestrial organisms. In a given volume of water less oxygen is present than in an equal volume of air and so aquatic organisms require a greater rate of ventilation. Movement of nutrients and toxic metabolites through water is extremely slow in comparison with air, and hence culture systems should provide uniform water distribution. Further, composition of water varies tremendously from one place to another requiring incorporation of local site and species characteristics into the culture environment

Environmental conditions of culture pond are regulated within the range required for survival and growth of fish. In low density culture practices, fish grows in a large water body, which is usually free from any factor affecting the water quality and production. However, aquatic productivity of these systems is rather low. The desirable condition of culture for increased productivity would be to modify the natural system, for which several management techniques are necessary. Intensification of aquaculture makes the culture system more complex and creates more impact on the environment. So one final question or challenge that emerges is to conduct aquaculture within the constraints of acceptable environment impact in the surrounding waters. Thus water quality management objectives become two-fold: one, to deal with water quality impacts on the aquatic organisms during culture, and the other, to prevent environmental impact of the farm effluent on the surrounding water bodies.

In general, the most widely recognized factors that are monitored and controlled within the pond system are: water temperature and pH, salinity, dissolved oxygen, carbon dioxide, hydrogen sulphide, ammonia, nitrite, nutrient concentration in soil and water, planktonic abundance, settleable solids, and pathogens. Several treatment methods have been designed with an objective to control and management of the water quality factors. Pollution by nutrients or pathogens in waste water effluent is currently recognized to be one of the main sources of impact on the environment. The four main components or water quality factors of environmental impact assessment are nutrients, biochemical oxygen demand (BOD), suspended solids and pathogens.

The whole text has been dealt in three parts: Water Resources for Coastal Aquaculture, Water Quality, and Water Quality Management. In the last part, the subject has been discussed in two sections: treatment of pond water for control and management of aquatic environment, and treatment of farm effluent to reduce the environmental impact.

Part I
WATER RESOURCES FOR COASTAL AQUACULTURE

2

Water Resources

2.1 Sources of Water

The various water resources of the world are enormous. It covers about 71 per cent of the entire earth's surface. On volume basis, 97.5 per cent of the world's water is saline with 99.99 per cent of this in the oceans, and the remainder making up the salt lakes. Only 2.5 per cent of the water is actually fresh water. Of this fresh water some 75 per cent is locked up in ice caps and glaciers, 24 per cent located as ground water, and about 1 per cent water is found in lakes, rivers, wetlands and the soil. This means only 0.01 per cent of the world water budget is present in lakes, rivers and wetlands, with another 0.01 per cent present in the soil. So while there is vast resource of ocean water in reality, very little freshwater is available for use in various purposes.

The total water volume in the world remains constant. Only its quality and availability change. The hydrological cycle shows (Figure 2.1) that the water is constantly moving, driven by solar energy. About 80 percent of rainfall or precipitation falls back into the ocean, and the rest falls on to the land. This water replenishes the soil and ground water, feeds the streams and lakes, and provides all the water needed by plants, animals and humans. The continuous cycle makes the water a renewable resource. The fresh water bodies are often closely inter-connected, forming part of a continuum from rain water to saline or marine water, and so may influence each other quite significantly. However, the chemical and physical natures of fresh water are subject to more rapid fluctuations than that of ocean waters, and are therefore more vulnerable to pollution. The quantity or availability and characteristics or quality of various water sources decide the use of water in an aquaculture enterprise in a particular locality in general. The various water sources can be compared with respect to these two factors: quantity and quality.

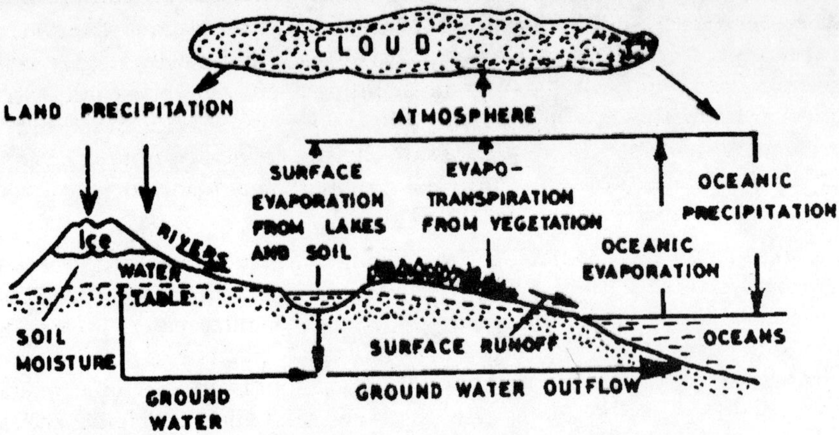

Figure 2.1: Hydrological Cycle Showing the Movement of Water

2.1.1 Surface Water

Surface water is described by any water body, which is found flowing or standing on the surface. It can be divided into four major sources: ponds, lakes and reservoirs, streams and rivers, and seawater. All these sources are fed by a combination of:

1. Direct precipitation or rainfall,
2. Surface run-off or rainfall flowing into the water body,
3. Interflow *i.e.* excess soil moisture constantly draining into the water body, and
4. Water–table discharge, *i.e.* water–table higher than an aquifer discharging water directly into the water body.

The quality and quantity of surface water depend on a combination of climatic and geological factors. In case of water bodies such as ponds, lakes and reservoirs water is collected and stored over a long period, and climatic condition such as recent pattern of rainfall is less important. In contrast, in rivers and streams water is in a dynamic state of constant movement and volume of water is very dependent on the preceding weather condition. Therefore, surface water quality and quantity can be discussed under the headings of ponds, lakes and reservoirs, and streams and rivers. The characteristic physical and chemical nature of sea or salt water and several difficulties for its use make it a surface water source for separate discussion.

1. Pond, Lakes and Reservoirs

In the lentic series lakes and ponds are the standing waters without current as a prominent feature. Bodies of water smaller and shallower than lakes are ponds. There are two types of natural ponds, those representing the pond stage of the previously extinct lakes, and those small and shallow basins remaining in the pond stage but not preceded by a lake. Ponds may be called as village ponds, homestead ponds, irrigation ponds, and so on (Jhingran, 1991). However, for aquaculture purpose ponds are mostly man-made, constructed by building up a dam, dyke or similar structure to impound water (embankment ponds) or by excavating soil from an area and filling with water (excavated ponds). The best topography for pond construction would be land surface with a reasonable slope in one or two directions and with a soil type of low permeability. Ponds are usually filled up by surface drainage or by ground water source.

Lakes and reservoirs are large surface water bodies, usually of area more then ten hectares. While the lakes are naturally formed, the reservoirs are man-made and mostly storage systems for river waters. A lake has been defined as a basin with standing deep water body and with no continuity with sea. However, brackish water lakes usually draw water from sea as well as rivers. Natural lakes are of diverse geological origins. Reservoirs are artificially created by impounding water across a stream with a bundh or dam. The purpose of building a reservoir is mainly for irrigation, hydroelectric power, flood control, fishery development, and so on.

Nature of lake bottom has a great role in determining its productivity. The lake bottom depends on the size, origin and nature of lake basin, and climatic conditions. Sources of bottom deposits are dead planktons; animals and plants of littoral zone; silt, clay and similar materials carried by inflowing waters; inorganic deposits such as marl secreted by animals or plants; and materials eroded from the shore. Bottom deposits may be homogeneous or heterogeneous in origin. Homogeneous deposits of inorganic origin are bed rock, boulders, rubbles, gravel, sand and clay. Homogeneous deposits of organic origin are detritus, fibrous peat (partially decayed plants), pulpy peat (finely divided plant remains in semi fluid greenish mass) and muck (finely divided black organic matter in advanced stage of decomposition). Heterogeneous may be of two or more kinds of homogeneous materials such as alluvial deposits (Biswas, 1996).

2. Streams and Rivers

In the lotic series creeks, brooks, streams and rivers are the flowing or running waters with a continuous current in one direction only. In the evolution process conditions continuously change, resulting in migration of environment. The river traverses through various ecosystems, from snow-bound mountains to vast alluvial plains and finally the estuarine reaches. These surface water bodies receive water from the runoff from direct rainfall, and snowmelt and glaciers during the dry weather. In India, the perennial rivers of the Himalayan region are fed by both the sources, whereas the peninsular rivers originating at lower altitudes are fed by the rainfall only. The flow is thus characterized by heavy discharge during the monsoon and very low or no discharge during the rest of the year. Stream discharge characteristics are important when the stream is the only source of water for aquaculture. However, volume of run-off waters depends on local conditions such as nature of soil, slope and vegetation; climate; volume and frequency of precipitation.

Rivers and streams flow into the coastal and deltaic areas, which have low elevations with respect to the mean sea level. In these areas large water bodies exist in the form of swamps, bays, estuaries and lagoons. The water bodies are mainly fresh and saltwater mixes, and their volume of flow depends on the stream characteristics of the rivers and streams, and the tidal condition of seawater.

3. Sea or Saltwater

Seas are the great natural water resources, which is about 97.5 per cent of the earth's total waters. About three quarters of the earth's surface lies under the ocean surface and only one quarter is exposed on the continent and on islands. About 65 per cent of the earth's surface is oceanic and 35 per cent is continental, which includes the shallow continental shelves. The sea or saltwater systems for use in aquaculture can, therefore, be better discussed as coastal water systems.

The ocean occurs along the coasts. The coast is buffeted by the waves generated by storms at sea, and as a result the rocks and sediments of the coastal region are shaped. The rivers debouch into the sea, influence circulation patterns along the coast and also provide sediments that respond to this circulation as well as to the waves and tidal currents. In the coastal area or zone, there are two distinct systems–the shoreland or coastal watershed and the coastal waters. The salt to brackish water systems operates from the coastal water regime to the shoreland or the coastal watershed.

The coastal watersheds drain fresh water directly into the coastal water basins. They include (*i*) all the water system including standing water such as ponds, lakes, (*ii*) natural detention areas such as bogs and marshes, (*iii*) drainage course such as streams, and (*iv*) all coastal flood plains, *i.e.* lower coastal areas flooded with tidal waters. The shoreland thus extends down to the upper boundary of the coastal water basin. These land areas strongly influence the coastal ecosystem through the discharge of land drainage waters. In turn coastal watersheds are considerably influenced by coastal waters.

The coastal waters can be categorized in three functionally distinct regimes: the estuarine, the nearshore and the oceanic regimes. The ecology of coastal water system has been dealt in several treatises on coastal ecology, environment and management (Odum, 1963; Clark, 1977; Tait, 1981; Boaden and Seed, 1985). In order to avoid repetition of the topic, only a brief introduction of the water bodies in the coastal zone has been discussed. An estuary, meaning in the dictionary, 'an arm of the sea', has been defined by Pritchard (1967), as a 'semi-closed coastal body of water which has free connection with the open sea and within which sea water is measurably diluted with fresh water derived from land drainage.' According to Clark (1977), estuarine water contains salinity greater than 0.5 ppt. The estuarine system includes mostly the water basins such as tidal river, bay, embayment and lagoon. A tidal river is the lower part of a river reaching the sea via an estuarine basin. Tidal rivers are usually flushed well by the opposing actions of fresh water and tide, but the freshwater part with less than 0.5 ppt salinity is not included even if it is tidally influenced. Bays are water bodies, quite open to the sea, and receive strong tidal flow and considerable circulation from fresh water inflow. They are usually larger water bodies well flushed through tidal exchange. Lagoons have restricted inlets to the sea and little freshwater inflow with sluggish water movement. Embayments are intermediate between bays and lagoons, but are shallower and smaller than bays with narrower inlets. Each of these estuarine systems is controlled by the combined circulation of tide, freshwater inflow, ocean current, wind or geological factors, which in turn govern the productivity of the ecosystem. The nearshore regimes include all marine waters seaward normally 3 miles distance from the shore in the territorial sea, and oceanic regime includes all waters seaward of the territorial sea boundary.

The coastal water system functions by intermixing of the two supply sources–the giant saltwater reservoir of the sea and the flows of the freshwater from the coastal watersheds. The ocean flow is modified somewhat by tide levels, storms and inlet size, while the freshwater flow from the land is much more variable due to seasonal changes in precipitation and runoff. Further, both the volume and the rate of freshwater flow depend on factors such as the size of the watershed, the landslope, the soil type and the interplay between rainfall and loss via evapotranspiration. Thus the water flow from the land applies more fluctuating amount of pressure and therefore influences more the characteristics of the estuarine environment.

2.1.2 Ground Water

Ground water is available in the underground as water trapped in the voids between the individual soil particles. Ground water availability depends on the two important properties: 'porosity' and 'hydraulic conductivity' of the soil which, in turn, decide its characteristics to store and transmit the significant quantity of water in the permeable layer of rock, sand, or gravel below the ground, known as 'aquifer' (Figure 2.2). In general, there are three types of ground water: springs, wells and depressions below the water table, depending on the type of aquifer. Ground water from the depressions below the

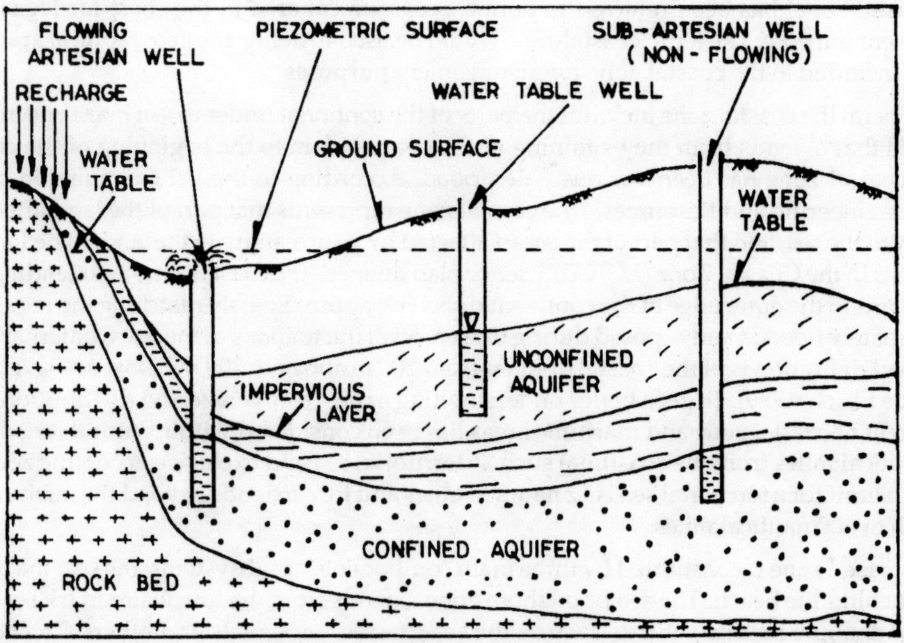

Figure 2.2: Different Types of Aquifers

water table is available in limited areas only, and is therefore not considered as a ground water source for aquaculture.

A crack or any other fault in the impermeable layer above a confined aquifer, or an aquifer exposed at the ground surface allows the water to escape to the surface as a spring. Most springs usually have sufficient discharge for use as aquaculture water source. However, type and location of spring, soil conditions, and topography are to be judged before a spring can be developed for aquaculture water source.

Three primary wells occur as sources of ground water: water table well, artisan or flowing well, and sub-artisan or non–flowing well. Water table wells draw water mostly from unconfined aquifers by pumping. They are sometimes referred to as shallow wells. In artisan or flowing well, water flows continuously through the well without pumping. In this case the ground level is below the 'piezometric surface', which is an imaginary surface passing through all points to which water will rise. Water begins to flow as soon as drilling is complete to the confined aquifer. Through the recharge area water can enter and flow by gravity to the aquifer. A non-flowing well will require pumping when the ground level is well above the 'piezometric surface' or recharge area. A detailed discussion on types of wells and their construction is given in Mal (1995).

2.2. Coastal Environment

2.2.1 Coastal Areas and Zones

Coastal areas or regions are commonly described as the interface or transition areas between land and sea. They are diverse in nature and function, and do not have exact natural boundaries like watersheds. For management purposes, a number of landwards and seawards boundaries, ranging from fairly narrow and definitive ones to much broader and vague ones are described. Usually the term 'coastal zone' has been referred to define geographical area in the coast for legislation and management purpose. The term 'coastal area' would be used to define the geographical area along the coast not included in the coastal zone for management purposes.

In general the coastal zone includes the parts of the continent under direct maritime influence. It is the land that extends from the beginning of the coastal plain to the beginning of the continental shelf. A coastal zone has been variously described. According to the US Commission on Marine Science, Engineering and Resources, "the coastal zone represents that part of the land affected by its proximity to the sea, and that part of the ocean affected by its proximity to the land". The Land-Ocean Interactions in the Coastal Zone (LOICZ) Science plan defines "the coastal zone as extending from the coastal plains to the outer edge of the continental shelves, approximately matching the region that has been alternately flooded and exposed during the sea–level fluctuations of the late Quaternary period". This second definition is of the coastal domain from 200 m above to 200 m below sea levels (LOICZ, 1998; Sa de Elrich, 1999). To get a better understanding of the relative size and distribution of coastal areas, extent of coastal zone and maritime areas has been considered within natural jurisdiction (up to 200 nautical miles from the coastline) such as territorial sea and exclusive economic zones (EEZ). Maximum limit for a territorial sea is 12 nautical miles and EEZ extends beyond the territorial sea up to a width of 200 nautical miles.

The coastal zone is constituted by three main components or subsystems: the sea, the beach, and the land behind the beach. The sea or offshore zone extends from the low water mark seaward and covers the shallow marine habitats such as the sea grasses, coral reefs and others. The beach zone extends from the low water mark to the seaward edge of the coastal vegetation. The land zone is the adjoining coastal land extending landward for same distance from the end of the beach. This distance is not clearly defined but varies according to each country and its environment management policy. The three subsystems of the coastal zone interact in many ways and the boundaries between them change. The coastal zone is not an isolated system. Various interactions from the rivers and waterways result from the inland activities and create impact on the coastal ecosystem.

According to Clark (1977), the major natural elements of coastal zone environment are hydrology or water system, soils, vegetation, topography and geology. Water is a most important resource element in the coastal environment because it plays a critical role in governing coastal ecosystems and in limiting its carrying capacity. Soils constitute a basic land suitability factor and a major control element on coastal water systems. Soil and hydrological factors together control the plant and plant community types, which, in turn, indicate the natural characteristics of many vital areas such as coastal marshes and mangrove swamps, freshwater wetlands, and submerged grass-beds. Topography such as the landforms of hills, mountains, ridges and valleys; water quality; and soil condition are also the closely associated factors of vegetation patterns.

The world's coastal regions are subdivided by physical rather than biological characteristics and categorized into a wide diversity of environments (Burke *et al.*, 2000):

1. *Near-shore terrestrial*–Dunes, cliffs, rocky and sandy shores; coastal xeromorphic habitats; urban, industrial and agricultural landscapes.
2. *Intertidal*–Estuaries, deltas, lagoons, mangrove forests, mudflats, salt marshes, salt pans, other coastal wetlands, ports and marinas, aquaculture beds.
3. *Benthic*–Kelp forests, seagrass beds, coral reefs, and soft bottom environments above the continental shelf, artificial reefs and structures.
4. *Pelagic*–Open waters above the continental shelf, freestanding fish arms, *e.g.* plankton blooms.

2.2.2 Ecological Divisions

Ecologically coastal areas are in general divided into three habitat zones: (*i*) maritime zone, the shallow-water zone overlying the continental shelf, which is the habitat of many terrestrial animals and plants, (*ii*) the sea itself, and (*iii*) the littoral or intertidal zone, which is the transitional zone between the land and the sea (Boaden and Seed, 1985).

1. Marine Environment

Ecologically the marine environment is divided according to depth, distance from land, and intensity of light penetration. The broad divisions are: *pelagic, i.e.* the water column or overlying water mass, and *benthic, i.e.* the sea bed or land mass beneath the water column. According to depth both the divisions are further subdivided. The pelagic division consists of *neritic* or coastal water zone, which is composed of all the waters above the continental shelf up to a depth of 200 m. The oceanic water consists of all the waters beyond the continental shelf.

The well illuminated part of the pelagic division, usually around 100–200 m, falls under the oceanic water zone and is called *euphotic* zone. The zone is a turbulent section of the sea and includes waters of shoreline, kelp beds, coral or rock reefs, etc. with a number of varied habitats. Below this level lies the *aphotic* zone where light dose not penetrate. There are further vertical divisions of the aphotic zone, not described here but are shown in Figure 2.3.

2. Rocky Shore

In the littoral or intertidal zone, there are shores composed of hard materials called rocky shores, which range from smooth uniform slopes to extensive boulder beaches and overhanging cliffs. Rock presents a stable, open surface for organisms to attach themselves or even bore into. The characteristic habitat of most rocky shores is tidepools, tidal rapids, crevices, boulders and algal epifauna. The tidal pools can occur at different depths in the intertidal zone and are subject to wide variations in physical factors unlike the sea. Tidal rapids are areas that occur in channels between adjacent land masses, and receive clean, fast flowing water, but are protected from wave action. Crevices in the rocks provide damp microhabitats. The upper surfaces of well-illuminated boulders support organisms similar to those on adjacent rocky beds. Micro-algae of many sheltered rocky coasts provide extensive surface for attachment of many organisms.

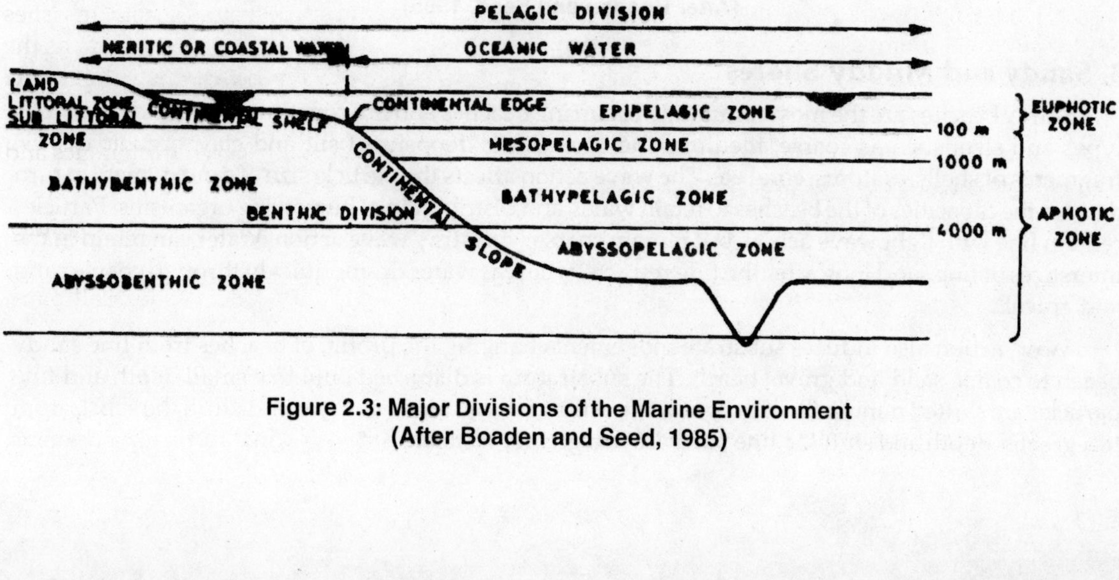

Figure 2.3: Major Divisions of the Marine Environment (After Boaden and Seed, 1985)

The intertidal zone is greatly inhabited by marine organisms, and hence can be termed a marine province. This abundance of marine organisms and periodical inundation by tides result in a vertical zonation. Based on tidal environment, the scheme divides the zone into littoral, supralittoral and sublittoral divisions (Figure 2.4). Littoral zone is essentially the intertidal area with extreme changes in water depths and turbulence, and in substrates from mud to sand and rock. The supralittoral zone ranges between high tide and the spray zone of the beach, whereas the sublittoral zone is the bottom portion of the neritic province. Considering the biological factors such as competition for limited resources, predation and grazing, the two other classification schemes have been introduced. According to Lewis's scheme, there are three zones: the littoral fringe, the eulittoral and the sublittoral. The littoral zone is a distinct biologically defined entity rather than physical like the intertidal zone. Littoral fringe zone, being transitional between land and sea, is relatively arid.

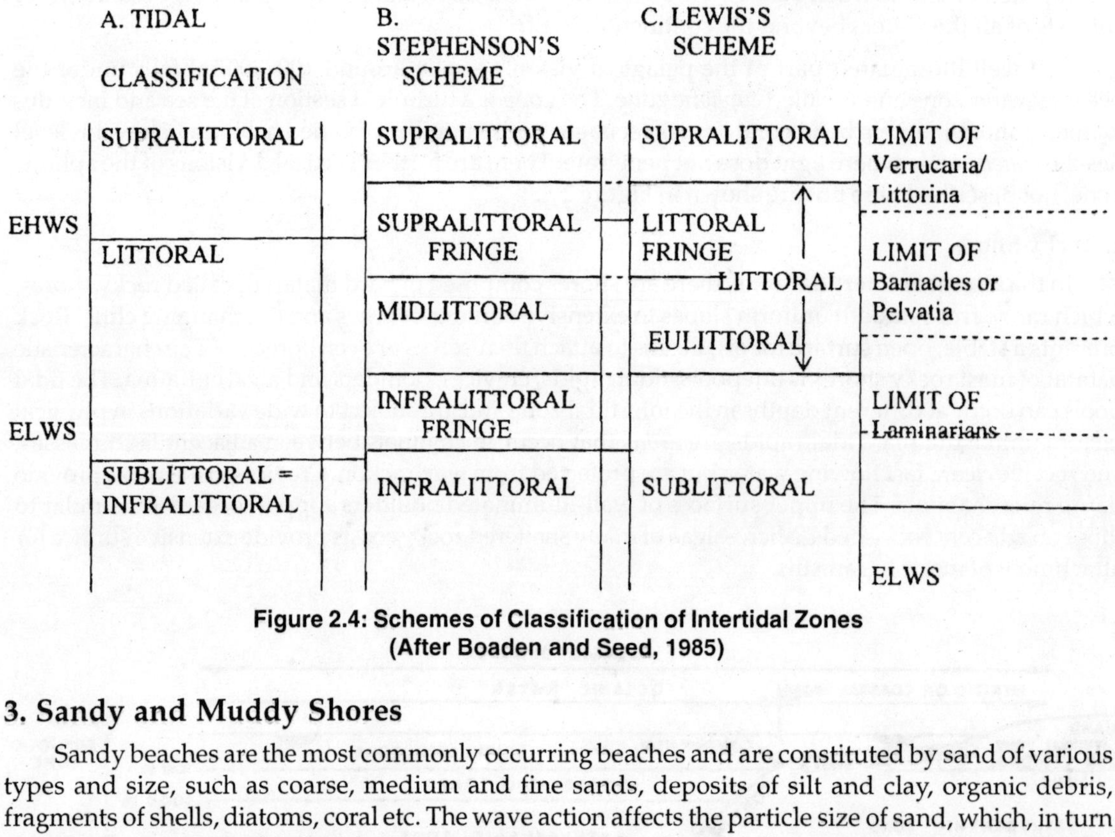

Figure 2.4: Schemes of Classification of Intertidal Zones (After Boaden and Seed, 1985)

3. Sandy and Muddy Shores

Sandy beaches are the most commonly occurring beaches and are constituted by sand of various types and size, such as coarse, medium and fine sands, deposits of silt and clay, organic debris, fragments of shells, diatoms, coral etc. The wave action affects the particle size of sand, which, in turn decides the capacities of the beaches to retain water, and distribution of burrowing organisms. Particles remain fine with light wave action, but remain coarse with stray wave action. Water can retain in the interstices of fine sand long after the tide retracts, whereas water drains quickly through coarse sand and gravel.

Wave action also induces substrate movement changing the profile of beaches from fine sandy beach to coarse sand and gravel beach. The substratum is disturbed only to a small depth and fine particles are shifted minimally with light wave action. Strong and heavy waves disturb the substratum to a greater depth and shift the fine particles to a greater distance.

When the sandy shores become more and more fine grained, and collect more organic matter, they become muddy beaches. Typical muddy shores are developed in areas with fine grained sediment particles and located notably in estuarine environment that are more protected from wave action. Due to minimum water movement, the muddy shores become more flat than sandy beaches, and are called "mud flats". These areas retain water much longer with poor interchange of tidal water, and therefore develop into anaerobic condition below the surface due to high internal load of bacteria and complete depletion of oxygen within the sediments.

4. Brackishwater or Estuarine Environment

Estuaries have been briefly discussed above as the sea or saltwater or coastal water systems, which include different types of water basins. Based on geo-morphological classification the four primary types of estuaries are (Boaden and Seed, 1985):

1. Coastal plain estuaries or drowned river valleys, the commonest type found along a coastline, and formed by the rising sea water level over the low lying coastal river valleys. They conform most closely to the definition of an estuary, which extends seaward up to the point of measurably diluted seawater and riverwards up to the measurable amount of saltwater.
2. Bar-built estuaries or semi-enclosed bays or lagoons, formed in low-lying areas by isolation of river water from the sea by offshore barrier, sand islands and sand spits running parallel to the shoreline.
3. Tectonic estuaries, formed as coastal indentures by faulting or local subsidence, and having an excess supply of freshwater inflow.
4. Fjords or drowned glacial valleys, formed as deep and narrow sill, having rivers entering at the head and showing estuarine features in the upper lays.

There are several other brackish water habitats such as estuarine realms, hypersaline lagoons, brackish seas, and brine seeps. Estuarine realms occur beyond river mouths, where the surface water is distinctly brackish due to mixing of large volumes of freshwater. Hypersaline lagoons are coastal water bodies with little or no fresh water input due to restricted inlets to the sea. Brackish seas are large semi-enclosed water bodies with waters permanently brackish but salinity lower than that of many estuaries. Brine seeps occur in areas where hypersaline water emerges subtidally on the sea floor.

5. Marshes and Mangroves

Salt marshes and mangroves swamps are the two most characteristic coastal environments prevailing in many places where soils and marine sediments meet mainly in the supralittoral and the upper littoral zones. Salt marshes occur mainly in temperate to subpolar climates and mangrove swamps in tropical to subtropical climates. Salt marshes are natural or semi-natural halophytic (salt-loving) grassland and dwarf brushwood on the alluvial sediments with average salinity in the range of 5-38 ppt. There are other vegetations such as reed und rush beds in the higher reaches of estuaries. The substratum of salt marshes is constituted by muddy and sediments of estuarine nature with high salt content.

Mangroves are described as an ecological group of dense terrestrial shrubs and trees growing up to 30 m heights. The term mangrove refers to both the tree and the forest. They are not taxonomically related, but are grouped together because of their tolerance of brackish to fully saline environments. To cope with extreme conditions of such as salinity, water saturation and dessication, and anoxia, mangroves have evolved a variety of coping mechanisms including prop roots, cable roots and salt glands. These roots provide a surface of attachment for marine organisms in a muddy or salty environment.

The mangroves occur along tropical shores, particularly in the tidal estuarine regions, never being totally isolated from the sea but growing also far inlands. They grow most in warm moist coastal belts and provide both fisheries and forest resources and goods.

Figure 2.5 shows a hypothetical coastal area with the various ecological divisions, and Figure 2.6 shows a sketch of the Hooghly-Matlah estuary of the two rivers falling in the Bay of Bengal.

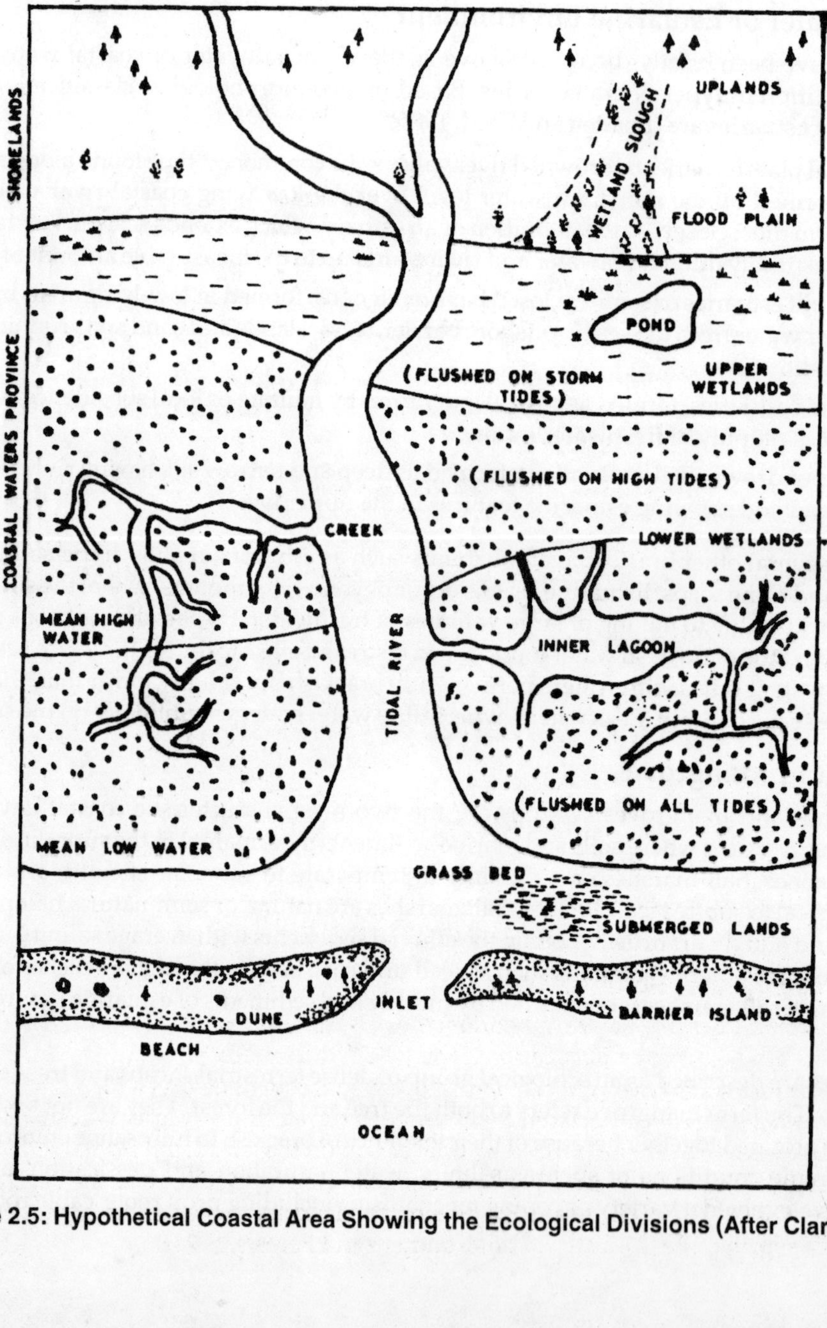

Figure 2.5: Hypothetical Coastal Area Showing the Ecological Divisions (After Clark, 1977)

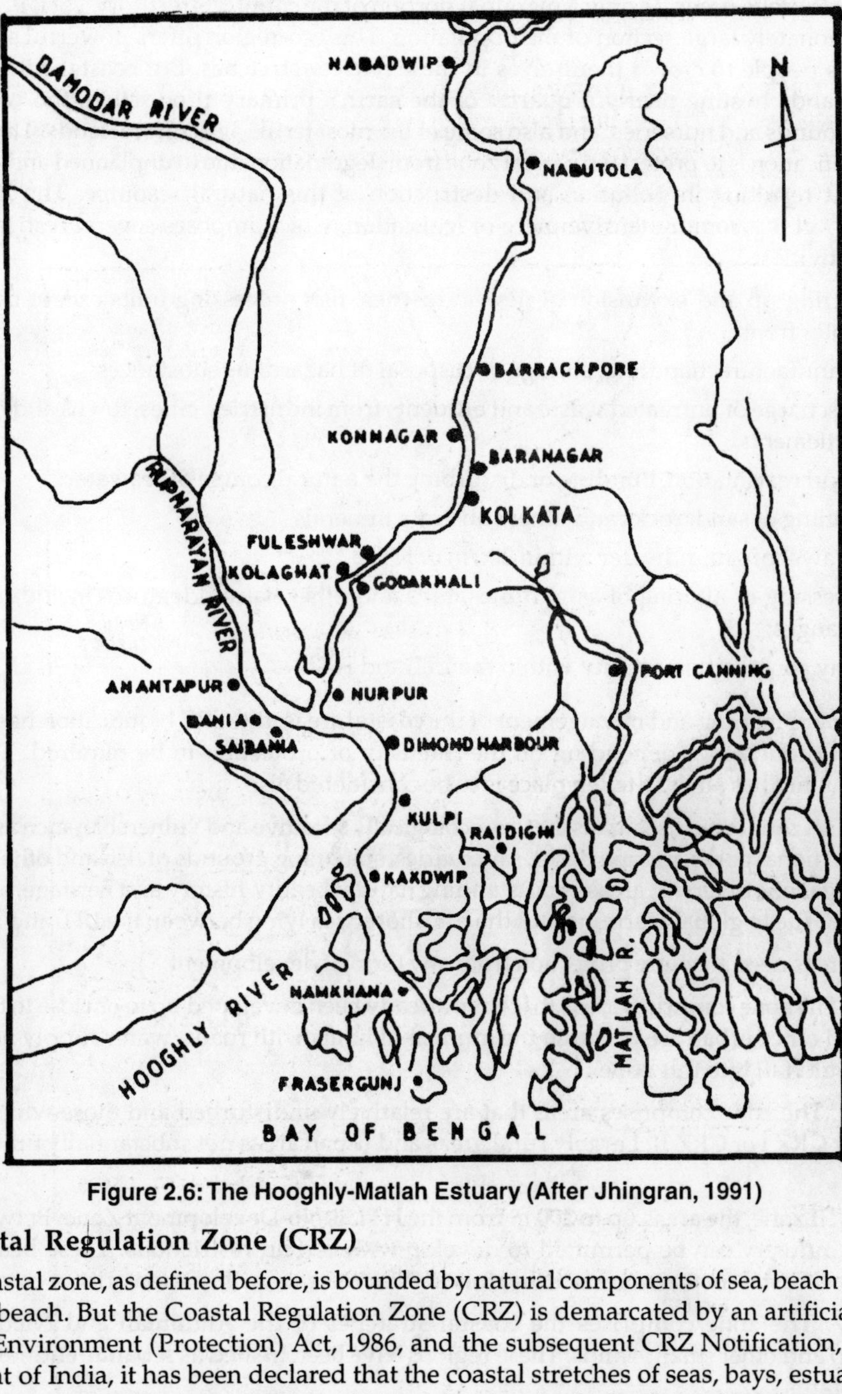

Figure 2.6: The Hooghly-Matlah Estuary (After Jhingran, 1991)

2.2.3 Coastal Regulation Zone (CRZ)

The coastal zone, as defined before, is bounded by natural components of sea, beach and the land behind the beach. But the Coastal Regulation Zone (CRZ) is demarcated by an artificial boundary. Under the Environment (Protection) Act, 1986, and the subsequent CRZ Notification, 1991, of the Government of India, it has been declared that the coastal stretches of seas, bays, estuaries, creeks, rivers and backwaters which are influenced by tidal action (in the land ward side) up to 500 m from the High Tide Line (HTL) and the land between the Low Tide Line (LTL) and the HTL as Coastal Regulation Zone (CRZ). The HTL means the line on the land up to which the highest water line reaches during the spring tide (Ministry of Environment and Forests, 1994).

The coastal zone occupies only a marginal portion of the country's territory, yet it is inhabited by a disproportionately large section of the population. This eco-region offers powerful scenic beauty, which drives people to crowd themselves in these narrow stretches. But coastal zones are highly productive lands hosting nearly a quarter of the earth's primary production, the world's major spawning grounds and nurseries, and also some of the most fertile agricultural lands. The objective of the CRZ Notification is to protect the coastal zone from degradation due to unplanned and/or excessive development resulting in pollution and destruction of this natural resource. The Environment (Protection) Act is a comprehensive piece of legislation, which imposes several restrictions on the following activities:

1. Setting up and expansion of new industries, fish processing units except that requiring water front.
2. Manufacture, handling, storage or disposal of hazardous substances;
3. Discharge of untreated waste and effluents from industries, cities, towns and other human settlements;
4. Land reclamation, bunding or disturbing the natural course of seawater;
5. Mining of sand, rocks and other substrata minerals;
6. Drawal of groundwater within 200 m of HTL;
7. Dressing or altering of sand dunes, hills and other natural features including landscape changes;
8. Any construction activity within the LTL and HTL.

For proper planning and management of the coastal areas, the CRZ Notification has categorized the areas into four zones depending on the intensity of protection to be required. The extent of development that has already taken place is to be considered also.

CRZ I: The zone comprises areas that are ecologically sensitive and vulnerable, such as mangroves, coral reefs, national parks, marine parks, sanctuaries, spawning grounds of fish and other marine life, areas rich in genetic diversity, areas of outstanding natural beauty, history and heritage, areas likely to be inundated due to global warming and the foreshore area lying between the LTL and HTL.

These areas need absolute protection from any form of development.

CRZ II: The zone comprises areas that have already been developed up to or close to the shoreline. All cities and other urban areas within the municipal limits with roads, water supply and sewerage facilities would fall into this zone.

CRZ III: The zone comprises areas that are relatively undisturbed and those which do not fall under either CRZ I or CRZ II. Largely rural areas and urban areas not substantially built up fall into this zone.

In CRZ III zone, the areas up to 200 m from the HTL is No-Development Zone. Between 200–500 m, tourism industry can be permitted to develop with certain restrictions. These beach areas are important for coastal communities of fisher folk and for recreation of vast majority of common people.

CRZ IV: The zone comprises the coastal stretches of the Andaman and Nicobar Islands, Lakswadeep and other small islands. These regions have been treated as separate entities with special protection status.

2.3 Characteristics of Water Resources

2.3.1 Environmental Characteristics of Coastal Water

Water, the most important component of natural resources, plays a critical role in governing the aquatic ecosystem. Any aquatic ecosystem, in the inland or in the coasts, is the basic geographic unit that encompasses all of the living and physical components involved with a distinct water basin. There are processes that "drive" the system, such as photosynthesis, water movement and nutrient recycling, which in turn limit the productivity of the system. There are variable factors that influence the condition of a water body or system and so limit the carrying capacity, for example temperature, available nutrients, dissolved gas content, and presence of pollutants. An understanding of these processes and factors is useful in characterizing the water resources particularly in the coastal area. A brief outline on the environmental characteristics of aquatic ecosystem and the basic governing principles has been given in the following pages. This will ultimately help in devising water resource management programs in the coastal area.

1. Carrying Capacity and Standing Crop

The biota of an aquatic ecosystem includes a large variety of plants, animal, fish and invertebrate organisms. The *carrying capacity* of an ecosystem, in general, is the limit to the amount of aquatic life (number or mass) that can be supported by that system. In narrow sense it is the potential number of individuals of a particular species. The actual number or mass of species (biomass) present in an area or volume at any time during measurement is termed as the *standing crop*. It is commonly expressed as mg carbon fixed per unit volume. The environmental limits existing in an ecosystem such as availability of food, space or breeding sites, disease or predator cycles, temperature, sunlight and salinity impose the ultimate constraint, *i.e.* the carrying capacity. In fact the carrying capacity measures the condition of the natural resource component–the water.

In a more general or non-ecological sense, the carrying capacity expresses the potential of the ecosystem to provide resources useful to human society. The carrying capacity should be maintained at the optimum level so that the resources of the ecosystem are capable to sustain at an optimally balanced resource base. The productive capacity of an ecosystem in coastal area should include a defined water basin, all shoreline areas and all shoreland watersheds. Each of these should be managed with respect to its relatedness and unity of the whole. This is essentially the goal of environment management of coastal waters.

2. Primary Productivity and Food Chain

Primary productivity relates to the capacity of an ecosystem to produce basic plant material. By means of photosynthesis the light energy is converted with the help of basic nutrients and carbon dioxide to plant tissues. Energy-rich organic molecules are synthesized from simpler inorganic molecules. The process is known as *primary production.* In contrast to standing crop, primary production is measured directly by the quantity of carbon (g) fixed per m^2 per day or per year. The indirect measurement is by the amount of oxygen evolved during photosynthesis. The total energy fixed is termed as *gross primary production*. Since some of this production is consumed by plants for respiration, the net primary production will be a lesser amount available to the consumers. Production, or carbon fixation rate is quite different from standing crop. Low standing crop may result not due to low productivity but due to high grazing, whereas high standing crop may not indicate high productivity.

The energy assimilated by plants is transferred through a series of lower to higher organism with repeated stages of eating and being eaten in a complex chain, known as *food chain or food web*. Animal life is nourished by plants that decay into detritus material in the water. Microscopic organisms act upon this material and reduce it into basic minerals. Some of the plant material is consumed directly by fish and shellfish, but more often it is first eaten by zooplankton, which in turn, become the food for fish. The flow and amount of available energy in a food chain controls life process and limits the carrying capacity of the ecosystem. Plant communities also provide the natural system "storage" of energy for supply against shortages. In coastal water system, salt marshes, mangrove swamps and grass beds are particularly important units as storage on which its food chain depends. The standing stock of grass, for example, provides a large reserve of nutrient matter as foodstuff for bacteria to fish. Marshes store large quantities of mineral nutrients in their soils. Thus the energy storage is highly capable for providing optimum ecosystem function.

3. Principles Governing the Coastal Water Ecosystem

The coastal water ecosystem is the basic functional geographical unit that includes all of the living and non-living components involved within a distinct coastal water area. There are a few simple principles and rules that govern the ecosystem in order to achieve its highest carrying capacity. Clark (1977) highlighted eleven ecological principles that would help in effective management of the system.

1. All factors that influence the function of the ecosystem of which the water basin is a part control the carrying capacity of the whole water basin.
2. The flow of energy and its availability within an ecosystem controls the life processes and are limiting conditions of its carrying capacity.
3. The natural system of energy storage must be capable to optimize the ecosystem function.
4. The patterns of water circulation within the aquatic ecosystem control its carrying capacity.
5. Nitrogen is the basic plant nutrient that limits the primary productivity and carrying capacity of the ecosystem.
6. Natural salinity levels normally optimize the carrying capacity of the ecosystem.
7. Dissolved oxygen supply must be sufficient for ecosystem function and maximum carrying capacity.
8. The natural balance of the temperature regime optimizes aquatic ecosystem function.
9. Increase of sunlight penetration increases the productivity and carrying capacity of an ecosystem.
10. A water body essentially links land and sea elements of the coastal ecosystem.
11. The natural rate, volume and seasonal pattern of freshwater flowing into coastal waters optimize the ecosystem function.

4. Aquatic Biodiversity

Aquatic biodiversity can be defined as the variety of life and the ecosystems that make up the freshwater, estuarine and marine regions of a country or the world and their interactions. It encompasses freshwater ecosystems such as ponds, lakes and reservoirs, rivers and streams, groundwater and wetlands. It encompasses marine and coastal ecosystems including oceans, estuaries, salt marshes, sea grass beds, coral reefs, and mangrove forests.

Aquatic biodiversity is the rich and wonderful variety of plants and animals. It is the number of different native species or species richness. It is the variability among living organisms from all sources including *inter alia*, terrestrial, marine and other aquatic ecosystems and the ecological complexes of which they are a part. This includes diversity within species, between species and of ecosystems. Many species of plants and animals live in water; some, like fish, spend all their lives underwater; whereas others, like toads and salamanders, may use surface waters only during the spring breeding season or as juveniles. Some aquatic creatures live their entire lives in the deep ocean, whereas others, like water striders, spend their life skipping along the surface of water.

These species have evolved and adapted to watery habitats over millions of years. Aquatic habitats provide the food, water, shelter, and space essential for their survival. The greater the diversity of habitats, whether in water or in land, the greater the biodiversity will be. Coastal estuaries and mangrove swamps *e.g.* are "edge" ecosystems that link salt and fresh waters and trap nutrients that allow them to support a rich diversity of aquatic plants and animals. Generally, the more complex and larger is the ecosystem, the greater is its biodiversity.

Aquatic biodiversity has innumerable economic and esthetic value and plays a critical role in maintaining and supplying overall environmental health. Overexploitation of species; introduction of exotic species; urban, industrial and agricultural pollution; habitat loss and alteration through dams and water diversion all contribute to the declining levels of aquatic biodiversity in both freshwater and marine environments.

5. Ecological Factors

Coastal water ecosystem is controlled by climatic, oceanic and terrestrial factors that influence the condition of its waters and bottoms. Primary factors are the pattern of water–shed drainage and freshwater inflow, and secondary factors are dynamic physical processes of tidal forces and currents. All these factors combined result into a specific pattern of water circulation. The biological demand of the water is supplied by its constituents such as nutrients, suspended solids and dissolved gases.

(a) Water Circulation

Specific pattern of water movement, or circulation is found in each coastal water. In the oceanic waters large scale forces of massive distant currents dominate, whereas in nearshore waters tidal action, wind, waves and runoff from the land control water circulation. The ocean currents are the major cause of horizontal circulation of water. There are surface currents due to air movement causing wind–driven circulation. In addition, vertical circulation is caused principally by temperature–salinity or density gradients. Further, there are phenomena like coastal upwelling, which forces up the bottom layer of water to move towards the shore. As a result water gets enriched with nutrients brought up from the bottom.

In bays, the dominant force in water movement is the tidal, varying greatly with tidal amplitude. With high tidal amplitude circulation forces tend to be greater and flushing rates better. In lagoons, the wind forces are dominant, whereas in embayment and tidal rivers water inflow may dominate.

Water circulation is a key ecological factor in coastal waters, because it does various useful work such as transporting nutrients, propelling plankton, spreading planktonic larvae of fish and shellfish, controlling salinity, flushing wastes, cleansing pollutant and so on. In fact, water circulation greatly influences the aquatic life and dynamic balance in the coastal water body.

(b) Light and Temperature

Sunlight being the fundamental source of energy for aquatic life, is the basic driving force of the aquatic ecosystem. Light penetration into water depends on the intensity of the incident light, which varies with the geographical location of the water body. It also depends on the weather conditions such as cloudiness, shade of the surrounding vegetation, and transparency of water. A portion of the solar radiation reaching the water body does not enter the water, but gets immediately reflected. About 50 percent of the incident light reflects from the water surface. Water absorbs light more strongly than does air. So the plants below the surface do not receive the light penetrating the water surface. The light penetration also depends on the transparency of natural water, which usually varies with turbidity. The term turbidity refers to the decreased ability of water to transmit light caused by a wide variety of suspended materials ranging in size from colloidal to coarse dispersions. The higher concentrations of suspended materials increase the turbidity, and thereby decrease the primary productivity. Estuarine waters are normally more turbid than ocean waters, being more mixed with silt and more rich in suspended organisms. However, water clarity is to be maintained to natural condition of the environment.

The solar energy from the sun heats the water body, raises the temperature of water, and provides energy for the aquatic organisms. The heat transfer between the atmosphere and the water determines the average temperature of a particular water body, and the heat exchange is dependent on various factors such as intensity of solar radiation, wind velocity, evaporation, water circulation and relative humidity of air. Temperature alteration by any external means is critical to estuarine water body in particular, and therefore, is to be avoided.

(c) Chemical Constituents

The chemical constituents of water fall into three classes: nutrients, trace elements and dissolved gases. The nutrients and trace elements are present as inorganic ions, organic substances and particulate substances, both organic and inorganic. The quantity of each ion in a particular water body is influenced by the chemical composition of surrounding rock formations, watershed characteristics, inflow and outflow of water, pH of water and others. Particulate inorganic matter consists of fine, suspended soil particles.

Organic substances in natural water are derived from incoming waters or from wave action on the shoreline, or organic materials blown into the water, and from organic materials produced in the water by living organisms. The organic substances are usually in the form of dissolved materials and suspended particles.

The most abundant gases in natural water are nitrogen, oxygen and carbon dioxide. The other gases present in minor quantities are unionized ammonia, hydrogen sulphide and methane. Among these gases only the oxygen is directly utilized by all the aquatic organisms. Carbon dioxide makes the aquatic environment fit for life. The other gases are either unnecessary for or toxic to fish.

Much of the animal life in coastal waters gets nourishment from the wetlands detritus consisting of plant particles and colonies of microscopic organisms. These are consumed by a variety of estuarine species such as oysters, shrimp, fish and zooplankton. Plants are nourished by nutrients, nitrogen and phosphorus compounds dissolved in water. Any reduction of the natural nutrient supply tends to alter the coastal ecosystem.

2.3.2 General Characteristics of Source Water

In the most basic classification water is separated into surface and ground waters, which are then subdivided into their various sources. The other very basic division of water, based on its most significant chemical property of salinity, separates water as fresh water, sea or saltwater and brackish water. Salinity of natural water varies from practically zero to more than 40 ppt. Sea or saltwater has a salinity of 33 to 37 ppt, with the average being about 35 ppt, and fresh water has essentially zero salinity. In estuaries seawater mixes with fresh river water and gradually increases the salinity of fresh water. This mixed water of intermediate salinity is called brackish water. For aquaculture industry the classification of source water based on these basic divisions is in vogue since long. Therefore, general characteristics of source water will be discussed based on both or either of these divisions.

An aquaculture farm takes water from ground water and/or surface water depending on its location. These source waters possess certain chemical, physical and biological characteristics suitable for animal growth. Each source water has advantages and disadvantages, and limitations for its use in aquaculture purpose. So the discussion of these characteristics with respect to the intrinsic quality of the various source waters will be pertinent. The quality characteristics can be categorized as water temperature and circulation, dissolved oxygen content, pH and carbon dioxide, nutrients and organic substances, and plant and animal community.

1. Water Temperatures and Circulation

The average temperature of a water body is determined by the heat exchange with its surrounding and water circulation pattern. These two phenomena vary considerably from a small and shallow pond through lakes and reservoirs to brackish water and seawater systems. In freshwater system there is variation between small and shallow ponds, lakes and reservoirs, and flowing water bodies. Strong thermocline rarely exists in shallow ponds for longer periods, where overturn due to seasonal temperature variation and surface winds creates circulation.

Water bodies with well-defined basins, *e.g.* lakes show a property of thermal stratification. This occurs when the thermal absorption is more rapid at the surface of water and the warm water, being less dense than cool lower waters, remains at upper stratum. In deep water bodies, wind action is not powerful enough to mix the two strata, and so a thermal stratification develops, with the difference in temperature being even several degrees Celsius. Three distinct zones of thermal stratification develop in lakes: (*a*) *epilimnion* or surface zone, a small portion of the total depth and relatively warm, (*b*) thermocline or *metalimnion*, with rapidly dropping temperature with increasing depth, and (*c*) *hypolimnion*, from below the thermocline to the bottom of the lake, with slowly decreasing temperature.

In flowing water systems of streams and rivers upstream history such as water source, season and geographical location, flow velocity, and volume and physical features of the stream channel decide temperature.

Saltwater systems of oceans under the influence of horizontal and vertical circulation, and high salt content create conditions highly different from other water bodies. Major currents owing to horizontal circulations create conditions for highly productive fisheries. The vertical circulations of the oceans are responsible for maintaining thermoclines deep and over a great depth in temperate and tropical oceans. However, thermocline extends over a much greater depth in the temperate than in the tropical oceans due to more gradual temperature decrease beginning at a shallower depth. In the oceans below the thermocline environment is considerably stable, and temperature is maintained within a few degrees over a vast area of the ocean.

Estuarine circulation pattern is influenced by the extremely complex and wide variation of their thermal characteristics. The two major factors are density of fresh and salt water mix, and tidal forces. Since water density is dependent on both temperature and salinity, brackish water shows maximum density at about 25 ppt and its freezing point. Therefore, estuarine water shows many characteristics similar to fresh water below about 25 ppt, and similar to salt water above about 25 ppt. Depending on the tidal velocity, river flow and physical dimensions of the estuary circulation pattern due to mixing of fresh and saline waters varies considerably (Figure 2.7). Where river flow is large compared to tidal flow, and where the width/depth ratio of the estuary is comparatively small, *highly stratified* or *salt wedge estuary* is formed. Seawater moves into the riverbed in the form of a salt wedge, over which less dense river water flows out in a gradually decreasing layer. Considerable mixing takes place at the interface of salt wedge and fresh water. Since mixing is not complete, the partially mixed water flows out over salt water. Mixing becomes more complete down the estuary. Limited mixing across the depth results in a marked discontinuity in the salinity profile or halocline.

Increased tidal flow compared to river flow increases vertical mixing breaking down the salt wedge and establishes a *moderately stratified* or *partially mixed estuary*. In these estuaries salt water moves upstream in the lower layers and less saline water moves out in the surface layers. When tidal flow increases more than in the moderately stratified estuary, all vertical stratification breaks down and a *vertically homogeneous estuary* results with no halocline, *i.e.* salinity is constant with the depth. Coriolis effect becomes more pronounced with increased flow velocity and deflects the water to one

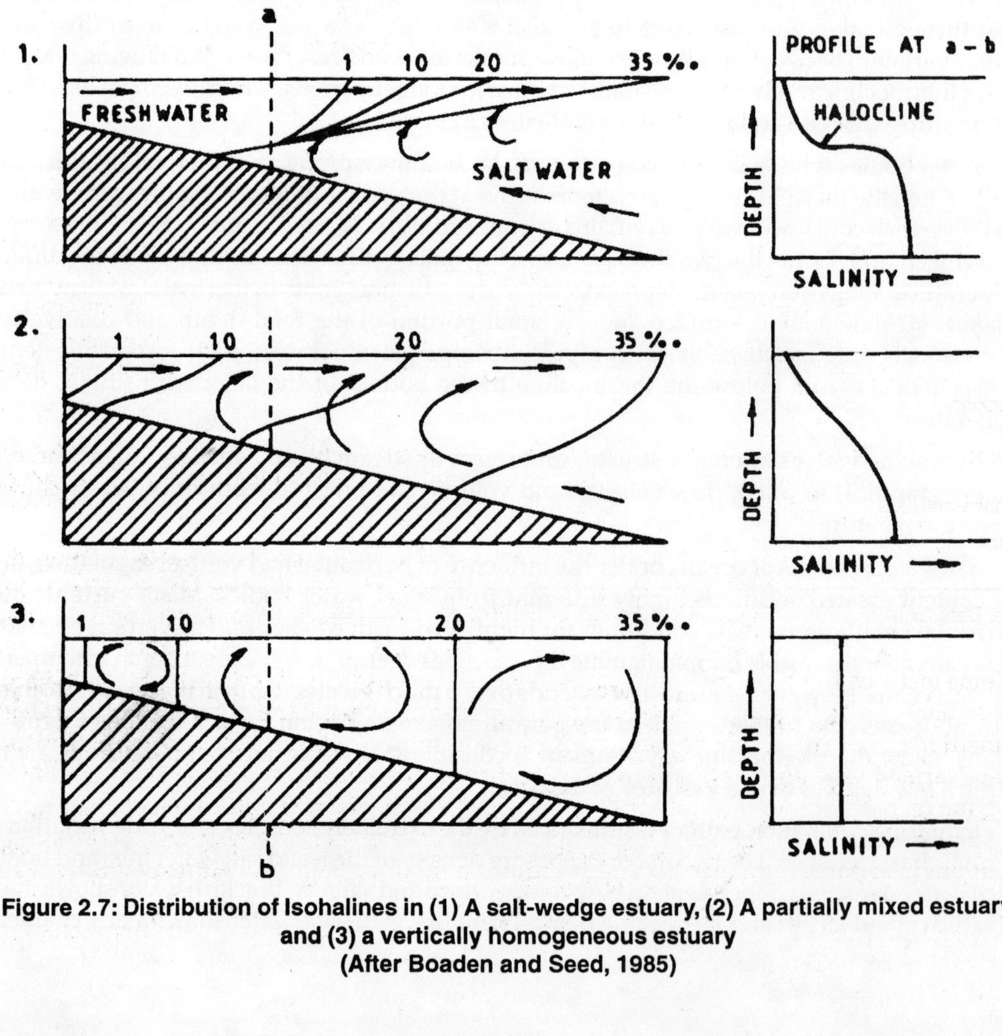

Figure 2.7: Distribution of Isohalines in (1) A salt-wedge estuary, (2) A partially mixed estuary, and (3) a vertically homogeneous estuary (After Boaden and Seed, 1985)

side of the estuary. An estuary may thus exhibit markedly different salinity levels at opposite banks and river water–seawater interface may become almost vertical. For detail information on water circulation in various water bodies the reader may refer Wheaton (1977).

Table 2.1 shows surface water temperature and salinity data of some Indian estuaries, lakes and brackish water ponds in the coastal belt. Temperature variations were recorded between pre-monsoon and post-monsoon periods or between April and December and salinity values were recorded according to seasonal variations (Jhingran, 1991).

Table 2.1: Surface Temperature and Salinity Variation of Estuaries, Lakes and Brackish Water Ponds in India

Water Body		*Temperature Range (°C)*	*Salinity (ppt)*	*Remark*
Estuary:	Hoogly-Matlah	19–32.5	1.5–32.8 4.5–31.6	Lower zones
		20–31	Trace to 22.2	Upper zones Post-Farakka period
	Mahanadi	19.4–34	Traces to 35.4	
	Godavari	29–35	Traces to 34	
	Adyar	25–32	19–31.5	
	Zuari-Mandovi	26–31	13.5–35	
Lake:	Chilka	17.5–32	0.13–36	
	Pulicat	24.6–30.6 26.1–31.7 28.6–35.0	11.2 – 50.5	Monsoon period Post-monsoon period Pre-monsoon period
	Kerala Backwaters	28.0–31 25.4–31.2	1.4–33.5 –	 At 9 m depth
	Mangalore Backwaters/1	27.2–35.1	0.36–26.43	Post-monsoon to pre-monsoon period
Brackish water ponds:	Bheri in West Bengal: 24-Parganas (N) Lower Sundarbans	 14.4–35 22.2–36.2	 3.5–34.3 2.2–22.4	
Canal water/2	Digha	2.0–36	1.0–25.0	
Creek water/3	Kandaleru, Andhra Pradesh	28.0–30.8	10–18 About 25	Dry season Other periods

Source: Data extracted from Jhingran, 1991. 1. Naik *et al.*, 2005; 2. Maity, 1990; 3. Joseph *et al.*, 2003.

Some features of estuarine circulation in the coastal waters of India can be highlighted here. The major estuaries in India are connected to the Ganga in West Bengal (Hooghly–Matlah estuary); the Mahanadi in Orissa; the Godavari and the Krishna in Andhra Pradesh; the Kaveri in Tamil Nadu; and the Narmada and the Tapti in Gujarat. All estuaries become practically fresh water dominated during the monsoon months of July to October because of complete nullification of flood tides and strong predominance of ebb tides. During the post–monsoon months the estuaries first exhibit high fluctuations in salinities and then become more or less marine dominated due to cessation of almost all fresh water discharge and dominance of flood tide.

The Hooghly–Matlah estuary (Figure 2.6) occupies the marshy deltaic area known as the Sundarbans. The Sundarbans at the Ganga–Brahmaputra delta is the largest single mangrove forest of the world and transcend the boundaries of India and Bangladesh. The main Hooghly estuary is a positive estuary of the mixohaline type, with a triangular wide mouth and great overall circulation. Salinity does not show any vertical stratification. After construction of Farakka barrage, the one-time salinity gradient zone has become virtually a fresh water zone due to flushing of Hooghly with the bypassed fresh water from the Ganga. Longitudinal temperature gradient does not essentially exist from the fresh water to the marine zone, but seasonal temperature fluctuations are observed.

An example of mixing and circulation in the Gautami–Godavari estuary is shown in Figure 2.8. Here tidal influence extends up to about 45 km from the mouth of the river. During the monsoon

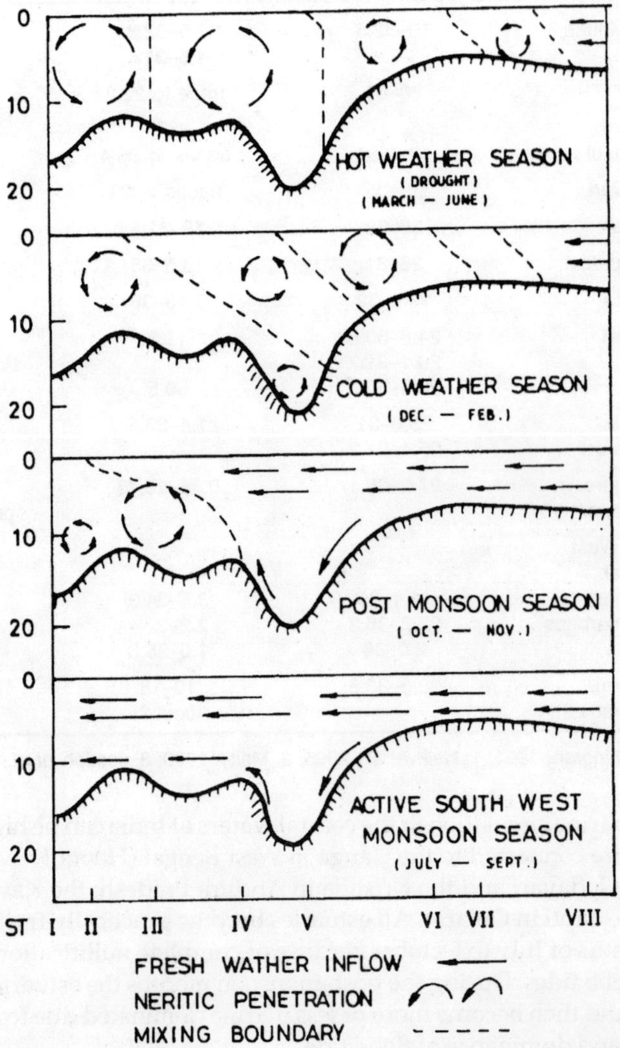

Figure 2.8: Mixing and Circulation Pattern in the Gautami-Godavari Estuary (After Ganapati and Sharma, 1965)

seasons a steep halocline preventing vertical mixing exists. From October neritic penetration predominates with diminution of fresh water flow. Vertical salinity gradient progresses in the direction of the estuary. The neritic domination prevails during the hot weather season. Intense vertical mixing exists, which however gradually reduces to very low values in the upper reaches.

On some major brackish water lakes and ponds, temperature and salinity variations can be exemplified with Chilka and Pulicat lakes, Vembanad lake in Kerala backwaters, Mangalore backwaters, Kandaleru creek in Andhra Pradesh, and ponds at lower Sundarbans. Chilka lake is a brackish water lagoon situated on the Bay of Bengal coast of India in Orissa. The lake receives several branches of Mahanadi river system. Being connected by a long outer channel with the Bay of Bengal through a single mouth, the lake contains a number of rocky islands, and a network of swamps and water ways (Figure 2.9). The important physico-chemical features are salinity and temperature variations shown in Table 2.1. Salinity fluctuates between November–December to April–June, but differently in different sectors of the lake. The extreme annual cyclic changes of salinity are between 0. 13 to 36.0 ppt.

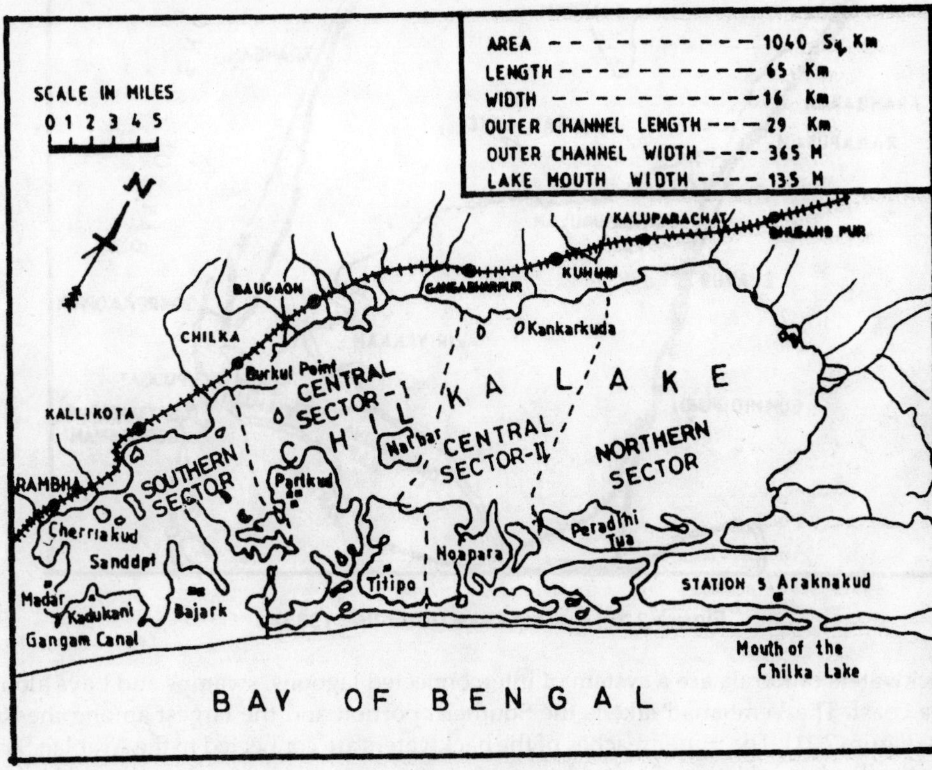

Figure 2.9: Chilka Lake
(After Jhingran, 1991)

The Pulicat lake lies in the two adjoining districts of Andhra Pradesh and Tamil Nadu (Figure 2.10). Kandaleru creek is a flood drain creek which has brackish water almost throughout the year except during the wet season. Water temperature fluctuations between pre–monsoon, monsoon and post–monsoon periods are shown in Table 2.1. Table also shows salinity variations from hyper saline conditions during pre–monsoon season to almost fresh water conditions during monsoons.

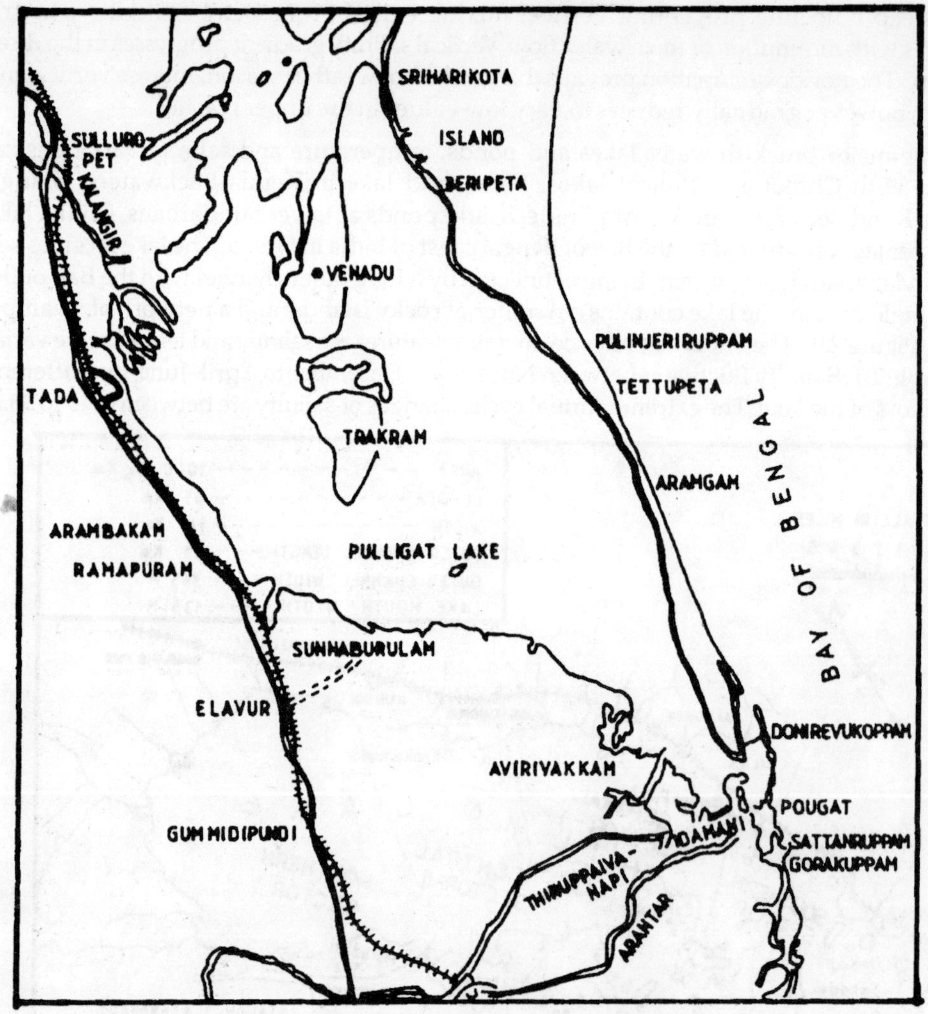

Figure 2.10: Pulicat Lake (After IIM, 1983)

The backwaters of Kerala are a system of interconnected lagoons, swamps and bays along the Arabian Sea coast. The Vembanad lake is the Southern portion and the largest among the Kerala backwaters (Figure 2.11). The upper reaches of the backwaters are connected to the Arabian Sea and flushed with the flood and ebb tides. The fresh water from two rivers and several rivulets and irrigation channels, flows through. The lower reaches are not connected with any river but gets water from its catchments area during monsoon. The hydrography during the pre–monsoon, monsoon and post-monsoon periods shows seasonal variation of temperature, salinity and related parameters. In the backwaters the surface water temperature ranges from 28 to 31°C with fluctuation of 3–4°C. Thermal stratification is absent in the summer months. But during monsoon months a steep thermal gradient develops in the estuary with a maximum difference of about 5° C at the surface and bottom. The temperature difference is due to fresh water discharge and inflow of seawater into the deeper layers of the backwaters. The surface salinity fluctuates between 1.4 and 33.5 ppt during the year. The salinity

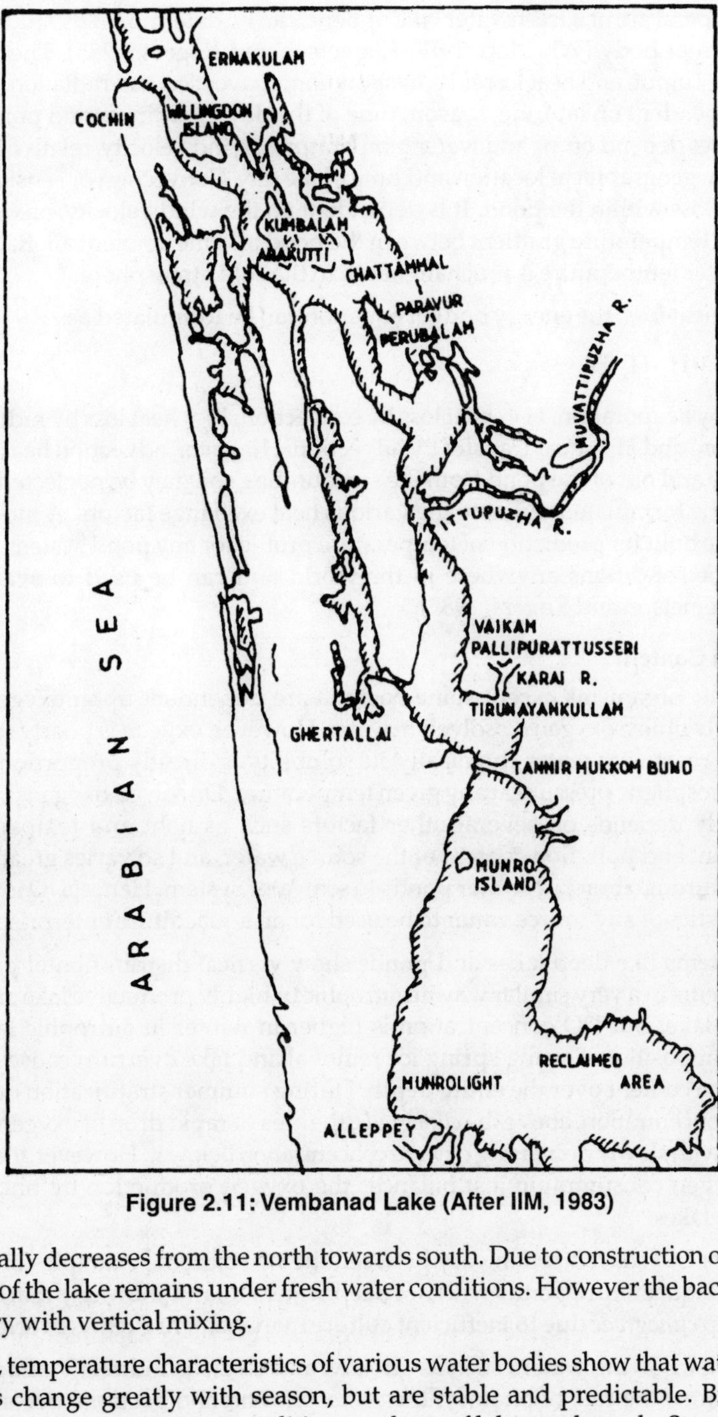

Figure 2.11: Vembanad Lake (After IIM, 1983)

of the lake gradually decreases from the north towards south. Due to construction of a bund across the lake, a large part of the lake remains under fresh water conditions. However the backwaters are a two–layer flow estuary with vertical mixing.

In summary, temperature characteristics of various water bodies show that water temperatures in lakes and ponds change greatly with season, but are stable and predictable. Brackish water and seawater have the same temperature conditions as those of lakes and ponds. Streams and river have normally wider seasonal and diurnal temperature fluctuations than any other water source.

The average temperature of a freshwater lake or pond can be determined by studying the energy relationship of the water body (Wheaton, 1977; Klemetson and Rogers, 1985). The various energy factors are solar energy input, and heat losses by evaporation, convection and radiation. The magnitude of solar energy is dependent on latitude, season, time of the day, cloudiness and pond surface area. Evaporative heat losses depend on air and water temperatures, wind velocity, relative humidity of the air, pond surface area, geographical location and time of the day. Convection or sensible heat loss is a major source of heat loss within the pond. It is dependent on the wind velocity, mixing of the water within the pond, and temperature gradient between the pond and the ambient air. Radiation loss can occur, if the pond water temperature is much higher than the cold atmosphere.

If H_{loss} is the net heat loss, the energy budget equation can be formulated as:

$$H_{loss} = H_e + H_c + H_r - H_s - H_a \quad (2.1)$$

where, H_e = heat loss by evaporation, H_c = heat loss by convection, H_r = heat loss by radiation, H_s = heat gain by solar radiation, and H_a = heat transfer by advection. However, advection heat transfer due to water movement into and out of the pond from the surrounding soil may be neglected. The total heat losses and gains can be formulated utilizing the various heat exchange factors. A model comprising all these factors can be built for prediction of temperature profile for any pond system. The model can be set up for climatic conditions anywhere in the world and can be used to evaluate potential aquaculture sites (Klemetson and Rogers, 1985).

2. Dissolved Oxygen Content

Nearly all aquatic organisms except some bacteria are dependent upon oxygen for their life processes. They mostly utilize oxygen dissolved in water. However, oxygen is poorly soluble in water, and since it does not react with water chemically, its solubility is directly proportional to its partial pressure, and the atmospheric pressure at any given temperature. Dissolved oxygen (DO) content of a particular water body depends on several other factors such as light and temperature, salinity, turbulence, silt content and pollution content of the source water, and so varies greatly from ponds, lakes and reservoirs through brackish water ponds to seawater system. Hence, DO is one of the very important characteristics of any source water to be used for an aquaculture enterprise.

Fresh water systems like deep lakes and ponds show vertical distribution of oxygen. Vertical distribution of DO occurs in a very similar way in eutrophic or highly productive lake and oligotrophic or low productivity lake. But DO concentration is higher in winter in eutrophic lake due to less respiration and decomposition. During spring ice removal and lake overturn cause a constant and relatively high oxygen content over the entire depth. During summer stratification of water column causes high oxygen concentration, above the *clinograde*, the area of rapid drop of oxygen concentration, and a mild *clinograde* with slowly decreasing oxygen concentration below it. However, the compensation level, where the oxygen consumption just balances the oxygen production by photosynthesis, is deeper in oligotropic lakes.

Characteristics of DO concentration vary between the two extreme examples of eutrophic and oligotrophic waters. Aquaculture ponds with not deep waters do not show any marked *thermocline* but may show a sharp *clinograde* due to inefficient culture management such as overfeeding.

The DO content of flowing waters such as streams and rivers is indirectly affected by several factors, which determine the rate of photosynthesis, respiration and bacterial decomposition, and extent of natural aeration. The daily oxygen cycle is largely dependent on temperature fluctuations and photosynthesis-respiration factors. The annual oxygen cycle is closely dependent on the temperature. The oxygen content is highest in winter and lowest in summer in most rivers. Along the length of a stream the DO concentration in the upper reaches is affected mainly by temperature and turbulence. In the lower reaches the DO concentration is mostly controlled by photosynthesis-respiration relationship.

Seawaters have DO concentration lower than that of freshwaters due to salinity of about 35 ppt. Saturation to supersaturation conditions prevails in areas near coral reefs. In general temperature variations cause DO in tropical surface waters to reach 6-7 mg/l, and DO in polar waters to reach more than 11 mg/l. From surface water down vertical oxygen distribution occurs. Oxygen concentration drops rapidly below the photosynthetic depths to a zone known as "minimum oxygen zone" varying between 150 m to 1000 m in different oceans. The DO concentration in the "minimum oxygen zone" varies between 0.14 and 1.4 mg/l. Below this zone the DO concentration increases with depth to a fairly high value in the deep layers.

Several factors influence the oxygen concentration in estuaries. Temperature and salinity create an oxygen gradient along the estuarine length. Upstream areas have highest oxygen levels during high stream discharge and fresh water influx. DO level decreases in dry periods due to increase of salinity. Stratification and estuary depth cause marked difference between the DO concentration in the surface and deep layers. Type of freshwater source, tides and vegetation influence the diurnal and seasonal variations in DO concentration. Tides affect the oxygen concentration due to inflow of seawater.

Table 2.2 shows data on DO variation in some estuaries, and brackish water lakes and ponds in coastal belt of India. DO variation have been observed due to decrease in salinity level with increase in freshwater inflow (Hooghly-Matlah estuary), extreme annual cyclic change in salinity (Chilka lake) and depth (Kerala back waters). One observation in a brackish water pond in Lower Sunderbans shows monthly variation.

3. pH and Carbon Dioxide

Most natural waters are buffered mainly by a carbon dioxide-bicarbonate system so that they maintain a relatively constant pH in the aquatic environment. Carbon dioxide is normally present in all natural waters. Being highly soluble in water, it goes into solution producing a weak acid (carbonic acid). Mainly salts of weak acids and strong bases contribute to the alkalinity of waters, and act as buffers. In most waters bicarbonates represent the major form of alkalinity due to their formation in considerable amounts from the action of carbon dioxide upon basic materials in the soil. However carbonates, ammonia or hydroxide may make a contribution to the total alkalinity of water.

The carbon dioxide–pH relationship of freshwater lakes and stream is largely affected by soil and rock characteristics of the area. The carbon dioxide level in water is normally very low in areas predominated by igneous rocks where a carbonate substrate is lacking and insoluble rocks limit the nutrient content of lakes and streams. Water tends to be slightly acid with pH at about 6.0.

On the other hand, high carbon dioxide content with pH 7.0 and above characterizes water flowing over areas predominantly composed of sedimentary rocks *e.g.* limestone. Rocks and soils of these areas are rich in carbonates and bicarbonates and make the water high in calcium and/or magnesium, commonly known as "hard" water. Hard water lakes and streams with sedimentary rock bottom release carbonate to the water and make the pH in the range of 4-6. High rates of bacterial decomposition provide high free carbon dioxide. These waters are characterized as "soft".

Table 2.2: Dissolved Oxygen, pH and Alkalinity Variations in Some Estuaries, Lakes and Brackish Water Ponds in India

Water Body		DO (mg/l)	pH	Total Alkalinity (mg/l)	Remark
Estuary:	Hooghly-Matlah System	5.0–10.8	7.6–8.5	68–195	All variations in a year
	Hooghly			2–7	
	Adyar	3.0–4.6	7.4–8.6	0–15	As carbonate
				64.0–152.5	As bicarbonate
	Zuari–Mandovi	0.9–7.76	–	–	In Mandovi river
		1.36–5.35			In Zuari river
Lake:	Chilka	2.6–15.6	6.8–9.7	25.8–157	
	Pulicat	0.84–6	8.1–8.7	45-120	
		4.0	–	–	Pre-monsoon
		6.8	–	–	Post-monsoon
	Mangalore backwaters/2	3.2–6.8	7.1–7.6		Post–to pre-monsoon periods
	Kerala	3.5–4.97	–		DO values at:
	backwaters	2.7–5.04	–	5.88–26.25/1	surface
		2.6–5.90	–	15.92–26.82	4 m depth
				21.2–27.05	9 m depth
Brackish water ponds	Bheri in West Bengal:				
	24-Parganas (N)	–	7.5–8.3	–	All monthly variations
	Lower Sundarbans	6.0–14.0	8.0–9.2	–	
Canal water/3	Digha	–	6.5–6.7	–	
Creek water/4	Kandaleru, Andhra Pradesh	3.8–4.9	7.9–8.0	168–200	

Source: Data extracted from Jhingran, 1991.1. All alkalinity values reported as mg C/m^3. 2. Naik *et al.*, 2005; 3. Maity, 1990; 4. Joseph *et al.*, 2003

In cooler temperate regions the deep lakes show seasonal variation in carbon dioxide concentration almost inverse that of oxygen concentration. In winter, near the surface the carbon dioxide content is low with slightly alkaline condition, but in deep waters bacterial decomposition lowers the pH with increase in free carbon dioxide and decrease in oxygen concentration. In warmer temperate zones, continuous circulation keeps carbon dioxide concentrations normally constant over the entire depth. Spring circulation produces a uniform pH over the depth of lake with loss of excess free carbon dioxide to the atmosphere. As summer stratification beings, carbonate reappears in surface waters of hard water lakes. Productive lakes with summer stratification shows that surface waters contain little free carbon dioxide, but considerable carbonates, whereas free carbon dioxide decreases with depth. This condition is also in inverse order of DO concentration in these lakes. The autumn overturn and summer stratification create the same conditions as with warmer temperate zone deep lakes.

Flowing water bodies like streams do not show any stratification and carbon dioxide gradient. pH of the stream is influenced by the chemical characteristics of the watershed. However, photosynthesis, respiration and decomposition have some influence on the carbon dioxide content of streams and river waters.

In oceans colder polar waters absorb carbon dioxide more from the atmosphere at lower temperature and transport it toward tropical region, where surface waters can hold less carbon dioxide at higher salinity and so diffuse it back to the atmosphere again. In general, the buffer system of saltwater is more effective than freshwater due to presence of an excess base or alkali reserve, and thus keep the pH values in surface waters between 8.1 and 8.3. Surface water pH increases because of photosynthesis, and then decreases beyond the photic zone because of respiration and decomposition. pH in the " minimum oxygen zone" may be around 7.7, which rises to about 8 in oxygen-rich depth of oceans.

In brackish water system, each estuary shows its specific pattern for carbon dioxide, which varies with depth, tidal and freshwater flow, temperature, salinity and biological processes. In the upstream reaches dominated by freshwater carbon dioxide levels vary under the influence of mainly temperature and biological activity. In the lower reaches, salinity increase decreases carbon dioxide saturation concentration. However the more effective buffer system of seawater can cause a decrease in free carbon dioxide concentration and an increase in the bicarbonate and carbonate forms. Thus pH increases gradually along the length of the estuary and reaches about 8.1 to 8.3.

Table 2.2 shows data on pH and alkalinity variations in some estuaries, and brackish water lakes and ponds in coastal areas of India. Alkalinity of the Hooghly estuary increased gradually from the fresh water zone to 7 mg/l, which was the highest in January-February month. Alkalinity was minimum at 2 mg/l in the monsoon. In Chilka and Pullicat lakes total alkalinity was highest in summer months.

4. Nutrients and Organic Substances

All aquatic organisms in the ecosystem are supported by a food chain with vegetation that is nourished by nutrient minerals dissolved in the water, particularly compounds of nitrogen (nitrates, ammonia) and phosphorus (phosphates). These nutrients are supplied from within the ecosystem continuously through natural cycle of storage-release-restorage system. Nutrients also include trace elements called micro–nutrients. However, nutrients continuously trickle out of the system by various processes and are replaced through land runoff and other sources. Organic substances ranging from simple molecules to very complex ones are present in natural water usually in the form of dissolved materials and suspended particles. These substances are supplied mainly from incoming waters or wave action and from decomposition and secretion by living organisms.

In fresh water system nitrogen is available in inert free state and in several combined forms such as nitrates, ammonia compounds and nitrite. The nitrogen compounds come from outside through rains, surface runoff, ground water and biological processes. Commercial fertilizers, manures and sewage are other sources of nitrogen in the water bodies. Organic nitrogenous substances found in natural waters are results of metabolic and other life processes and decomposition of dead organisms. These compounds are mainly simple amino compounds, plant and animal proteins, and uric acids released from animal wastes.

Inorganic nitrogen compounds are normally found in low concentrations in unpolluted water. Ammonia is available in equilibrium with ammonium ion (NH_4^+), both concentrations being dependent on pH. Unionized NH_3 in highly toxic to moist aquatic organisms, but both NH_3 and NH_4^+ ion concentration is usually about 1mg/l or less in unpolluted water, which is below the toxicity limit. Ammonia, being a product of decomposition, varies in concentration over the depth of a lake. It is usually low in the surface layers but increases in concentration in the hypolimnion due to decomposition in summer. Fall and spring overturn causes an even distribution by mixing.

Nitrite and nitrate are both products of bacterial decomposition of ammonia and closely follow in seasonal variation in lakes. In late summer both are at or near their minimum concentration owing to the use of nitrates for all green plants. However, in early winter the ammonia and nitrate concentrations decrease but nitrate concentrations increase. At maximum nitrate concentration it varies vertically within a eutrophic lake, being strongly influenced by productivity of the lake.

Nitrogen concentration in streams and rivers depends on runoff water volume and its characteristics, drainage basin activity, stream turbulence and biological activities in the stream. Drainage basin activities may include farming in the natural land cover, which results in fertilizer runoff or sewage disposal from large cities.

Unlike nitrogen, phosphorus occurs in a number of forms such as soluble inorganic and organic phosphorus, particulate phosphorus, and as metal compounds of calcium, magnesium, sodium, and aluminium. Further the phosphorus chemistry in water is highly variable and complex. Concentration of phosphorus in freshwater systems depends primarily on several factors: (1) basin characteristics like volume, stratification and circulation; (2) chemical characteristics of the watershed; (3) biological processes within the system; (4) pH, oxygen content and water temperature; (5) concentration of metal ions; (6) loss of phosphorus in the sediments; and (7) seasonal variation. In eutrophic lakes with summer stratification, phosphorus concentration in the hypolimnion increases greatly following oxygen depletion, probably due to formation of ferric phosphorus complex with oxygen. Sulphur is present in significant quantities in some lakes, normally as sulphate (SO_4^{2-}) and hydrogen sulphide (H_2S), the primary sources being rain and sulphur–containing geological formations. Silicon is found in all natural waters in the low concentrations. Silica (SiO_2) is very important for diatoms and some sponges.

Iron is present in natural waters as both ferrous and ferric forms. In absence of oxygen ferric iron is reduced to ferrous iron. Oxidation of ferrous iron present as dissolved form in water results in formation of ferric hydroxide, $Fe(OH)_3$, which can be taken up easily by phytoplankton. Calcium and magnesium ions in carbonate forms are the most predominant in freshwater. Their concentration primary depends on chemical composition of the watershed, rock and soil. Availability of calcium and magnesium is influenced by photosynthesis, respiration and decomposition processes through the CO_2 buffer system. Sodium and potassium are of limnological importance, but are present in low concentrations. Sodium occurs frequently as Na^+ ion, but also as borate and sulphate. Potassium occurs as K^+ in natural waters.

There are many trace elements available as dissolved or suspended in low and undetectable concentration. However, there are a group of elements, commonly present in surface waters, in easily detectable form, such as manganese (Mn), aluminium (Al), zinc (Zn), copper (Cu), chromium (Cr), titanium (Ti), molybdenum (Mo), iodine (I), lead (Pb), barium (Ba), bromine (Br) and stronsium (Sr).

In saltwater systems phosphorus, nitrogen and silica are the most important and most limiting nutrients. The concentrations of these nutrients as phosphates; ammonia, nitrate and nitrite; and silica vary with depths in most oceans. The deep areas are the large reservoirs of the nutrients, which make the areas of upwelling highly productive. The surface layers contain low and uniformly distributed nutrients, and in between the nutrients vary in increasing concentrations. Nitrate is found in all depths of oceans in greater concentrations than other compounds. Ammonia concentrations are high in the ocean surface waters, varying from about 0.025 mg/l during fall and winter to almost nil during spring. Nitrite concentrations are very low even in the surface layers.

Vertical distribution of phosphate in the various oceans shows that phosphorus is distributed throughout the depth. The phosphate distribution is similar to that of nitrate, but its turnover rate is faster than that of nitrate. Very little organic phosphate is found below 300 m due to quite rapid phosphorus turnover rate in the surface layers. Silica concentrations vary largely in the oceans and its vertical distribution is somewhat different from that of phosphate and nitrate.

Iron and manganese are found in oceans in very low concentrations. Potassium, calcium, magnesium that are necessary for plant growth are abundant in the oceans. A large number of elements is available in seawater in variable concentrations, but their necessity to plants is not understood.

Table 2.3 gives some data of nutrient variations in estuaries and brackish water lakes in India. Although the range of values of phosphorus is very low in the Hooghly-Matlah estuary system, phosphorus concentration in Hooghly estuary showed maximum (above 4 mg/l) during summer months and traces during monsoon months. Silicate values showed a decreasing trend from the fresh water zone to the marine zone and about 500 mg/l during January and February. In Chilka and Pullicat lakes higher values of the nutrients also occurred during monsoon months.

Table 2.3: Variations of Some Nutrients in Estuaries and Brackish Water Lakes in India

Water Body		NO_3-N (mg/l)	Phosphate (mg/l)	Silicate (mg/l)	Remark
Estuary:	Hooghly-Matlah system	Trace–0.22	Trace–0.61	3–27	
	Adyar	–	–	40–60	
			83–304 (inorganic)	1.6–14.8	
Lake:	Chilka	Trace–0.19	Trace–0.18	0.1–6.0 (as SiO_2)	
	Pulicat/1	–	0.10–0.90 (inorganic)	1.0–24.4	Average of lake mouth, southern and northern sectors
	Mangalore backwater/3	2.32–9.76	0.36–1.64	29.8–114.8	
	Kerala backwater/2	0.811–39.97 1.349–39.46	0.363–1.268 0.531–2.108	6.42–59.71 4.79–35.73	At surface At 9 m depth (Average of monthly variations)
	Kandaleru creek/4	0.16–0.24	0.04–0.08		

Source: Data extracted from Jhingran, 1991. 1. As mg P/l; 2. As µg/l; 3. Naik *et al.*, 2005; 4. Joseph *et al.*, 2003..

5. Plant and Animal Community

Neritic or Coastal Waters

In the marine environment two major categories of pelagic organisms are plankton and nekton. Planktonic organisms include phytoplankton (plants), zooplankton (animals) and bacterioplankton (bacteria). Classification of plankton is usually based on size and life-history characteristics. Nektons are the larger pelagic organisms such as fish, cephalopods and mammals. In general, planktons, having limited powers of locomotion, are drifted by the water currents. Nektons can counteract water current and swim.

The two main groups of phytoplankton are diatoms (non-motile unicellular algae) and flagellates (motile form). Some forms of flagellates produce toxins, which when transmitted through molluscs or fish, cause paralytic shellfish poisoning. The phytoplankton is of relatively limited taxonomic range, whereas the zooplankton includes representatives of almost all the animal phyla either as adults or as eggs and larvae. Copepod crustaceans are an important group of zooplankton. Pelagic bacteria form a highly important source of primary production in pelagic water.

The neritic and oceanic waters are populated by different plankton communities. The differences in water circulation and sedimentation patterns and effects of solar radiation play a major role. Phytoplankton are largely restricted to euphotic zone because of availability of dissolved carbon dioxide and major nutrients like nitrogen, phosphorus and silicon in this zone. But the nutrients are temporarily depleted as photosynthesis is a surface phenomenon. A turbulent vertical mixing or upwelling due to wind and other forces restores the nutrients continuously to the euphotic zone, and thus makes the coastal waters highly productive.

There are geographical as well as seasonal variations of planktonic growth. In temperate waters marked variations in the abundance of plankton occur due to variations in light and nutrient levels in the euphotic zone. The phytoplankton growth is essentially prevented by poor light conditions and by vertical mixing of plant cells to greater depths in water, but is effectively maintained within the euphotic zone with increase in solar radiation in the spring. During summer phytoplankton growth is restricted by the grazing zooplanktons due to severe shortage of nutrients in the surface, and cells continually sink out of the euphotic zone. In the tropical waters phytoplankton growth is inhibited by the low nutrient content of surface waters in spite of well illumination and thermal stratification throughout the year. In addition to the seasonal changes, considerable changes in species composition of phytoplankton also occur. This change is most pronounced in areas with marked temperature changes, but certain changes also occur in polar and tropical waters. Zooplankton communities generally grow in abundance after growth of phytoplankton, and this herbivorous grazing causes decline in phytoplankton bloom. Copepods are the most important grazers and contribute to 70–90 percent of the herbivores. Grazing effectively varies seasonally as well as spatially. In coastal waters with large variation of environment conditions grazing does not occur for a prolonged period, whereas it appears to be very predominant throughout the year in more open water conditions. In the tropics grazing continually limits the phytoplankton growth to a low level throughout the year.

The distribution of plankton is often characterized by patchiness, which is more pronounced in coastal waters due to variable conditions of temperature, salinity and grazing. Patchiness occurs within the water column. Many zooplankton organisms show diurnal vertical migration, at times more than 100 m. During the day time they occur deep in the water column, but at dusk come up to the surface to feed on the phytoplankton.

In the neritic or coastal waters zone food chain is associated through five trophic levels in the oceanic regions, three trophic levels in the shelf and one or two trophic levels in the upwelling regions. The length of the food chains varies with difference in size of the dominant phytoplankton, the primary producers and the nektonic consumers. Since a considerable amount of energy (about 90 percent) is always lost between successive trophic levels, shorter chains are preferred in the naturally unstable areas like the coastal waters.

Life in the sea is measured in term of primary production, which is mostly through the phytoplankton in the euphotic zone. Below this zone bacteria are the most important primary producers. In the inshore areas algae, mangroves and sea grasses are the primary producers. A global primary productivity varies between 25 and 200 g C/m^2/year, with a value of 50 g C/m^2/year as the average. In the open seas it is generally low, but in the regions of upwelling it is very high (300 g C/m^2/year). The shallow coastal waters are usually nutrient–rich and highly productive, but estuaries yield less productivity due to high turbidity and reduced depth of euphotic zone.

Rocky Shores

Rocky shores exhibit the most diverse plant and animal life, and are most densely populated with microorganisms. The organisms are exposed to extreme physical conditions such as strong water current, high temperature, high light intensity and dessication. Therefore, most species show one or more adaptive features. The substratum of rocky shores is the chief factor determining the distribution and adaptive characteristics of the organisms. Strong attachment devices like the algal holdfasts, cementation by oysters and barnacles, devices for boring into the rock surface, use of rock crevices, and formation of dense aggregates are the examples of adaptive factures of the organisms to resist strong water movement.

Barnacles with tightly fitting operculum plates and a non-porous shell, snails with a horny or calcareous operculum, anemones ejecting mucus and limpets returning to the rock contours are the examples of organisms with adaptation to prevent water loss. Some tropical molluscs having pale, heat reflecting shells; some other shells with ridges, spines and the like as heat radiating surfaces; intertidal barnacles and limpets evaporating water stored excess in mantle cavity; and many crabs and snails migrating to take shelter in crevices or weed beds during low tide are the organisms counteracting high temperature. Retreat of photosynthetic pigments from the cell surface in some algae, masking substances protecting chlorophyll against photolysis and bright pigments in many high shore lichens are some examples to protect from damaging effects of high light intensity.

Many rocky-shore organisms show feeding adaptation. Suspension feeders utilize gills (bivalves), setae (barnacles), tentacular crowns (tube worms) to collect particulate materials. Some gastropods and chitons use a file like radula, and sea urchins use a complex series of teeth to scrap the rock. Many anti-predator devices like spines of urchins, thickened shells with narrow openings of gastropods and calcified tissues of some algae are structural adaptation of these herbivores prevalent in the tropics.

The intertidal zone, being biologically defined, is inhabited mostly by marine organisms and a few terrestrial species that can tolerate a short period of tidal submersion. In vertical zonation each species is typically most abundant in a zone of favorable condition. The zonation scheme shows the limits or lines above or below an optimal zone of abundance (Figure 2.4). Few species such as small snails and blue green prokaryotes live in the littoral fringe, which is relatively arid. Barnacles, mussels and some oysters live in the eulittoral zone. The sublittoral zone, exposed during spring tides is typically rich with flora and fauna. Major organisms are kelps in all regions, corals in tropical and turfs of red algae. Net primary productivity in kelp beds is ground 1000 g C / m^2/year, which is comparable to highly productive tropical coral reefs. Kelp fronds are rich source of organic matter, both particulate and dissolved, which enters the detrital food chain. The most sea urchins and some gastropods are the herbivores in kelp beds. A diverse population of suspension feeders like sponges, mussels, and predators like starfish, lobsters and fish constitute the trophic structure.

Sandy and Muddy Shores

As the two types of shores are the two extremities of a continuous spectrum, the flora and fauna change according to their typical habitat in these shores. In both shores dominant macro fauna encountered are polychaete worms, bivalve molluscs and crustaceans. In sandy shores large plants and animal like barnacles and mussels are absent. However, mud flats are unique in coastal environment by occurrence of two separate layers–the surface layer with photosynthetic green plants and a deeper layer inhabited by chemosynthetic autotrophic bacteria. Diatoms, micro algae and sea grasses grow on the surface layer of mud flats. Bacteria in the deeper layers are mostly sulphur reducing type.

Zonation of Stephenson's scheme occurs in most sandy beaches. Scavenger excavation burrows like amphipod crustaceans in the temperate zone and fast moving ghost crabs in the tropical zone are the inhabitants in the supralittoral fringe of sandy shores. The broader midlittoral zone is inhabited by variable groups of isopods, crabs and similar animals. The lower most infralittoral fringe contains the highest number of organisms like large surf clams, sand dollars, polychaete worms, crustaceans and large carnivorous snails. On the contrary, the muddy shores do not show any well-defined zonation. Crabs, clams and polychaete worms are the inhabitants in mudflats.

In sandy shores there is virtually no primary productivity as large multicellular plants are lacking. Hence macroscopic herbivores do not grow in sandy shores. Most organisms are either suspension or detritus feeders. The trophic structure of mudflats is built up from a detritus bacteria base of deposit and suspension feeders, and the other one based on microscopic diatoms. The productivity of mudflats increases through huge population of bacteria in the sediments and macro fauna in the surface. However there are only few herbivores. The carnivores of muddy shores are the fish and the birds.

Estuarine Environment

Estuaries are relatively poor in diversity and richness of species in comparison to the adjacent marine and freshwater environments. Some groups like the cephalopods and echinoderms are completely absent. Fluctuating and stressful condition, lack of food and relatively recent origin of estuaries contribute to the rather poor diversity of the estuarine biota. The benthic organisms move further ashore than the pelagic forms due to the stratified nature of estuaries. More or less fully saline water in the mid-upper shore at high tide and freshwater flow down the shore line at low tide enable the marine forms to reach a higher intertidal zone. However, the interstitial water of less variable conditions than overlying water mass causes a buffering influence to the infaunal species.

Estuarine water encounters the three major faunal groups of marine, freshwater and brackish water (Figure 2.12). The marine group is the largest among these, with the steno–haline species being restricted to the lower reach of narrow salinity conditions of the estuary. The euryhaline species tolerating much wider salinity fluctuations occur further up the estuary. The small group of brackish water species occurs in the middle reaches with highest salinity fluctuation. Areas with 5–8 ppt salinity appear to represent a critical salinity range with minimum species richness. Freshwater species rarely tolerate salinities over 5 ppt at the head of the estuary. The polychaetes, bivalves and crustaceans dominate the cooler water, and gastropods and crabs in tropical and subtropical waters. Migratory species including fish, birds and decapod crustaceans use the estuaries as a feeding and / or nursery grounds. Numerous others pass through estuaries *en route* to breeding grounds in rivers or sea.

Estuarine flora is also limited. Muddy sediments are unsuitable for the attachment of macro-algae, and high turbidity with poor light conditions restricts the growth of phytoplankton. However, abundant rooted plants form an important component of micro–flora. High organic content also contributes to the rich growth of microorganisms and meiofaunal species.

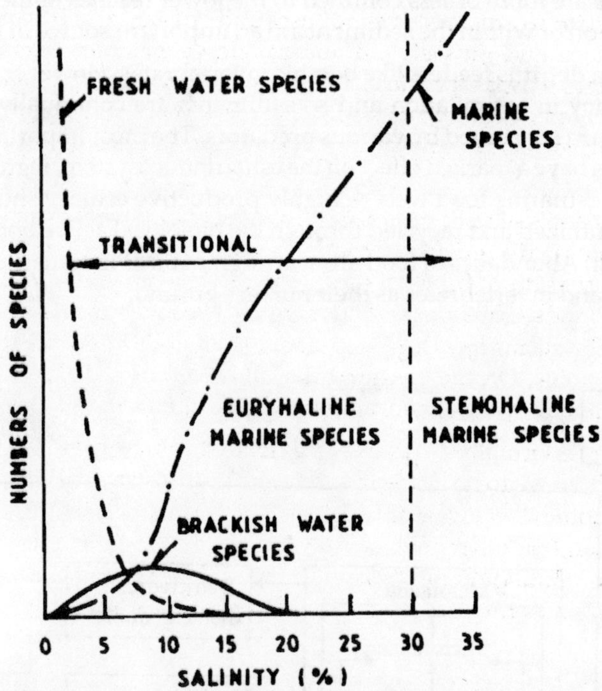

Figure 2.12: Number of Species in Three Major Faunal Groups in Estuaries (After Bose *et al.*, 1991)

Many marine species exhibit varying degrees of osmoregulation and ionic regulation in order to adapt very quick change in salinity gradients present in the estuaries. Many higher crustaceans are strong osmoregulators. Most bivalves are osmoconformers, as they are unable to regulate either their cell volume or its ionic composition. They close their shell in case the salinity falls far below the tolerance level. Some organisms burrow into the sediment, or migrate to avoid wider salinity changes. For example, many crustaceans like blue crab migrate for breeding. Conversely, species like mullet and flounder use estuaries as their breeding grounds. Some adaptive features occur in order to avoid loss of a large number of planktonic larvae. Examples are spionid polychaetes with ability to swim in the water column in the flood tide and to remain at or near the bottom at ebb time; benthic nature of some larvae; direct hatching of eggs of sand prawn into miniature adult, etc. Some morphological and anatomical peculiarities develop in the typical estuarine organisms, *e.g.* partially sterile nature of freshwater worms that inhabit brackish water and lower rates of reproduction of the original marine forms.

Estuaries are highly productive, with phytoplankton, benthic algae and rooted plants all contributing to primary production. Although phytoplankton are at a disadvantage in many estuaries due to high turbidity, net transport of surface water to the open sea and competition with macrophytes for nutrients, they grow substantially in larger estuaries with restricted interchange with open sea. Usually zooplanktons are not very efficient grazers of phytoplankton and about 50 percent of the net phytoplankton production may be available to benthic forms as living cells or as detritus. Similarly the large proportion of the primary production as rooted plants is mostly utilized through the detritus pathway. The seaweeds are more or less confined to the lower reaches of the estuaries having high salinity. Diatoms living on or within the sediment are an important source of flood.

Estuaries are rich in detritus feeders like benthic invertebrates. However, many detritus feeders are having low efficiency in assimilation and so sediments are continually being reworked. The detritus feeders are in turn consumed by various predators. The most important groups of predators like crabs, fish, and birds have a marked effect on the estuarine ecosystem. Figure 2.13 shows the major interaction in a typical estuarine food web. A highly productive estuary shows how efficiently the available nutrients are utilized and recycled through the biological active bottom sediment with the help of water movement. Abundance of food attracts migrant birds in estuaries as feeding grounds as well as varieties of fish and invertebrates as their nursery ground.

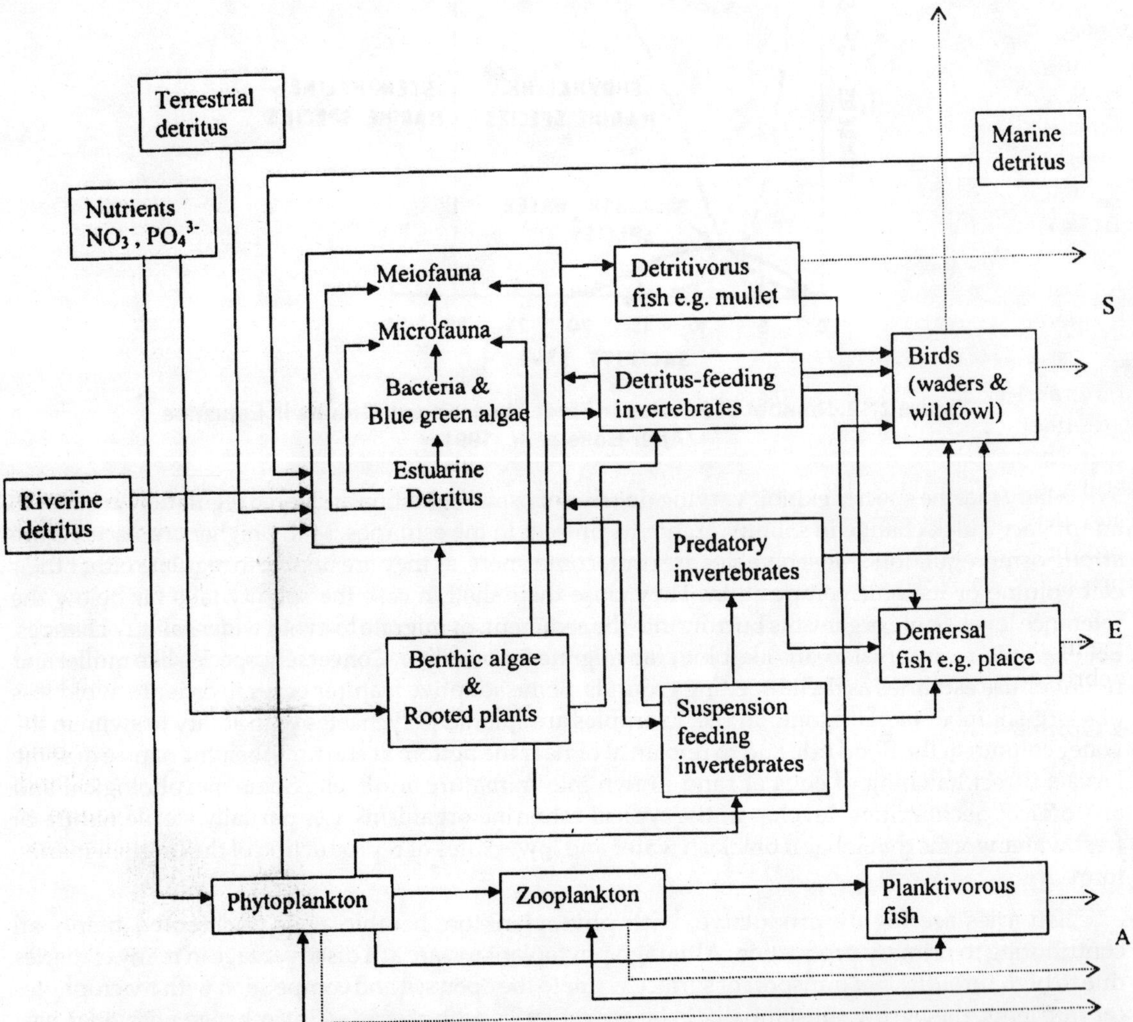

Figure 2.13: Schematic Representation of a Typical Estuarine Food Web (After Barnes, 1984) Dotted lines indicate 'losses' from the estuarine system.

Marshes and Mangroves

Terrestrial rooted plants are originally the dominant organisms in salt marshes, such as grasses of several genera. Areas with finer sand particles admixtured with silt favor growth of protists, prokaryotes and filamentous algae in the surface. Both the sediment and the overlying water are highly populated by bacteria. Marine animals found in salt marshes are various crabs, mussels, some snails and smaller crustaceans like the amphipods. Terrestrial animals like insects are permanent inhabitants. Many animals like duck and geese species, rats, moles, rabbits and hares also graze during low tide on the salt marshes and effect on the vegetation. At high tide marine and estuarine organisms inhabit the area.

A basic zonation pattern of species occurs in salt marshes that extends from the tidal creek to the terrestrial vegetation. The creek bank is the lowest zone, without any macrophytic vegetation, but containing mainly estuarine and marine animals and many snails. A border zone between the creek bank and the main marsh land comprises the creek side marsh. Next comes the main part of the marsh, which can be further subdivided to grasses of different genera. The primary production due to emergent marsh plants and various micro-algae is higher than in estuarine waters and mud flats. The major primary producer is the different species of *Spartina*. Average annual net productivity of marshes in American east-coast and Europe has been shown to be within 200-400 g $C/m^2/y$. Most of this is not directly consumed but utilized through detritus pathway, which ultimately forms the basic food chain for the different marsh animals. The high nutrient content of salt marshes makes them effective nursery grounds for many crustaceans and fish.

The dense growth of evergreen mangroves shelters a variety of terrestrial animals like insects, birds and arboreal mammals. Their complex root systems and thick silty sediments of the substratum are suitable habitats for many marine invertebrates dominated by various mollusks, crustaceans and fish. The mangrove snail feeds on the mud surface during low tide. Tropical land crabs, fiddler crabs and ghost crabs can strain organic detritus from the mud through their mouth parts. Seeds of mullets and penaeid shrimps also are carried by the tides into mangrove areas. Mangroves are highly productive with a net primary production of 350 to 500 g $C/m^2/y$. Plant debris contribute to a major proportion of this production. Microorganisms rework this debris and provide food for the detritivorous organisms. Snails and higher carnivores enter the food chain at a later stage. The overall food web in the mangroves is controlled by tidal flushing, riverine input, accumulation and decomposition of debris and faunal behavior.

6. Ground Water Characteristics

Ground water can be characterized by a number of factors:

1. The nature of rain water, which can vary considerably in terms of acidity due to pollution, and effect of wind–blown spray from the sea in the coastal areas;
2. The nature of ground water already existing since time immemorial;
3. The nature of the soil through which water penetrates; and
4. The nature of the rock comprising the aquifer.

All these factors influence the ground water quality. However, avoiding the detail discussion on the influence of each factor, it can be said that the quality required of a ground water supply depends on its purpose: for drinking, industrial and irrigation. For use as aquaculture water source the quality characteristics already set would be followed.

Ground water temperatures tend to remain relatively constant throughout the year. Only shallow ground water displays any appreciable fluctuation in temperature due to variation in solar energy received at the earth's surface. Therefore the water temperature in shallow ground approximates the mean temperature of the area.

The most ground waters contain dissolved gases derived from natural sources. The atmospheric gases carbon dioxide, oxygen and nitrogen come through the normal geochemical cycle of ground water. The other gases such as methane and hydrogen sulphide are derived from underground biochemical processes. These gases are the most common undesirable components of ground water. All ground water contains salts in solution. The type and concentration of salts depends on the environment, movement and source of the ground water. Usually, ground water contains higher concentration of dissolved constituents than the surface water because of the greater exposure to soluble salts in geological strata. The major anions are carbonate (CO_3^{2-}), bicarbonate (HCO_3^-), sulphate (SO_4^{2-}) and chloride (Cl^-) as salts of calcium, magnesium, sodium, potassium and to a lesser extent iron and manganese.

Table 2.4: Comparison of Basic Quality Factors of Various Water Sources

Water Quality Factor	*Lakes, Ponds and Reservoirs*	*Streams and Rivers*	*Springs and Wells*	*Estuaries and Oceans*
DO content	Slightly lower in water drawn from below the thermocline	High due to turbulence	Low concentration	Usually lower than fresh water, but varies with salinity and temperature conditions and depth
pH	More constant than stream water	Determined by type of soil of area through which the stream flows	–	Buffer system is more effective than fresh water
Temperature	Change markedly with season. More stable and predictable than that of streams	Not predictable	Constant	Same as lakes and ponds
Nutrients (dissolved and suspended)	More constant than stream water	Determined by geographical conditions, stream depth, width and turbulence	A number of major ions are present. Some wells have undesirable gases or chemicals	Rich in essential nutrients, and also in many non-essential ones. Concentration varies with depths
Organisms	Almost same as in streams	Various types are present, *e.g.* predators and competitors	–	Enriched most with a variety of plant and animal life, and so with high productivity

The precipitation falling on the earth contains only small amounts of dissolved mineral matter. The soluble products of soil weathering and of erosion by rainfall and flowing water add salt to the water passing through soils. Excess irrigation water percolating to the water table may contribute considerable salt concentrations. In addition, soluble soil materials, fertilizers, and selective absorption of salts by plants modify salt concentrations of percolating waters. Soils and ground water of arid climates and poorly drained areas may contribute high salinities. In coastal regions, airborne salts originating from the air-water interface over the sea are an important source of salinity in ground water. In arid regions with small surface runoff and large evapotranspiration, airborne salt deposition becomes highly intensified in ground water.

7. Summary

Water is obtained usually in sufficient quantities for aquaculture purpose from surface and/or ground waters. These waters are available normally from the source with the quality characteristics existing in the natural resources. We have discussed in the previous pages the general characteristics of the different water resources. The basic quality factors of the source waters that an aquaculture farm can rely on remain more or less same unless pollution load and the related environment hazards make this water considerably unsuitable for the animal life. A comparative statement of the basic quality factors of natural waters is given in Table 2.4.

3

Water Resource Use in Coastal Area

Industrialization and urbanization are the two most important development activities in any country that place resource demands on land and sea. For decades, the growth of Indian economy has made a rapid shift from a predominantly rural based population to one that is now increasingly concentrated in large cities mostly near the coast. India has a coastline of 7516 km including that of Andaman and Nicobar Islands, and Lakshadweep. Out of the three mega cities with population more than 10 million, Mumbai and Kolkata are the coastal cities. The population density is also much more in coastal areas than the national average. A study on land use changes in coastal towns revealed that the east coast lying towns changed at the rate of 435.67sq km/25 years and the west coast lying towns at the rate of 210.78 sq km /27 years (Ramachandran, 2003). About 20 percent of the Indian population lives in the coastal areas, and about 20.6 percent of the world population lives within 30 km of the coast. Coastal population is growing at a rate faster than the world population and within the next 20–30 years the coastal population is estimated to almost double (UN, 1994).

Coastal development activities create tremendous impact on the coastal resources and its environment. The coastline of any country supports a wide variety of activities such as human settlement, industries, ports and harbors, agriculture and forestry, fishing and aquaculture, tourism, etc. A list of such activity is presented in Table 3.1. Perhaps the oldest use of coastal resources is that of the sea for transport. Even today the transoceanic shipping remains of great importance in world trade. Many coastal villages, even though started for agriculture, fishing and trading, have grown into major settlements demanding resources of land and water. Fishing of near-shore waters and farming of coastal low land are the major economic activities in the coastal villages. The fisheries sector often accounts for a significant proportion of GDP in the economies of many coastal countries. Catch from coastal fisheries contributes a large proportion of the total fish production of these countries. On the other hand, coastal aquaculture has been growing fast in a number of industrialized and developing countries, particularly in parts of Southeast and East Asia and parts of Latin America. The number of people engaged in fisheries and aquaculture activities worldwide as their source of income and food has grown tremendously. While in 1970-80 it increased by 26 percent, in 1980-90 the growth became 72 percent. This has occurred in majority of the developing countries, where the number of people engaged in the agricultural sector is from low and middle-income groups (Sampath, 2003).

Table 3.1: Use of Coastal Resources

Harvesting of aquatic animals and plants	
Aquaculture or farming of fish, crustaceans and algae	
Agriculture and Forestry	
Use of coastal land	– Industry
	– Urban development
	– Recreation
	– Military activities
Use of water space	– Enclosures of estuaries for farming, docks, fresh water storage, causeways for roads or railways
	– Port installations
	– Industry
	– Recreation
	– Transportation
	– Military activities affecting water space
Extraction of mineral resources	– Sand and gravel
	– Shell, sand and merl
	– Coastal mines
Energy and Power	– Oil and gas
	– Tidal and wave
Waste disposal	– Sewage and domestic refuse
	– Refuse and oil from ships
	– Industrial waste
	– Radioactive waste
Miscellaneous	– Introduction of exotic species
	– Natural processes like erosion, siltation of estuaries, storms, etc
	– Coastal defense and maintenance
	– Scientific studies and education

Source: Adapted from Boaden and Seed, 1985.

When land becomes limiting in the coastal areas, estuarine and other wetland habitat is filled in or dredged out. In this way intensive changes occur in some areas with loss of mangrove swamps, tidal flats and coral reefs. Many industries are sited on the coast because of better access to transport.

Electricity generation requires a large cooling water supply, which is readily met from the lower reaches of rivers or from the sea. Mangroves constitute a significant portion of the coastal wetland in India. The Report of the Ministry of Environment and Forests (1994) has shown that about 37 percent of the mangrove forests in India have been lost between the years 1988 and 1994. Indiscriminate use of timber and firewood, conversion of forest areas to residential and industrial estates, extension of harbors and docks all contribute to the degradation of mangroves. In recent past development of coastal aquaculture, especially shrimp culture has caused concern in the coastal environment. The two southern states, namely Andhra Pradesh and Tamil Nadu have been significantly affected due to unplanned development and lack of scientific management of aquaculture. Loss of agricultural land and labor, salination of soils and ground water, and discharges of untreated farm effluents affected the ecosystem characteristics.

Urbanization and industrialization in the coastal zone inevitably lead to problem of waste disposal. The surrounding seas and coastal wetland are often used as a repository for waste discharges of all kinds. As a result coastal and marine habitats degrade and create environmental impacts on tourisms, fishing and/or agriculture. Discharges of industrial wastes to municipal sewerage systems compound this problem.

3.1 Coastal Fisheries and Aquaculture

3.1.1 Coastal Fisheries

1. Types of Fisheries

The fisheries, in general, depend on natural resources that are found in a great variety of environments ranging from high seas through coastal waters to inland water. Accordingly, and for historical reasons also, the fisheries have been categorized as marine (exploiting open sea and coastal stocks), inland fresh water (capture fisheries of rivers, lakes and estuaries), and aquaculture (cultivating fish as livestock) (FAO, 1995). The fisheries are the common property resource from which aquatic organisms can be exploited with or without appropriate licenses. Fisheries are usually not considered as a homogeneous unit of a production 'system', as these consist of various activities and do not take into account the differences in the basic production processes, the types of ecosystem, harvesting methods and the economics.

Aquaculture is an activity of farming of aquatic organisms, including fish, molluscs, crustaceans, and aquatic plants. Farming implies some form of intervention in the rearing process to enhance production as well as ownership of the stock being cultivated. The production enhancement techniques are regular stocking, feeding, protection from predators, etc. Aquaculture is far more akin to farming and animal husbandry than to fishing from open or common access resource, as it involves the rearing and management of living aquatic resources in a restricted environment. Aquaculture is dependent on the inputs provided by the farmer. The catch from capture fisheries is usually dependent upon virtually uncontrollable, natural production processes of the water bodies that are affected by their environmental conditions (FAO, 1999).

In general, the capture fisheries pertain to the rivers, estuaries, reservoirs and lakes in the inland sector, and coastal and marine fisheries in the marine sector. However, capture fisheries in the marine sector are particularly dependent on the coastal area, as these are mostly based on coastal stocks, and others exploit offshore stocks that spend part of their lives in inshore waters as a nursery or feeding area. Similarly coastal aquaculture is also highly dependent on the coastal water for resources. There are four other categories of fisheries (FAO, 1999):

1. Culture based fisheries *i.e.* the capture fisheries, which are maintained by stocking with seeds originating from aquaculture.
2. Enhanced fisheries *i.e.* the activities of supplementing or sustaining the recruitment of one or more aquatic organisms and raising them beyond a level, which is sustainable by natural processes.
3. Recreational *i.e.* fisheries conducted by individuals primarily for sport but with a possibility of capturing fish for domestic consumption.
4. Wild fisheries *i.e.* the fisheries based on natural production and recruitment.

Table 3.2 shows the classification of capture fisheries systems. In the following discussion the reservoir fisheries and deep-sea fisheries have been left out, as these production systems do not strictly come under coastal fisheries sector.

Table 3.2: Various Capture Fisheries Systems

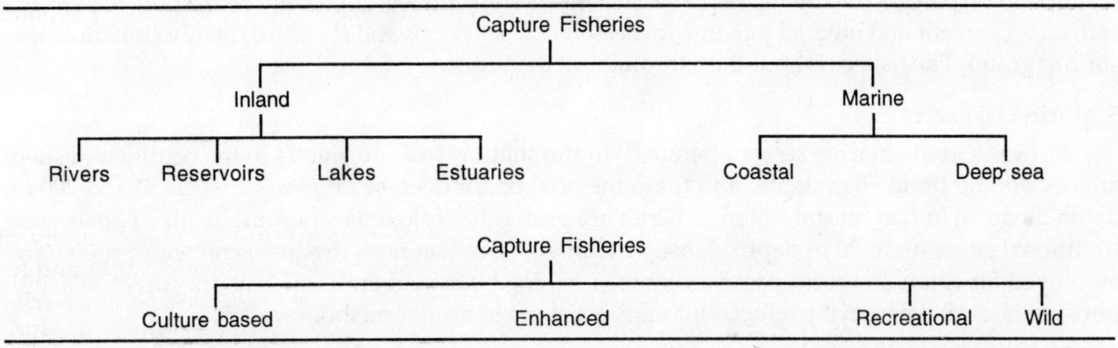

2. Inland Capture Fisheries

In coastal areas the inland capture fisheries operate in rivers and streams, lakes and estuaries. The riverine fisheries, in general, pertain to the fresh water zones in the upstream sections of the rivers, whereas the fisheries in lakes and estuaries operate in the brackish water areas. The estuarine capture fisheries can be discussed with an outline of fishery resources of major estuaries and brackish water lakes of India. The Hooghly-Matlah estuary is dominated by marine species with an increasing number of species towards the lower zone. The most commercially important are the clupieds: *Hilsa ilisha*; Perches: *Lates calcarifer*; Mullets: *Mugil Persia* and *M. tade*; Threadfins: *Polynemus paradiseus* and *P. indicus*; and *Pama pama*, *Sillaginopsis panijus*, and shrimps. A system of so-called migratory fishing is practiced during winter in the lower zones of the estuary. Bulk of the total landings is usually carried by bag-nets. It has been found that most of the areas of the Hooghly estuary are not suitable for bottom trawling due to uneven riverbed contour consisting of loose mud.

Clupieds, mullets and shrimps also dominate the Mahanadi estuary in Orissa. In the Gautami-Godavari estuary in Andhra Pradesh shrimps form an important fishery, the *Metapenaeus monoceros* being the most abundant, particularly in the mangrove swamps and backwaters. Drag nets and stake nets are the common gears used in the fishery. Clupieds and mullets also form a sizeable proportion of the catch. In general, clupeids, mullets and shrimps form the major catches in almost all the estuaries in the east coast.

The fishery potentials of the important estuaries in the west coast of India are little known. Hilsa ilisha forms an important fishery in the Narmada estuary. *Sillago sihama* and *Mystus gulio* are the dominant species and crustaceans are of limited resource in the Nethravathi and other adjacent estuaries.

Fisheries of the Chilka lake in Orissa are exploited (*i*) throughout the year by means of net fishing, locally known as 'bahani'; and (*ii*) seasonally by means of large impoundments of split bamboos in shallow areas, locally known as 'janos', and by traps. The important species of the lake's catch are mullets: *Mugil cephalus* and *M. macrolepsis*; perches: *Lates calcarifer* and *Gerres setifer*; threadfins: *Polynemus tetradactylus*; and penaeids such as *Penaeus semisulcatus, P. indicus, Metapenaeus monoceros*, and *M. dobsoni*. Crabs constitute a minor proportion of the lake fishery. In Pulicat lake of Andhra Pradesh shrimps dominate the catches with *P. indicus* followed by *P. semisulcatus, P. monodon, M. monoceros* and *M. dobsoni*. Mullets form the next group with *M. cephalus* as the major species. Among the gears used is stake net, drag net and shore seine. In Kerala backwaters the composition of fish and shrimps is shrimps 60–70 percent (including *Macrobrachium rosenbergi*), mullets 11 percent, pearl spots 10 percent, catfishes 9 percent and others 1 percent. *M. dobsoni, M. monoceros* and *P. indicus* mainly constitute the shrimp group. The fishery is abundant throughout the year.

3. Marine Fisheries

Fisheries in the marine sector operate (*i*) in the shallow inshore waters in the continental shelf area extending up to 50 m depth, and (*ii*) in the offshore or deep-sea region between 50 and 200 m depth range. In Indian coastal waters fisheries are primarily exploited by non-mechanized boats with traditional gears up to 20 m depth. Mostly nylon gill nets, bag nets, dragnets and shore seines are employed for catch. In recent past mechanized fishing boats and crafts such as shrimp trawls and purse seine etc have been introduced for marine fishing in deeper offshore waters.

The coastline of about 6000 km running through several maritime states with characteristic features and ports is shown in Table 3.3. India has total estuarine areas of 3.9 million ha and backwaters of 3.5 million ha. Mangroves constitute about 0.4 million ha. The island coasts of Andaman and Nicobar, with coastline as reported from the existing data have been devastated in the recent Tsunami disaster in December, 2004. There have been severe changes in the ecological pattern of these islands. The average width of shelf area from the shore base varies from about 32 km off the coast of Andhra Pradesh to about 175 km off the coast of Maharashtra. The fisheries potential in the marine sector have been categorized traditionally as pelagic, midwater and demersal, with further separation of demersal categories as crustaceans and/ or cephalopods, which are treated as shellfish (Table 3.4). The major share of the marine catch, about 80 percent of the total in all the regions, comes from the inshore fisheries up to 50 m depth, which is contributed mostly by non-mechanized boats. However, contribution by mechanized boats gradually increased since the 70's. The Indian fisheries resource data of the inland and marine sectors have been collected from IIM (1981, 1983) and Jhingran, 1991), which can be consulted by the reader for details.

3.1.2 Coastal Aquaculture

1. Types of Aquaculture Production Systems

From ancient times till date a number of aquaculture techniques have evolved for a large variety of aquatic organisms in marine, brackish and fresh water environments. Since fish farming system is usually country-specific and fish fauna is distributed mainly in fresh water and marine environments, aquaculture practices have grown from very much local and traditional operation to scientific and sophisticated technologies with economy ranging from small rural-based output to large scale diversified production. Therefore it becomes pertinent to outline the various types or classes of aquaculture and/or fish farming practices or systems prevalent in different countries in the world. Table 3.5 shows various types of aquaculture systems practiced in coastal waters. Although the term "coastal aquaculture" covers primarily land-based and water-based brackish water and marine aquaculture practices, fresh water aquaculture is also practiced in coastal areas.

Table 3.3: Maritime States of India with Coastline Length and Continental Shelf Area

Sl.No.	*States*	*Length of Coast Line (km)*	*Continental Shelf Area Above 200 m Depth (km²)*	*Characteristic Features*	*Ports*
1.	Gujarat	1640	99,370	Mangroves and coral reefs at Gulf of Kuchch, extensive continental shelf area and shallow coast, sandy intertidal zone with muddy or sandstone areas	Kandla, Porbandar, Okha
2.	Maharashtra	720	1,04,760	Rocky coast with small bays, creeks, fringed by islands No major rivers Marine sanctuary at Malvan	Mumbai
3.	Goa	110	9,980	Coastline formed by many rivers, with estuaries and creeks	–
4.	Karnataka	270	25,470	Straight coastline, some part rocky, with rivers, rivulets, creeks and bays Wildlife and mangroves at Gokaran	Mangalore
5.	Kerala	600	35,940	Numerous brackish water lagoons and backwaters along the coast Beautiful beaches	Kochi
6.	Tamil Nadu (including Pondichery)	1000	41,420	Narrow belt of sand dunes, low-lying beaches, plains mostly formed by rivers, chank and pearl beds, coral reefs and mangroves Wildlife sanctuary at Vedaranyam	Chennai
7.	Andhra Pradesh	970	31,040	Long, smooth coastline with inundations, deltaic regions of Krishna and Godavari with marshy and muddy coasts	Vishakha-patnam
8.	Orissa	480	25,000	Depositional coasts, formed by deltaic regions of three rivers	Paradip
9.	West Bengal	200* 65	20,000	Large intertidal deltaic regions of Ganga and Brahmaputra Massive sedimentation in the Hooghly mouth, innumerable tidal creeks and estuaries Sunderbans–largest mangrove forest, swamps and backwaters	Haldia, Kolkata
10.	Andaman and Nicobar Islands	–	16,060		
11.	Lakshadweep	–	4,340		
	Total	6050*	4,33,780		

* Higher figure implies inclusion of coastlines of several islands in the Sunderbans.

Source: IIM, 1981; Jhingran, 1991; SAC, 1992.

Table 3.4: Marine Fisheries Potentials in Different Maritime States in India

Sl.No.	*Fisheries*	*Species*	*Fisheries potential*
1.	Pelagic	Oil Sardines, other Sardines, *Hilsa ilisha*, other Hilsa, other Clupeids, Flying Fish, Mackerel and Tunnies	Richest in Goa, Karnataka and Kerala up to 50 m depth.
2.	Midwater	Catfishes, Chirocentrus, Anchovilla, *Thrissocles*, Bombay Duck (*Harpodon neherius*, Lizard Fish *Saurida* and *Saurus*), Ribbon Fish, Carangidas family, White Fish *Lactarius*, Pomfrets, Seerfish *Sphyraenea*, Mugil, Belone, *Hemirhamphus*	Rich in all states up to 50 m depth, with Gujarat, Maharashtra, Orissa and West Bengal ranking first. Between 50 and 200 m depth it is the richest in Goa, Karnataka and Kerala, Gujarat and Maharashtra ranking second.
3.	Demersal	Elasmobranches (Shark, Rays), Eels, Perches, Red Mullets, Treadfins Polynemids, Croakers Sciaenids, Silver Bellies *Leignathus Gazza*, Soles, and miscellaneous.	Rich in all the states up to 50 m depth with ranking Orissa and West Bengal as first, following Gujarat and Maharashtra; Tamil Nadu, Pondichery and Andhra Pradesh; and Goa, Karnataka and Kerala in that order. Between 50 and 200 m Goa, Karnataka and Kerala rank first and Gujarat, Maharashtra, Tamil Nadu, Pondichery and Andhra Pradesh rank second.
4.	Crustaceans	Penaeid Shrimps, Non-penaeid shrimps, Lobsters, Crabs, other Crustaceans	Rich in Gujarat and Maharashtra waters up to 50 m, and in Goa, Karnataka, Kerala, Orissa and West Bengal between 50 and 200 m depth. Cephalopod fishery has a greater potential in Orissa and West Bengal than in other states.

Source: IIM, 1981, 1983.

Land and Water-based Aquaculture

Mainly aquaculture can be categorized as: land-based and water-based systems. Land-based systems comprise largely ponds, built on dry land. Ponds, the most common of all aquaculture systems, are primarily two types: embankment and excavated ponds. In embankment ponds water is impounded by building up a dam, dyke or similar structure. Excavated ponds are constructed by excavating soil from an area to form a depression and filling it with water.

Aquaculture, particularly hatchery operations, is also carried out in tanks, made of concrete or plastic materials. While ponds are larger and of earthen construction, tanks are smaller and constructed from a variety of materials. Tanks can be constructed of various sizes and shapes such as rectangular, circular and oval. Raceways are ponds of flowing water. They are rectangular in shape, much longer than their width. Raceways are of series and parallel designs. They can be easily constructed and maintained like ponds, but require large water flows particularly in parallel operation.

Water-based systems include enclosures, pens, cages and rafts and are usually sited in sheltered coastal or inland waters. Enclosures are formed from bay by closing the open water access by a solid, net or mesh barrier. The natural bottom of the water body forms the enclosure bottom. Pens and cages are enclosed structures, made from poles, mesh and netting. Pens are supported on the bottom of a water body, but cages are hung from poles or rafts, which float on the water surface. In contrast to the enclosures pens and cages are entirely man-made.

Table 3.5: Various Coastal Aquaculture Systems

- Land-based systems
  - Earthen ponds
    - Embankment
    - Excavated
  - Tanks
    - Concrete
    - Plastic
    - Rectangular
    - Circular
    - Oval
  - Raceways
    - Series
    - Parallel
- Water-based systems
  - Enclosures
  - Pens
  - Cages
- Inland aquaculture systems
  - Fresh water
    - Ponds
  - Brackish water
    - Ponds
    - Estuaries
    - Lakes
    - Swamps
- Marine aquaculture systems
  - Enclosures
  - Pens
  - Cages
- Culture Practices
  - Traditional
  - Extensive
  - Semi-intensive
  - Intensive

Inland Aquaculture

Inland aquaculture is broadly divided into freshwater and brackish water culture practices. Fresh water culture is practiced mainly in ponds, whereas brackish water culture is carried out in various water bodies from ponds through lakes and swamps to estuaries. In some coastal areas salinity in ponds is reduced naturally by rainfall and fresh water runoff, whereas in estuaries, lakes and swamps brackish water with salinity in the range of 15–30 ppt is utilized for farming.

Marine Aquaculture

Marine aquaculture, or 'mariculture' in the shortened form, means organized culture of marine organisms in sea water, *i.e.* rearing of marine animals and plants in the marine environment itself. The terms 'marine cultivations', 'marine farming' and 'sea farming', similar in meaning, have also been used under this aquaculture system. Usually the habitat selected for mariculture is the edge of the oceans, its inshore bays, inlets and the like, and the open sea as well.

Extensive to Intensive Aquaculture

Similar to livestock production systems that range from extensive ranching to intensive feedlot systems, aquaculture also is practiced at different intensities. Extensive farming usually involves unsophisticated technologies relying on natural food organisms of the pond and has a low input to output ratio. Fish juveniles procured from wild stocks are introduced into the ponds and production inputs, in the form of feed and fertilizers, are provided occasionally, if at all. With increase in production intensity, fish stocks are purposely increased and natural food supply is enhanced by more production inputs such as fertilizers and supplemental feed in the form of oil cakes, rice bran and similar agricultural by-products. The system is then called semi-intensive farming, which produces the bulk of aquaculture products in ponds.

Intensive systems produce more output from a given production unit through higher levels of technology and greater degree of management. Hatchery-produced seeds; higher stocking densities in smaller, well-designed facilities; use of compound manufactured feed pellets and carefully controlled water quality are the major factors of intensive systems.

Shrimp farming systems of *Penaeus monodon* under different culture intensities including traditional aquaculture in India is given in Table 3.6.

The dominant aquaculture production systems followed in Asia and Europe are shown in Table 3.7. The growing system in Asia is extensive, semi-intensive and intensive land-based ponds, and water-based cages and pens for fish and crustaceans, but predominantly extensive for plants (beds, rafts) and for molluscs (rafts, stakes and longlines). In Europe fish are reared only in intensive tanks, cages and pens, and mollusks are farmed in trays, rafts and ropes.

2. Species Cultured in Coastal Waters

The following categories of marine organisms are usually cultured in coastal areas in different countries of the world:

1. Finfish
2. Shellfish, viz crustaceans and molluscs
3. Turtles
4. Miscellaneous invertebrates, *e.g.* sponges and bait worms
5. Aquatic plants, *e.g.* seaweeds

Table 3.6: Shrimp Farming Systems for *Penaeus monodon* in India

Shrimp Farming System	*Main Characteristics*
Traditional	A variety of polyculture systems with a large component of miscellaneous fish and a small component of shrimps. Ponds are filled with tidal water with no control over quality and quantity of stocking. The most well known systems are *bheries* of West Bengal and the paddy-cum-fish culture in Kerala, Goa and Karnataka, with pond sizes from 2–200 ha.
	Average production is low and ranges from 200 to 500 kg/ha/year of mixed sizes and species.
Improved traditional	Traditional ponds with supplementary stock of with shrimp seed. Most of the limitations of the production system remain; the entrance of predators and competitors, insufficient pond depths, and full dependence on natural food in ponds, with size 2–5 ha.
	Overall yields increase to about 300 to 600 kg/ha/year, with one-third of the crops being shrimps.
Extensive	Ponds, 1–2 ha, are usually squared with excavated walls of some 1.5 to 2 m high. Water is pumped from canals or creeks. Stocking is at rates of 2-5/m², with one to two crops a year. Pond-side prepared feeds from a variety of ingredients are used.
	Crop yields are about 300–700 kg/ha, with large variations between different ponds.
Modified extensive	Lay out same as extensive, but involving pond preparation with tilling, liming and fertilization, and higher stocking of 5–10/m². Combination of locally prepared or imported pellet feed is often used.
	One or two crops of 600–1100 kg/ha can be harvested per year.
Semi-intensive	Ponds are 0.25 to 1.0 ha in size, with regular water supply and drainage of controlled water exchange, and stocking densities of 15–30/m². Use of imported pellet feeds, and drugs and chemicals are common.
	Average yields of 2200 kg/ha/year with an average of 1.2–1.5 crops per year.
Intensive	Ponds are 0.25–0.5 ha in size with 4 aerators per pond, a central drainage system to remove accumulated sludge, and stocking of 30–80/m².
	Yields of over 8000 kg/ha are possible, but the actual production in India is about 4500 kg/ha with 1.6 crops per year. Although very common in Thailand and Taiwan, it is not frequently practiced in India.

Source: FAO/NACA, 1995; Hein, 2001.

Table 3.7: Dominant Aquaculture Systems in Asia and Europe

Aquatic Species	*Asia*	*Europe*
Finfish	Extensive, semi-intensive ponds (land-based), intensive cages, pens (water-based)	Traditional and semi-intensive ponds, and intensive tanks (land-based) cages (water-based)
Crustaceans	Traditional, extensive, semi-intensive, intensive ponds	
Molluscs	Extensive: beds, stakes, rafts	Extensive: bottom (trays) and suspended (rafts, ropes) systems
Aquatic plants	Extensive: beds, stakes, rafts	–

Source: Primavera, 2002.

The major aquatic species from Asia are fresh water and marine fish, crustaceans, molluscs and seaweeds, and from Europe salmonids and molluscs. Figure 3.1 shows the ecological pyramid of 1999 global aquaculture production according to species group and feeding habit (FAO, 2001). Carnivorous fish and crustaceans comprise only 10.4 per cent of production volume, but more than treble in value due to high market price of these commodities. In contrast, plants and lower trophic level molluscs and herbivores contribute up to 45.8 per cent of total production by volume, but only 27.5 per cent by value.

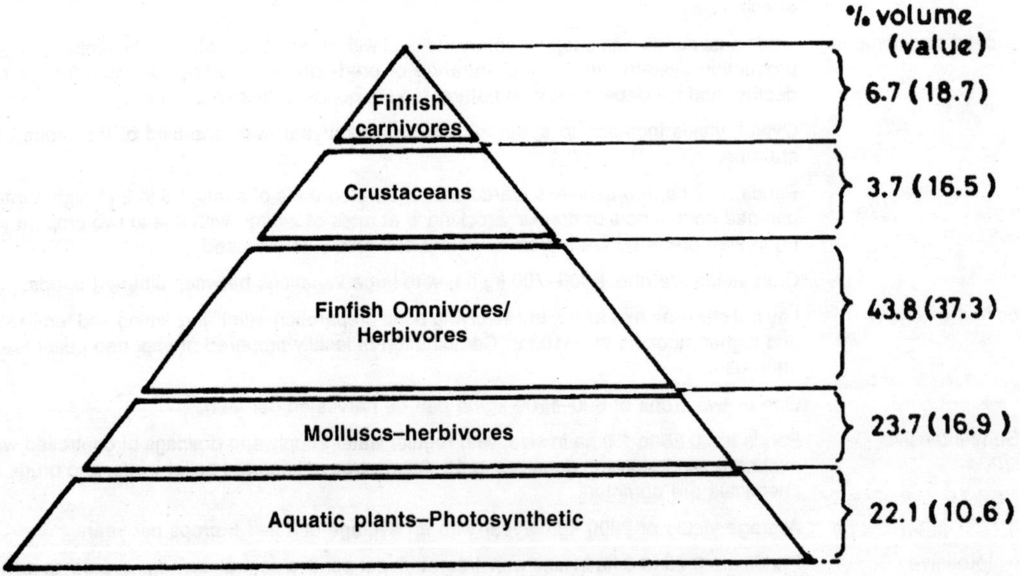

Figure 3.1: Ecological Pyramid of 1999 Global Aquaculture Production According to Species Group and Feeding Habit (After FAO, 2001)

In the following pages, a brief outline of the various categories of marine organisms cultivated in coastal area has been given. For detail reference the reader may consult Milne (1972), Bardach *et al.* (1972) and Iversen (1976). For culturable species of Indian coasts reference is made to Jhingran (1991).

Finfish

Table 3.8 shows distribution of some important fish species cultured throughout the world. Mullets are probably the most important culturable brackish water species. They occur in the sea, in estuaries, and some even migrate into fresh water. Their fry are collected in shallow coastal waters. The Hooghly-Matlah estuary in the Bay of Bengal coast is rich in fry and/or fingerlings of *M. persia* and *M. tade*. *Chanos chanos* fry is also available extensively in the Bay of Bengal coast. Fry of betki (*L. calcarifer*) are available in the Hooghly Matlah estuary, the intertidal pools with grassy vegetation forming the main source of collection. These are also available in Chilka in Orissa. The pearl spot breeds in shallow confined waters throughout the year and the fry are available in Adyar in Tamil Nadu, Chilka and Kerala backwaters. However, culture of finfish (*Lates* species, *Chanos* species and mullets) is done in isolated patches due to non-availability of seeds.

Table 3.8: Some Important Fish Cultured in Coastal Water

Group	*Common Species*	*Feeding Gabit*	*Distribution*
Mullet	Grey or striped mullet: *Mugil cephalus, M. persia, M. tade, Rhinomugil corsua, Liza persia, L. tade*	All herbivorous Omnivorous	Almost throughout the tropical and subtropical waters
Milkfish	*Chanos chanos*	Herbivorous	Tropical and subtropical waters bordering the Pacific and Indian oceans
Yellow tail	*Seriola quinqueradiata*	Carnivorous	Widely cultured in Japan
Eel	*Anguilla japonica, A. anguilla*	All carnivorous	Japan Europe, Israel, UAR, Russia
Pearlspot	*Etroplus suratensis*	Herbivorous	India, Pakistan, Sri Lanka
Perch	*Lates calcifer*	Carnivorous	Indo-pacific region
Tarpon	*Megalaps cyprinoides*	Carnivorous	Indo-pacific region
Rainbow trout	*Salmo gairdnerii*	Carnivorous	Norway, Scotland, Australia
Salmon	*Salmo salar, Oncorhynchus* spp.	Carnivorous	Norway, Scotland, USA
Red sea bream	*Pagrus major*	Carnivorous	
Sole	*Solea solea*	Carnivorous	Britain
Sea bass	*Morone labrax*	Carnivorous	

Crustaceans

Table 3.9 shows distribution of some important culturable shrimp and prawn species in the world. *Penaeus, Metapenaeus* and *Parapenaeopsis* are the three genera of shrimp species cultured throughout the world. All varieties of shrimps are available all along the Indian coast from November to May on the west coast and from December to August on the east cost. Tiger shrimp (*P. monodon*) is the largest number of penaeid shrimps available from offshore catches. Adult shrimps are bottom dwellers at depths of 20-70 m offshore areas. After spawning and hatching out of larvae in the sea, the post-larvae or fry move towards the shore and encounter the coastal waters. They enter mangroves and other brackish water areas for their nurseries or feeding grounds, and grow to juveniles, post-juveniles and adult stages. Tiger shrimps spawn throughout the year and post-larvae are available in estuaries. Similar is the case with most of the marine shrimp species and their post-larvae are abound in tidal creeks and coastal shallow waters during specified season. Wild seeds of *P. monodon* and *P. indicus* are still collected for culture in West Bengal, Orissa and Andhra Pradesh in India.

Macrobrachium rosenbergii, called the giant fresh water prawn, is widely distributed in the Indo-Pacific area. Though a fresh water species it migrates into the estuaries for breeding and spawns in water with salinity fluctuating between 5 and 20 ppt. It migrates up the estuary to fresh water habitats at the post-larval stage. The other *macrobrachium* species also behave in the similar pattern.

Lobster, one of the highly valued crustacean seafood, is cultured in the USA and European countries. The two important species, American lobster, *Homarus americanus* and the European species *H. vulgaris*, occur in the North Atlantic. The spiny lobsters, which supply the major portion of the lobster tail in the market, are abundant in tropical waters. There are a number of commercially important species of spiny lobster that are cultured in different countries. In India the peuruli of the spiny lobsters, *Panulirus homurus, P.Ornatus* and *P. Polyphagus* are collected from inshore waters and cultured on trial basis.

Table 3.9: Some Important Shrimps Cultured in Brackish Water

Species	*Feeding Habit*	*Distribution*
Penaeus indicus (white shrimp)	Omnivorous	India, Bangladesh, Indonesia, Thailand, Singapore, Philippines, Ecuador
P. monodon (Tiger shrimp, grass shrimp)	Omnivorous	India, Bangladesh, Indonesia, Thailand, Singapore, Philippines, Taiwan
P. semisulcatus	Carnivorous	India, Singapore
P. japonicus (Kuruma shrimp)	Omnivorous	Japan
P. durorarum, P. aztecus, P. setiferus		USA
P. plebejus, M. benettae, M. macleayi		Australia
Metapenaeus dobsoni	Omnivorous	India
M. monocearos	Omnivorous	India, Bangladesh, Indonesia, Thailand, Philippines, Taiwan
M. affinis	Omnivorous	India
M. brevicornis	Omnivorous	India, Indonesia
Parapenaeopsis styliferus, P. sculptilis		India
Macrobrachium rosenbergii, M. malcomsoni	Omnivorous	India

Majority of edible crabs of commercial importance are swimming crab (*Portunidae*), mud crab (*Xanthidae*) and cancer crabs (*Cancridae*). *Scylla serrata* of the portunid group is the product of brackish water pond culture in south east Asia and the Philippines. This is also cultured in a bay near Tuticorin, Tamil Nadu. The highly valued blue crabs of Portunid group and mud crab are distributed along the Atlantic coast of the USA, Mexico and Cuba.

Molluscs

Mollusks are commercially important cultrable species and a number of varieties are cultured throughout the world (Table 3.10). There are three taxonomic classes: bivalves, gastropods, and cephalopods. The edible oysters, pearl oyster, clam and mussel are all bivalves. The categories of *Ostrea* (flat) and *Crassostrea* (cup) oysters are cultured throughout the world. The Japanese pearl oyster *Pinctada fucata* is the most common species occurring on large sea beds in Indian seas, which are more extensive and productive on the east coast than on the west coast. The species thrive on coarse sand, gravel or rock bottoms of inner bays with clean water and moderate sea conditions.

Mariculture of two types of clams–hard and soft is largely practiced in the USA In Japan also several species of clams are cultured. Mussels are in great demand in Europe, Spain being the leading producer. They are often easier to culture than oysters. *Mytilis edulis* is the most common edible mussel. It usually forms thick carpet like growth over submerged rocks. Sea scallops are being cultured by the long line method in Japan's bays. The species is the cold water animal occurring at 10–30 m depths in sea beds of sand with mud or gravel.

Many marine gastropods such as abalones are food delicacies. Large numbers of abalone species are distributed worldwide. About 10 species of temperate water are commercially exploited. They are found in small crevices of rocks or under stones, but grow in deeper zones. There are many important species of shell and chank available along the Indian coasts. They are utilized for food to a limited extent.

Table 3.10: Some Commercially Important Culturable Molluscs

Sl.No.	*Type*	*Species*	*Distribution*
	Bivalve		
1.	Edible oyster	*Ostrea edulis* (European flat oyster)	Britain, France, Norway, Spain
		Crassostrea virginica (American oyster)	USA, Canada
		C. gigas (Japanese oyster)	Britain, Canada, USA, Philippines
		C. angulata (Portuguese oyster)	Britain, France
		C. madrasensis (backwater oyster)	East and west coasts of India
		C. cucullata (rock oyster)	Intertidal rocky coasts of India
		C. discoides (disc oyster)	Littoral zone of coastal areas of India
		C. gryphoides	Muddy creeks in Indian coasts
2.	Pearl oyster	*Pinctada fucata* (Japanese pearl oyster)	Indian coasts
3.	Clam	*Mercenaria mercenaria* (hard clam)	Atlantic coasts
		Mya arenaria (soft clam)	Labrador to North Carolina
		Andara granosa (haigai) and several others	Japan
		Meretrix meretrix (bay clam) and several others	Indian estuaries and backwaters
		Cockle clam	Mud flats of south Indian coasts
4.	Mussels	*Mytilus edulis*	Spain, Italy, France, Holland, South-eastern and western coasts of India
		M. viridis	Along both the coasts of India
		M. galloprovincialis	Italy
		M. smaragdinus (green mussel)	Philippines
		M. edulis planuties	Australia
5.	Scallops	*Patinopecten yessoensis* (sea scallop)	Japan
	Gastropods		
1.	Abalones	*Halistis discus* and others	Japan, China, South Africa, New Zealand, Australia, Mexico and pacific coasts of USA
2.	Conches	*Trochus niloticus* (top-shell)	
		Umbonium vestiarium (bottom-shell)	
		Turbo marmoratus (turban-shell)	All in Indian coasts
		Turbinella pyrum (sacred chank)	
		Several other shell and chank varieties	Caribbean sea and Atlantic ocean coasts
		Strombus gigas (Queen conch)	
	Cephalopods		
	Squid	*Sepioteuthis lessoniana*	Japan
		S. sepioidea	Miami
		S. arctipinnis and several *Loligo* species	South Indian coast

Cephalods are molluscs and include squid, cuttle fish and octopuses. However, they are supplied from commercial fisheries. Few species of squid have been experimentally cultured in Japan and Miami. Several species are available in the south Indian coasts.

Turtles and Miscellaneous Invertebrates

Sea turtles are particularly suitable for culture. Their omnivorous feeding habit makes it easy for cultivation. The green turtle *Chelonia mydas* is the commercially important species available in India. But this and the Olive ridley (*Lepidochelys olivacea*) have been taken up for conservation measures for protection. Most of the marine turtle populations in the Indian region are in decline due to deliberate human predation by large scale netting. Recently in the U.S.A. the turtle farming has been found to be successful. The sponges are lower invertebrates commercially used for surgical purpose and cleaning. Their farming is simple and is common in the Mediterranean Sea and in the Florida, U.S.A. Some bait worms, *e.g.* blood worms (*Glycera dibranchiata*) and sand worms (*Neanthes virens*) are the earthworm-like dwellers of the intertidal zone and are very popular. Their spawning and feeding habits make them easy to cultivate

Sea Weeds

Sea weeds are collectively the species of marine algae. They form one of the most important aquatic resources of any country, as they can be used as food, fodder, manure and pharmaceuticals. The anchored sea weeds are found in the photic zone, all around the sea coast. The marine algae without true roots, stems and leaves grow attached to rocks or other plants. The class of *Rhodophyceae*, the red algae is the most commercially important food algae. The most popular one, *Porphyra* are cultivated in Japan, Europe and America. This and the *Gracilaria* grow in the Indian coasts, particularly in the south. Among the green algae, *Monostroma* fetches the highest price in Japan. The brown algae, collectivity known as kelp, are cultured in Japan and the Philippines. Few species of green and brown algae also grow in the Indian coast. However, commercial cultivation of sea weeds in not significant. Some commercially important culturable sea weeds are given in Table 3.11.

Table 3.11: Commercially Important Culturable Seaweeds

Type and Species	*Distribution*
Rhodophyceae (Red algae)	
Porphyra ('Lever')	Europe and America
P. yezoensis ('Nori'), *P. tenera*, *P. angusta*, and others	Japan Taiwan
Gracilaria gigas and some others	Taiwan, India
G.lichenoides and some others	India
Gellidiella accrosa	Canada, Ireland, Scotland and mediterranean countries
Rhodymenia palmate ('Dulse')	
Chlorophyceae (Green algae)	
Enteromorphia, Monostroma	Japan
Ulva fasciata	India
Phaeophyceae (Brown algae, Kelp)	
Undaria pinnatifida	Japan
Caulerpa, Eucheuma, Hypnea	Philippines
Sargassum, Padina	India

3. Operation of Coastal Aquaculture Farms

Farming Areas and Farm Facilities

The site of any farm depends on the topography of the coastline, the climate, water supply, natural food organisms available in the water and the species to be cultured. Among the various coastal zones, the six zones that have been successfully adapted for farming are the shore, intertidal, sublittoral, surface-floating, midwater and seabed (Figure 3.2). The type of facilities to be constructed for farming in the various zones depends on the species to be cultured, and the environmental control over the species. A small overview on the farm sites and facilities in coastal zones is presented below. For detailed information the readers may consult Milne (1972).

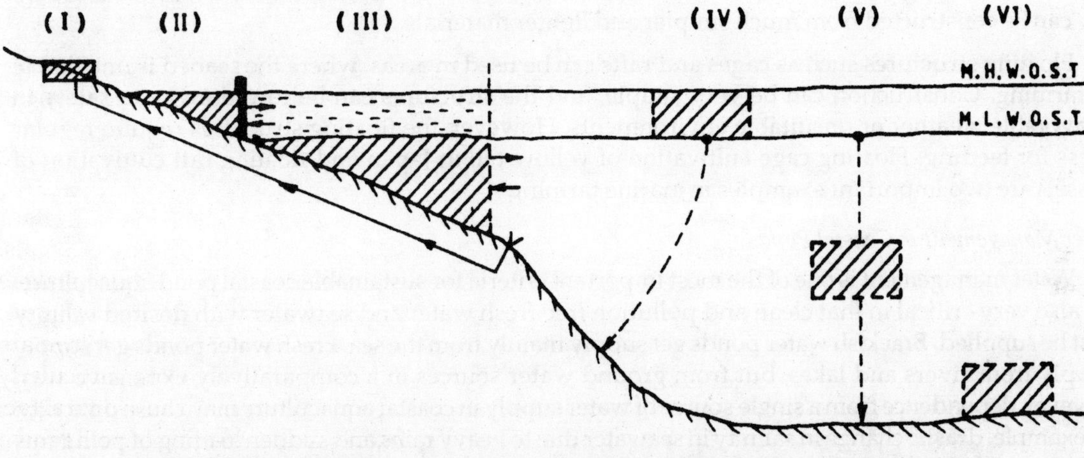

Figure 3.2: The Six Possible Zones for Aquaculture in Coastal Waters: (I) Shore, (II) Intertidal, (III) Sublittoral, (IV) Surface floating, (V) Mid-water, and (VI) Seabed; MHWOST: Mean highest water of spring tide, MLWOST: mean lowest water of spring tide (After Milne, 1972)

The shore facilities required for aquaculture of fish and shrimps would be ponds, tanks, raceways and hatcheries, which are very similar to those existing for inland freshwater aquaculture throughout the world. Intertidal farming techniques require impermeable pond barriers with sluice gates to hold the animals because of water movement and circulation between high and low tide. Impoundment of sea water by a bund or barrage facilitates water circulation in the pond during each tidal cycle and requires no pumping, if designed properly with the tidal water level. Salinity reduction in the pond water due to rainfall is also considered between each water replenishment period. However, culture of organisms such as molluscs, do not need such coastal constructions and can adapt racks and frames in the tidal seaways. The intertidal zone is traditionally used for pond culture of fish and shrimps in mangrove swamps and for mollusc culture by monitoring the hydrographic conditions.

In the sublittoral zone enclosures are constructed in natural bays by using either embankments with mesh sluices or net barriers. The enclosure can be constructed (a) in the middle of a bay with no foreshore, or (b) with a portion of the foreshore extending into deep water, or (c) as a sea bay enclosure at the entrance. The enclosures are relatively cheep in construction but with additional maintenance cost to keep the nets clean to preserve the water circulation. The embankment method is used for oyster and yellow-tail culture, and the barrier method for yellow-tail, rainbow trout and pompano culture. In chilka lagoon in India, mud embankment or bamboo net enclosures are used for "pen culture" of marine shrimps. The culture technique is locally known as *gherybundh* (Biswas, 1996). However, the impoundment technique is not used for green mussel culture in the sublittoral zone in the Philippines.

The seabed below the low water spring tide mark has been used traditionally for culture of molluscs and crustaceans in Europe, the USA and South Africa. The selection of sites with a hard bottom and strong tidal currents prevents silting of the beds. Depths of 1–12 m water are used for oyster culture in the USA, whereas scallops are found to inhabit water depths of 0.5–25 m on sand or silty sand bottoms in Russia. While oysters and mussels are planted in the seabed for cultivation, seabed cages are used for retention of the lobsters. Usually, if the seabed is very soft and silty, it is not suitable for farming. Since the cages rest on the seabed, out with the range of wave and wind action, they can be constructed from much simpler and lighter materials.

Floating structures such as cages and rafts can be used in areas, where the seabed is unsuitable for farming. Construction can be very simple, and the structures can be towed away to safety in unfavorable weather or unsuitable environments. However, the floating structures require regular access for feeding. Floating cage cultivation of yellow-tail in Japan and floating raft cultivation of mussels are two important examples in marine farming.

Water Management in Coastal Ponds

Water management is one of the most important criteria for sustainable coastal pond aquaculture. It is also very critical in that clean and pollution free fresh water and seawater with desired salinity must be supplied. Brackish water ponds get supply mainly from the sea. Fresh water ponds get supply cheaply from rivers and lakes, but from ground water sources in a comparatively expensive way. However, dependence from a single source of water supply in coastal aquaculture may cause difficulty. For example, drastic change in salinity in seawater due to heavy rains and sudden loading of pollutants in river water make the water unsuitable for fish growth. Another important criterion is perfect draining of effluent water from the pond. A natural drainage outlet is necessary to minimize contamination of the water source. A site close to the coastline should have good access to seawater so that effluent water can be discharged from the pond.

On the basis of water supply, an ideal brackish water farm can be sited in the following locations:

1. In a coastal zone nearby a river estuary,
2. In a coastal zone away from rivers,
3. In the upstream areas of a river having tide water influence, and
4. Along the periphery of a lagoon, creek or semi-closed bay.

A comparative analysis between these locations is given in Table 3.9, and Figure 3.3 shows the simplified sketches of the systems. At any location the water intake point may be the most important factor of farm design. Usually the intake opening in surface water should be located near the water surface in order to draw surface aerated water. But care should be taken to avoid suction of floating pollutants. In case of seawater intake, it should be located as far offshore as possible to avoid salinity fluctuations and pollution problems. The intake point should be located some distance below the surface, but not near the bottom to avoid suction of low DO and high sediment water (Upadhyay, 1994). Shrimp farms in Taiwan located near river or estuary utilize high and low tides for water supply, supplemented with pumping. Direct pumping of seawater without filtering, or through pipes buried into beach sand, or from a dug well in the beach is also followed for water supply (Chen, 1976).

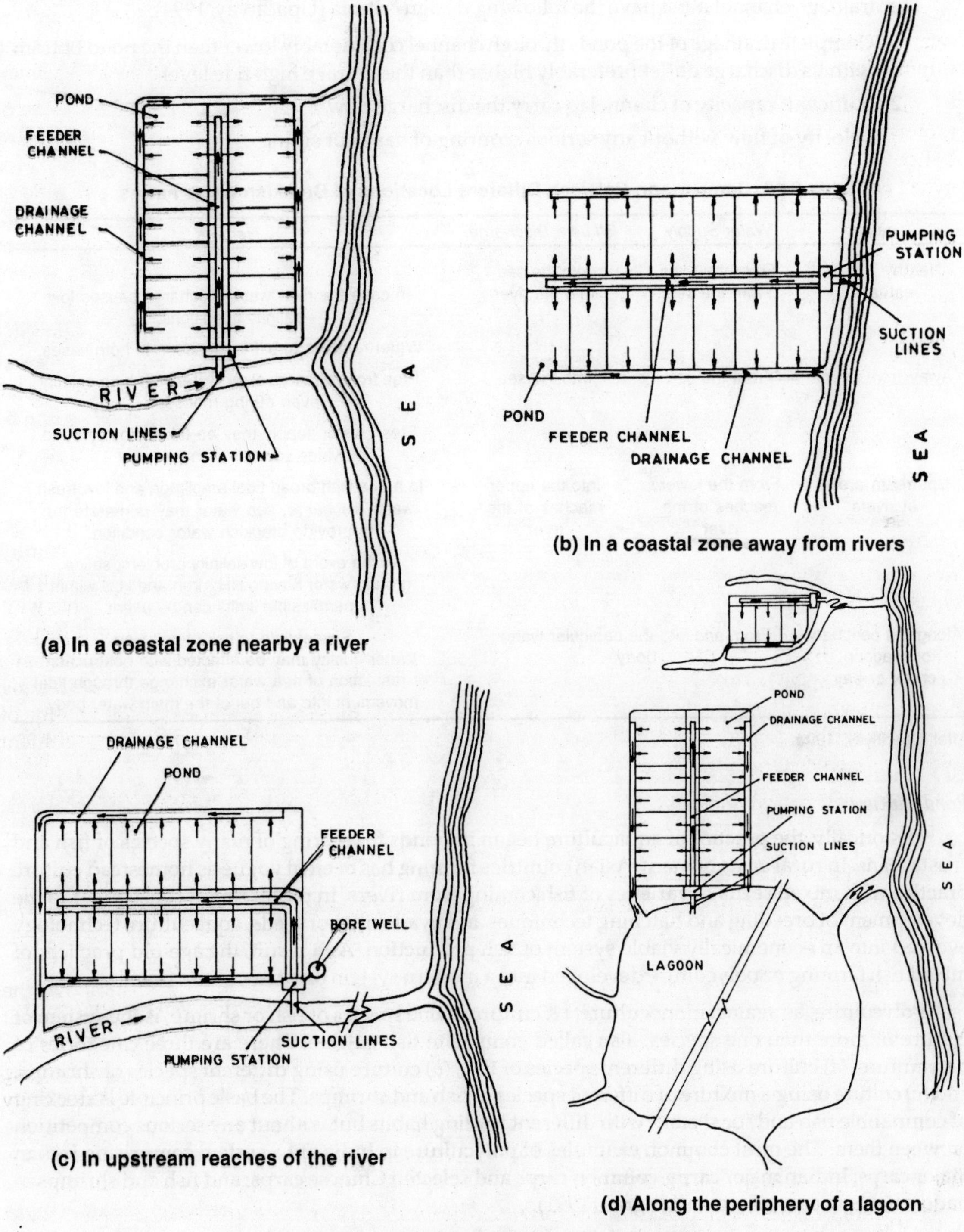

Figure 3.3: Location of Aquaculture Farm Site (After Upadhyay, 1994)

The drainage channel must have the following design criteria (Upadhyay, 1994):

1. Complete drainage of the ponds through channel considerably lower than the pond bottom, with its discharge outlet preferably higher than the average high tide level.
2. Sufficient capacity of channel to carry the discharge flow.
3. Velocity of flow without any serious scouring of banks or silting.

Table 3.12: Comparison Between Different Locations of Brackish Water Farms

Location	*Water Supply*	*Effluent Discharge*	*Remark*
Nearby a river estuary	From the river From the sea	Into the sea Into the river	In case the river water discharge causes low salinity conditions. Water remains turbid in monsoon in both cases.
Away from rivers	From the sea	Into the sea	Free from siltation, clear water with high salinity even during monsoon. Fresh water supply may be taken from ground water sources, if necessary.
Upstream areas of rivers	From the lower reaches of the river	Into the upper reaches of the river	In areas with broad tidal amplitude and low fresh water discharge, sea water may penetrate far to provide brackish water condition. In the event of low salinity problem, saline ground water having NH_3, iron and H_2S within permissible limits can be used.
Along the periphery of a lagoon, creek or bay	From and into the particular water body		Susceptible to risk of pollution. Water quality may be affected with obstruction or restriction of sea water exchange through tidal movement into and out of the main water body.

After Upadhyay, 1994.

Pond Culture

Historically, the practice of aquaculture began in ponds for rearing of many species of fish and crustaceans. In rural areas of many Asian countries farming has been in vogue as homestead culture practice with mixed seeds of varieties of fish coming from rivers. In recent times, consequent to the development of breeding and hatching techniques and availability of seeds, aquaculture technology evolved into an economically viable system of fish production. As a result, the age-old practices of mixed fish farming or polyculture developed into a modern system of monoculture practices.

Polyculture, as against monoculture, *i.e.* culture of one species of fish or shrimp, is the system of culture of more than one species, also called composite fish culture. There are three categories of polyculture: (*a*) culture using different species of fish, (*b*) culture using different species of shrimps, and (*c*) culture using a mixture of different species of fish and shrimps. The basic principle is stocking of compatible fish and/or shrimp with different feeding habits but without any serious competition between them. The most common examples of polyculture in India are a system comprising Indian major carps; Indian major carps, common carps and selected Chinese carps; and fish and shrimps in paddy fields in coastal areas (Jhingran, 1991).

Pond culture practices followed in brackish water areas particularly in Asian countries can be categorized as traditional and modern.

Traditional Culture Practices

There are three principal types of culture practiced in India.

1. *Impoundments, locally called Bhasabadha or Bheri*–The bheris constitute one of the important fishery resources in estuarine wetlands of the Sunderbans in West Bengal. They form a major fish and shrimp production system in an area of about 33,000 ha with low to high salinity zones of mudflats, swamps, marshes and paddy fields. The *bheris* are irregular in shape and range in sizes from 2 to 200 ha, with average size in the range of 15–34 ha. In *bheris* of perennial nature water is let in and impounded during high tide through improvised or permanent sluice gates, and a wide variety of fish and / or shrimps seeds carried along with water are cultured. The water starts filling up from February–March with tidal fluctuation between 1.1 to 4.5 m. Large-scale fishing is done between September and November. Selective stocking with compatible species of fish and shrimp and nursery rearing of fry are rarely practiced in *bheris*. No capital investment except for construction of embankments and sluices is required. These brackish water fisheries, in general, use reclaimed lands situated on the verge of flowing estuarine water. The lands are usually unfit for agriculture due to high saline nature of the soil.
2. *Paddy fields*–Paddy fields are utilized for culture of shrimps and fish usually in West Bengal and Kerala. The practice followed in West Bengal uses the irrigation canal if any lying in the vicinity of paddy fields, and in low salinity regions of upper Sunderbans. During the southwest monsoon in June–July paddy cultivation is started. When the water in the canal increases its level with accumulation of rainwater and decreases in salinity, bundhs surrounding paddy fields are cut at some places to allow fish fry enter the fields. The fish are grown and harvested just before the paddy harvest. In some practices seeds of shrimp, which tolerate low salinity are stocked also. The practice in Kerala utilizes low-lying paddy fields (locally known as *pokkali* fields) for shrimp culture after harvesting paddy crop by October. The bundhs are strengthened and through sluices the fry are let in during high tide with the incoming water. The fry are trapped in conical bag net during the escape of water at neap tides. The stocked fry are allowed to grow undisturbed in paddy fields till December and then harvesting begins and continues till April. Traditional culture systems also exist in *ghaznis* in Karnataka and *khazan* lands in Goa.
3. *Small shallow coastal lagoons*–Small low-lying lagoons which remain disconnected with sea for most part of the year but are liable to be periodically inundated during the monsoon are used for culture. These lagoons are usually poor in productivity due to poor quality soil, meager nutrient content and hyper saline conditions.

In Indo-Pacific region such as Indonesia, the Philippines and Taiwan, brackish water ponds are arranged along a flat shore of salt marshes, which have been reclaimed from the sea by filling in the ponds and eliminating the salt. The ponds are known as *Tambaks* in local language. Ponds are surrounded by dykes, and are filled with water by tidal flow through a channel linked with the sea. A high tide of about 1.5 m is just sufficient. A sluice-type device is used for drainage of water. An area of about 10 ha is considered suitable for a *tambak* farm.

Modern Pond Culture Practices

A modern brackish water aquaculture farm mainly consists of a system of ponds, water supply and drainage channels, control structures and support facilities. A system of fish ponds may have a nursery pond, transition pond, production or rearing pond, and catching pond, as shown in Figure 3.4. The nursery pond, small in size and occupying about 1–4 per cent of total production area, rears the fry before transferring into a larger pond.

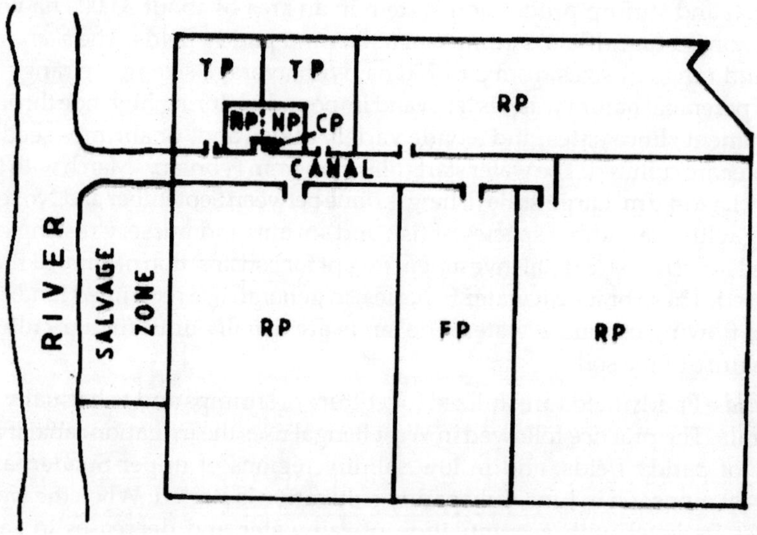

Figure 3.4: A Conventional Modern Pond System (TP: Transition pond, NP: Nursery pond, RP: Rearing pond, CP: Catching pond, FP: Feeding pond) (After Alcantara, 1982)

A transition pond, occupying about 10 per cent of the production area, rears the fingerlings for varying periods before stocking them in a rearing pond. The rearing pond occupies about 80 per cent of the total area and grows the animals to the adult size. The catching pond, intended for harvesting of fish, is constructed adjacent to the gate inside a big pond (Alcantara, 1982).

The main gate for the water supply connects the main channel of the pond system to the source of water. The main channel usually runs through the central portion of the farm. In modern farms each pond has individual water supply and drainage outlets to make them independent of each other. Usually a single gate per pond connected to a channel provides passage for tidal inflow and outflow. In a flow-through system, two gates at opposite ends of the ponds are used. A separate drainage channel is usually provided and located at the other end of the pond, opposite and parallel to the supply channel.

Brackish water farms are usually run as self-contained units with their complex of nursery, rearing and stocking ponds, along with the breeding operation in a hatchery complex as an additional facility. Most of the fish cultivated are known to breed naturally in the sea. Their seeds are available in coastal waters. Mature spawners of fish and shrimp are collected from the sea, and induced spawning or breeding technique is carried out to produce fertilized eggs. Subsequently hatchlings are produced in stages of operations, and finally larval rearing to fingerlings of proper size make them ready for release in outdoor ponds. All these operations are carried out in controlled conditions of water temperature, salinity, DO concentration and proper feeding.

Examples of Pond Culture

Grey mullet culture–Grey mullet is usually cultured in brackish water ponds along with other varieties. Similar practice is often followed in fresh water also. Artificial propagation and culture practices of grey mullet, developed in Taiwan, follow the procedure outlined below (Chen, 1976).

The breeders of about four years age are caught from the sea and kept in an indoor concrete tank with a provision of continuous supply of sea water and adequate aeration. Spawning is induced by hormone injection to the female, and eggs are fertilized by milt collected from the males. The fertilized eggs are then introduced into hatching tanks, made of plastic or concrete, where water quality is monitored continuously with change of water and aeration. Continuous circulating motion of water keeps all the eggs in suspension. Sea water used for hatching is filtered, irradiated and treated with antibiotics to control marine bacteria. During hatching, water temperature is maintained at 23–24.5°C and salinity between 30 and 33.8 ppt.

The small hatchlings are reared and developed into fingerlings of about 1.5 to 2.0 cm, which are ready for release in outdoor ponds for stocking. Polyculture system is practiced with several other fishes such as grass carp, silver carp, tilapia, etc. Usually 1000 to 2000 fingerlings per ha are stocked in polyculture ponds. They grow to the weight of about 300 g after one year, 1.2 kg after two years, and 2 kg if left for three years. Yield of 2500 to 3500 kg/ha has been reported in intensive ponds.

Marine shrimp culture–The shrimp culture technique has developed in countries like Taiwan, Japan and the Philippines in the late 1970s and also in the Pacific coasts of Latin America like Ecuador. In India shrimp culture based on the marine shrimp *P. monodon* attained the status of take-off stage in mid-eighties in the coastal zones in the east and west coasts. About 70 per cent of marine shrimp are produced in the east and 30 per cent in the west coast. The estimate of 2000 year shows that about 50,000 ha are under traditional or improved traditional culture, some 90,000 ha under extensive or modified extensive systems, about 20,000 ha under semi-intensive farming, and some 1000 ha under intensive farming (FAO/NACA, 1995).

The marine shrimp culture involves several steps: maturation of adult shrimp, spawning, hatching, larval rearing and growing to harvest size. The operations of first four stages are carried out in indoor tanks, and the last stage of culture is practiced in outdoor grow-out ponds. Discussion of these stages is beyond the scope of the book. Only a brief outline of larval rearing as an example of tank culture is given in the next section. The grow-out culture of marine shrimp is mostly carried out in earthen ponds of size 0.5–10 ha. The ideal average water depth, without aeration or water circulation, is 1.2–1.5 m, with minimum operational depth of 1.0 m. Intensive shrimp ponds in Taiwan use about 1.8 m depth of water. Stocking of larval shrimp directly into grow-out ponds, or rearing in nursery ponds and transferring the juveniles to the final grow-out ponds is usually the practice. In a typical example of practice as developed in Taiwan (Chen, 1976; Liao, 1986) shrimp fry is stocked in grow-out ponds of the size 0.1–0.5 ha with aeration equipment, and well-developed water distribution and drainage systems. Stocking densities as high as 30–40 PL/m^2 were practiced, and survival rates of 80 per cent and 8.4–11.2 tons/ha/crop production have been achieved.

Tank Culture

Aquaculture is practiced in tanks, usually made of concrete or plastic materials and in size smaller than ponds. Tanks are commonly used for small operations such as spawning, hatching and larval rearing but usually not for commercial fish production. Tank culture system is involved mostly in hatchery operations in brackish water areas. The ideal tank culture has several advantages such as self-cleaning characteristic of tank design, and provision of maintenance of high quality water for the cultured species (Wheaton, 1977). With land cost rise and scarcity of good quality water in several areas the tank culture system has become more attractive.

Circular indoor tanks are commonly used for maturation, spawning and larval rearing of shrimp species. Larval rearing tanks with sea water supply system and pretreatment of water can be described as a typical tank culture practice. Two-ton capacity wooden tanks (Figure 3.5) with conical bottoms have been used in Taiwan for larval rearing. Pre-filtered seawater, obtained from off-shore sub-sand wells, is first settled in reservoirs. Two-stage pressurized filters remove particulate matters, phytoplankton and certain hydrocarbon toxins, and in-line UV germicidal lamps finally sterilize the water. Tanks are filled with one ton sea water and then stocked with 50,000 to 100,000 nauplii per ton. The volume of water in the tanks is increased by daily addition of about 200 liters of water. Water management practice depends on density of diatoms feed, amount of particulate matter, concentration of metabolites, infections, etc. Larvae pass through several stages with adequate feeding, to grow into post-larva PL_5, which can be harvested for stocking into nursery ponds. For a 10-tank hatchery capacity maximum consumption of seawater is 25,000 liters per day, and compressed air requirement is 1430 liters per min. (Bose *et al.*, 1991).

Raceway Culture

Raceway culture units, which utilize flow-through water supply system, are widely used in hatchery operations. The reasons are that water quality and water quantity management are easier in

Figure 3.5: Schematic View of Larval Rearing Tank Showing the Strainer and the Supply Line for Water and Air (After Bose *et al.*, 1991)

raceways than in ponds. Flowing water flushes waste from the culture area and also forces the fish to swim thus increasing their survival rate when stocked into the wild (Wheaton, 1977). Raceway culture is also traditionally practiced for salmonids and other species in the USA.

A simple raceway design is shown in Figure 3.6. Pretreatment of incoming water by sedimentation and/or aeration and filtration, and pretreatment of effluent water before discharge into the environment may require (Lawson, 1997). Construction is mostly earthen or of concrete. A commercial raceway system for channel catfish in the U. S A. consists of 108 concrete raceways, each 11 m long by 5 m wide with a water depth of 1.2 m. Water flows by gravity from overhead tank with 6–7 minutes water retention time in each section. The system of culture is typically of inland fresh water species.

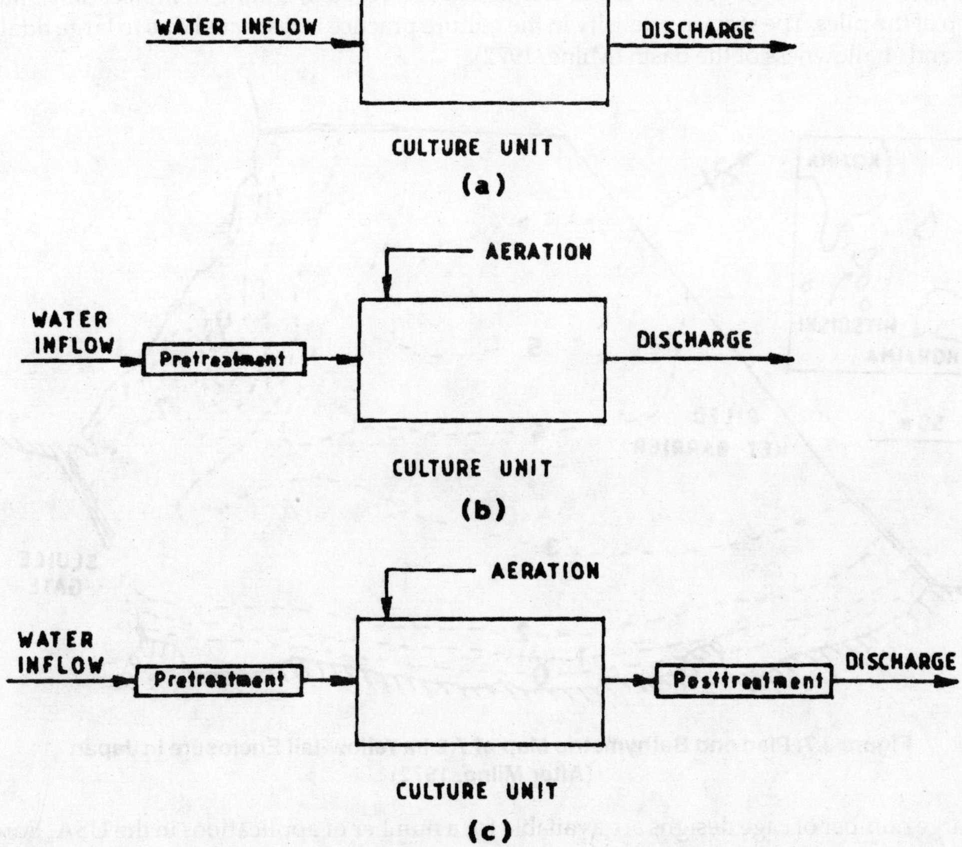

Figure 3.6: A Simple Raceway Design: (a) No treatment, (b) With pre-treatment, and (c) With pre- and post-treatment (After Lawson, 1997)

Culture in Enclosures, Pens and Cages

The technique of culture in the sublittoral zone involves the confinement of fish within structures like enclosures, pens and cages. Culture is practiced in open water bodies but with modification in that the organisms, being confined within the structures, grow in the vertical water column. In this way the production per unit surface is greatly increased (Wheaton, 1977). There are certain advantages of such culture system like less complex management than land-based systems and leasing of water bodies by non-land owning community. However, there are certain drawbacks such as of cage culture. Since cage facilities are built in public or multi-user water area, water pollution and deterioration of water quality conditions cause serious damage in culture. Stormy weather conditions like high wind, tides and waves may damage the cage installations. Further, negative impact on environment due to culture waste generation can introduce disease in the organisms (Beveridge, 1984).

Sublittoral enclosures for sea farming of oyster, yellow-tail, salmon and rainbow trout have been widely used in Japan and several European countries. Figure 3.7 shows plan of 7.2 ha yellow-tail enclosure in Japan. The channel between a large and small island was sealed off to provide a 7.2 ha enclosure. 350 m long net barrier is stretched between the two islands by hanging from a framework of concrete piles driven into the seabed. The 40-mm mesh, vinyl coated, is hung from steel cables attached to the top of the piles. The stocking density in the culture practice is restricted due to large tidal range of 2.4 m and shallowness of the basin (Milne, 1972).

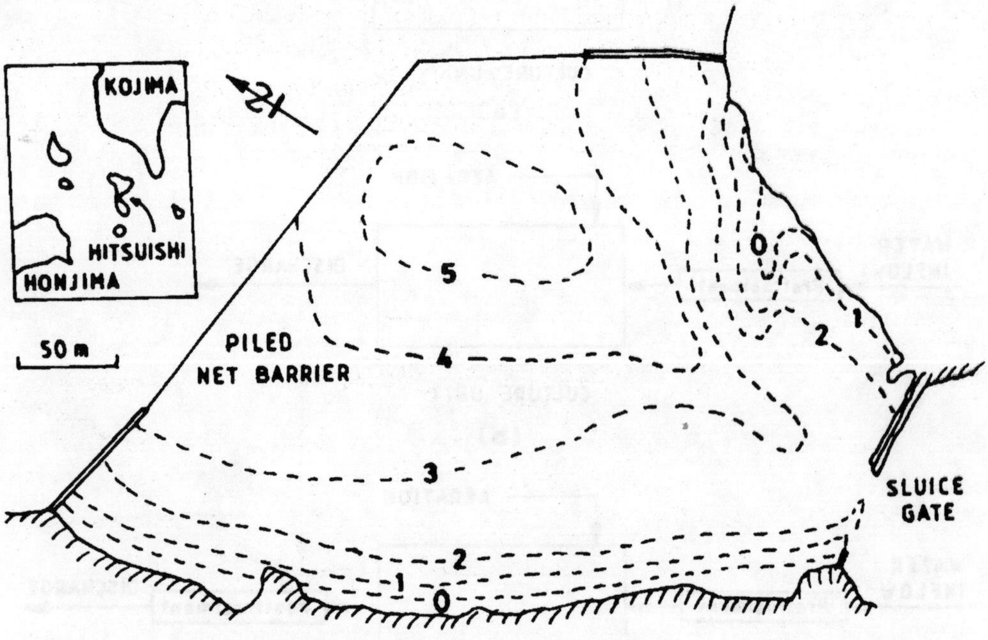

Figure 3.7: Plan and Bathymetric Map of 7.2 ha Yellow-tail Enclosure in Japan (After Milne, 1972)

A large number of cage designs are available for a number of applications in the USA, European and Far East Asian countries. The designs of the four basic types: fixed, floating, submersible and submerged are in use in many variations and combinations (Lawson, 1997). Floating net cages are the most popular method of yellow-tail culture in Japan. A single net cage, joined together to form a group of 4–10 is the most common. The cages originally constructed out of bamboo and oil drum floats are being replaced with Styrofoam cylinders. Bag nets of synthetic mesh, 10 m square by 3–5 m deep, suspended from the floating collars, are anchored at each corner (Figure 3.8). The group of these nets forms a large raft. The cages are commonly used for finfish culture. The red sea bream culture practice

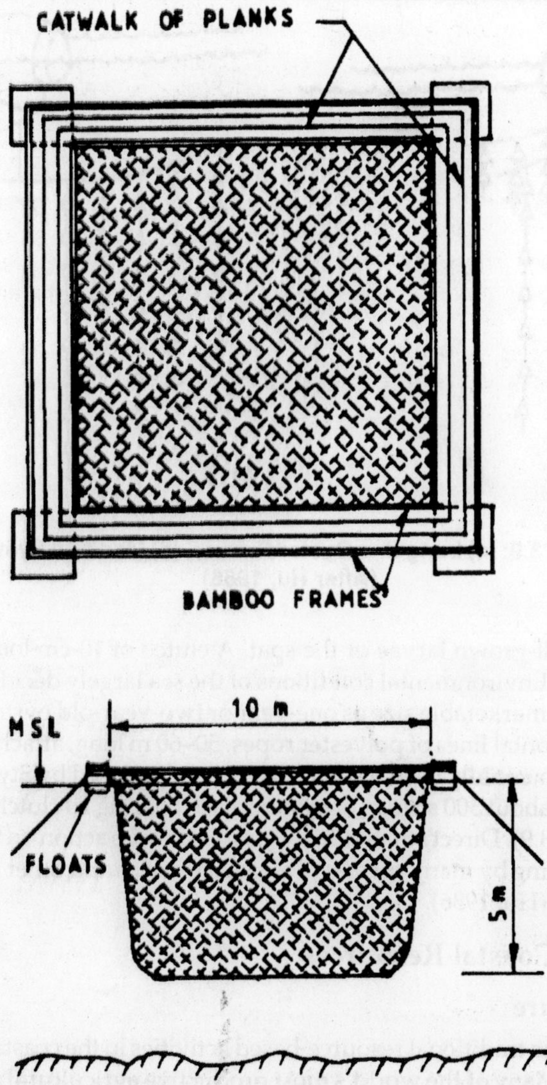

Figure 3.8: Floating Net Cage with Buoyancy Collar for Farming Yellow-tail in Japan (After Milne, 1972)

in Japan is followed in floating net cages. The cages are fitted in areas where the water current is between 5 and 15 cm/s. Water temperature fluctuates between 10 and 29°C in a year, the optimum temperature being 20–28°C. The nets are protected from rough seas, red tides and pollution. The juveniles of the fish, 30–100 mm in length, collected from the sea are placed inside the cages and immediately feeding started. The fish get accustomed to the new feeding within 12 hours.

The off-bottom culture techniques like floats, rafts, racks, trays and strings are employed for culture of molluscs such as oysters, mussels, clams and abalones. The edible oyster culture practice developed in Japan, Taiwan and other countries uses racks of bamboo and pine logs attached with strings as seed collectors in shallow areas of the seas. The strings contain clutches of oyster or scallop

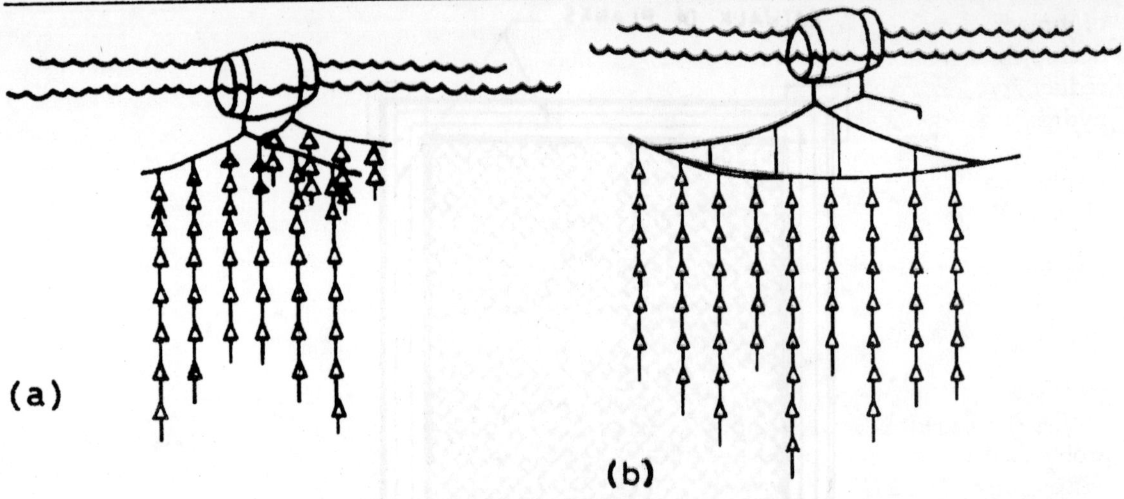

Figure 3.9: (a) Long-line Oyster Culture, (b) Modified System (After Hu, 1986)

shells for attachment of full-grown larvae or the spat. A clutch of 10-cm long shells usually gets an average of 25 oyster seeds. Environmental conditions of the sea largely decide the seed attachment to the strings. For growing to marketable size as one-year or two-year-old oysters, the long-line culture practice uses parallel horizontal lines of polyester ropes, 50–60 m long, attached to rows of floats. The oil drums coated with pitch initially used as floats have been replaced by Styrofoam boxes and balls. Each line, 4–5 m long with about 500 strings of oysters, each bearing 20 clutches, is anchored by 60 kg anchors at its ends (Figure 3.9). Direct stress due to wind and wave action on the strings, predation by oyster leeches, and bio fouling by many species of algae, barnacles and other organisms cause serious problems in oyster culture (Hu, 1986).

3.2 Multiple Uses of Coastal Resources

3.2.1 Coastal Agriculture

Agriculture is one of the traditional resource-based activities in the coastal areas besides fisheries, aquaculture and forestry. Many of the world's most productive agricultural areas are located in river deltas and coastal plains. Such areas generally have deep, relatively flat, fertile soils and benefit from a substantial supply of water, from surface and/or subsurface sources. Further, most coastal areas have a milder and more humid climate than the interior as a result of the moderating influences of the sea, particularly where favorable sea currents occur. In many developing countries the majority of the population is dependent on coastal agriculture for their livelihoods. In many small islands and in many countries such as Bangladesh and Egypt agriculture is concentrated on the coastal plains or in fertile river deltas and hence coastal agriculture production makes an important contribution to the local economy or to national agricultural production.

The crop-growing environment in the coastal area in India is categorized as tidal wetlands with four sub-divisions as perennially fresh water, seasonally or perennially saline water, acid sulphate soils, and peat soils. The lands are located in low-lying plains along rivers, seacoasts and in deltas. Under lowland submerged condition, several physico-chemical changes take place such as depletion of molecular oxygen, decrease in redox potential, increase in specific conductance, increase and decrease in pH depending on the soil reaction, reduction of Fe^{3+} and Mn^{4+}, denitrification and sulphate reduction. Flooded soil conditions also generate carbon dioxide, methane, toxins, organic acids and hydrogen sulphide. The most important change that takes place under anaerobic conditions is accumulation of ammonia in the crop root zone such as of rice plant. For phosphorus, reduction of insoluble ferric phosphate to more soluble ferrous one is the dominant process. Water-soluble phosphorus increases appreciably depending mostly on soil pH (Raghavaiah *et al.*, 1995).

1. Constraints Affecting Coastal Agriculture

There are some constraints that affect agriculture and agriculture development in coastal areas (Scialabba, 1998):

Proximity of the Sea

Many coastal agricultural lands are low-lying and subject to severe water logging and drainage problems due to periodic ingress of sea water during high tides and storm surges, and stagnation of rain and runoff water. There are extensive low-lying areas in the rice growing coastal belts of eastern and southern India, where poor drainage seriously affects crop production during the monsoon months.

Soil salinity problem is also caused by the seawater intrusion, and high, more or less, saline water table in the coastal areas. Saline and alkaline soils are commonly met in the coastal alluvium of littoral origin in India, where backwater flow or intrusion of seawater takes place. Ground water quality is also affected in wells in some coastal areas of the maritime states of India. Adverse effects of saline water on soil properties and crop growth increase with the concentration of salts and sodium, bicarbonate and boron content. The suitability of irrigation water is mostly judged on the basis of its salt concentration, proportion of sodium to other cations, bicarbonate concentration and boron content. Penetration of seawater far inland in the lower reaches of most streams and rivers during high tides also causes periodic increases in river salinity that may complicate its safe use for irrigation.

Some more consequences of the proximity of the sea are physical damage from wind storms or tidal waves, shoreline retreat and flooding as a result of coastal erosion, subsidence or rise in sea level due to global climatic change, and unfavorable climate in some areas.

Upstream Activities

Availability of surface water to coastal agriculture is affected by the water use in the upstream river flow. Dams and irrigation schemes, inland or near the coast, reduce the quantity and quality of surface flow. Silt removal and flood regulation activities affect fertility of coastal alluvium. Irrigation by extracting ground water, and cropping activities in coastal areas may lower the water table, and increase the salt intrusion. On the other hand, increased encroachment of forested land by agriculture, practices of slash and burn in the forests, overgrazing and inappropriate cultivation practices may increase runoff and erosion in catchment areas and as a result sedimentation in rivers, lowering of dry season river flows and increased flooding take place. Upstream discharges of industrial and domestic effluents also affect the quality of surface water available to coastal agriculture.

Limited Land and Water Resource

Coastal areas face problems of limited land and water resources for extension of agriculture further from the shore, estuary or port. Urbanization intensifies competition for land and water use. Agricultural activities of small holders like grain crops and grazing require extensive amounts of lands, and thus tend to be marginalized by those requiring higher capital inputs and less land for higher-value outputs such as livestock raising or horticulture. Similarly irrigation of agricultural lands using high volumes of water may face growing competition for water from residential and industrial uses, and from other types of agriculture owing to increasing water scarcity.

Reclamation of coastal saline lands for cultivation has been practiced in many counties by construction of suitable embankments with one way sluice gates to stop ingress of seawater but allow the monsoon rain water to drain out. Thousands of hectares of land have been thus protected in the coastal states in India. But most of the area still remains uncultivated due to lack of proper drainage and leaching.

2. Crop Selection for Salt-affected Soils

In general, salt-tolerant species or varieties should be preferred. Rice cultivation, being adapted to a variety of agro-hydrological conditions, is also grown with high yielding varieties mostly in the coastal area. In lowland, the rice crop is not irrigated but the soil can remain flooded, for at least a portion of the crop to a sustained water depth. In the coastal belt of West Bengal, soil is rich with pH 6.5–7.5 and total soluble salt content 3.0–18.0 mmhos, and medium to late duration varieties with one crop in the *kharif* season is grown. Short or medium duration high yielding varieties can be grown successfully in areas, where drainage facilities could be made by construction of drains and cross-drains (Mukherji, 1978).

Sorghum and *bajra* are semi-tolerant in saline conditions. Turnip, spinach and radish are suitable crops for such soils. Improved varieties of wheat, barley and *bajra* are more suitable than others for areas irrigated with saline waters (Paliwal, 1972).

3.2.2 Coastal Forestry

Technically forest is defined as an area set aside or maintained under vegetation for any indirect benefits, viz. climate, protective or environmental and/or for production of wood and non-wood products. FAO Forestry Department defines forests as 'vegetation formations with a minimum of 10 percent crown cover of trees and/or bamboo with a minimum height of 5 m and generally associated with wild flora, fauna and natural soil conditions.' Forest resources (including wild life) of coastal areas vary widely from their inland counterparts. Mangroves and tidal forests have no parallels in terrestrial uplands. In many countries, coastal areas such as beaches, dunes, swamps and wild lands, even without any trees, are officially designated as 'forested' lands. Forests constitute the largest, most complex and most self-generating of all eco-systems.

1. Types of Coastal Forests

Usually the inland forests and woodlands in each climatic regions of the world extend to the sea, thus forming a part of the coastal area. In addition to such forests, special forest communities are found in extreme water regimes of coastal areas and along inland rivers. These are mangroves, beach forests, peat swamps, periodic swamps (tidal and flood plain forests) in the coastal area, and permanent freshwater swamps and riparian forests along inland rivers (Scialabba, 1998).

Mangroves

Mangroves are the most typical forest formations in the coastlines of the tropics and subtropics. Trees and bushes growing below the high water level of spring tides are called mangroves, although the term itself also applies to the ecosystem. The root systems are usually submerged below the saline water, which is periodically diluted by freshwater surface runoff.

There are some 16 million ha of mangrove forests in the world (FAO, 1994). These are distributed in two distinct biogeographic zones: those of West Africa, the Caribbean and America; and those on the east coast of Africa, Madagascar and the Indo-Pacific region. Mangroves of the Indo-Pacific region are richer with 40 tree species. Mangroves form an evergreen closed high forest 30 m or more in height of *Heritiera* or less commonly *Bruguiera* species. There may be an underwood of same species or of others such as *Ceriops*. This two-storeyed forest type is of common occurrence due to the prevalence of trees whose maximum height is only 5-10 m mixed with others capable of further development. On areas flooded only by spring tides there is more varied growth of *Pandanus* spp., canes, ferns, etc. *Rhizophoraceae* replace *Heritiera* nearer the sea. The flora is a poor one and besides the genera of *Rhizophoraceae*, which are almost confined to this type, there are a few other genera, each usually with several species particularly adapted to the unusual conditions, such as *Heritiera, Carapa, Sonneratia, Avicennia, Excoecaria, Cerbora, and Cynometra*. Shrubs are few, *Acanthus* spp. being the most common.

Mangrove forests shelter most of the animal species found in the other environments like beaches, rivers, fresh water swamps or in other forests near water. Mangroves represent the last refuge of a number of rare and endangered animals such as the proboscis monkey in Borneo, and the royal Bengal tiger and the spotted deer in the Sunderbans in the Bay of Bengal coast. Mangroves are an ideal sanctuary for birds, the number of species in each of the regions ranging from 150 to 250. Some 65 species, listed as endangered or vulnerable, live in the mangrove forests.

Beach Forests

This type is in general found above the high-tide mark on sandy soil and dunes and may extend into agricultural land or upland forests. The vegetations are mostly scrub type comprising stunted tree growths. These forests are adapted to difficult growing environments of extreme soil or climatic conditions such as strong winds, salinity, and lack or excess humidity. They are very sensitive to modifications of the ecosystem and may die with slight changes in the groundwater level.

Swamp Forests

There are various types of swamp forests, which are mainly defined by habitat rather than on structure and physiognomy. Peat swamp forests were formed by the sea level rise in the early geological era in parts of Sumatra, Malaysia, Borneo and New Guinea. Tidal swamp forests are periodically flooded by fresh or brackish river water and are found on higher elevations than mangroves. They are characterized by more diverse natural vegetation than mangroves. Floodplain forests are seasonally flooded by rainwater rather than by tides, and are of natural vegetation types. Fresh water swamp forest is permanently wet and characterized by its eutrophic richer plant species.

Riparian Forests

This type, also called riverine or gallery forests, grows adjacent to or near rivers and are extremely dense and productive with large number of climbing plants. These forests have important indirect use values in preserving water quality, controlling erosion and as wildlife refuges. There are several other coastal forest formations and plantations, which are under severe threat of extinction or degradation by overexploitation and conversion to agricultural land.

Coastal forests in India fall under the category of littoral and swamp forests on coastal sand, tidal swamp forests on estuarine sand, fresh water swamps on wet alluvium and riverine fringing forests. Among the important species are *Casurina, Heritiera, Manilkara, Rhizophora*, etc. Canes and ferns form the undergrowth (Agarwala, 1985).

2. Socio-economic Values of Coastal Forests

Economic valuation of natural ecosystems has some shortcomings. They require much time for collection of diverse and extensive data, and environmental goods cannot be evaluated in monetary terms rather on value judgment. Ecosystems are, therefore, valued according to (*i*) direct use values in terms of commercial use such as timber and other forest products, (*ii*) indirect use values comprising all the ecological functions within a system such as coastal protection from flood control, (*iii*) option values considered as extra insurance against the risk of good and services, and (iv) non-use values as intrinsic and inherent characteristics of a system for future generations. The option and non-use values are difficult to convert into monetary terms and are defined qualitatively.

Table 3.13 illustrates the values related to coastal forests. Direct uses in the form of products obtainable from coastal forests range from timber, poles and post to food, beverages and drugs. Many direct uses imply social functions such as habitat to forest dwellers and opportunities for education, research, recreation and tourism. Indirect use values are illustrated by the prominent environmental role of the coastal forests in nutrient cycling in the mangrove-marine food web (Figure 3.10). The contribution of mangrove forests in feeding, breeding and nursery rearing of numerous commercial fish and shellfish is well-established. The value is not, however, directly attributed to mangroves, as many of these fish and shellfish are caught offshore. Other important indirect use values include valuable physical habitat for wildlife species; buffering action against wind, waves and water currents; and sediment trapping for protection of sea grass beds, reefs from siltation and for reducing water turbidity. The option values are in the form of premium to pay for preservation and use by future generations.

Figure 3.10: Mangrove-Marine Food Web (After Scialabba, 1998)

Table 3.13: Values Related to Coastal Forests (After Scialabba, 1998)

Values			
Use values			**Non-use values**
Direct uses	Indirect uses	Option uses	Objects of intrinsic values
Timber	Nutrient cycling (including detritus for aquatic food web)	Premium to preserve future direct and indirect uses	Endangered species
Non-timber products (including fish and shellfish)	Watershed protection		Threatened or rare habitats
Recreation and Nature tourism	Coastal protection		Cherished landscape
Genetic resources	Air and water pollution protection		Cultural heritage
Education and Research	Microclimate function		
Human habitat	Carbon store		
	Wildlife habitat (including birds and aquatic species)		

3. Special Characteristics of Coastal Forestry

Interaction with Terrestrial and Marine Ecosystem

Mangrove-marine food web, as illustrated in Figure 3.10, is an example in arid tropics, where leaf litter and detrirus brought from the terrestrial zone by tidal flushing of coastal mangrove areas is the only source of nutrients during dry season. Many wildlife species in coastal forests are also dependent on other ecosystems such as mammals, migratory birds, and marine shrimps and salmon..

Natural or Human-incurred Risks and Uncertainties

Natural risks such as strong winds, hurricanes and typhoons, floods and droughts cause considerable damage to coastal forests. Competition for land, water and forest resources incur many threats to coastal forests such as deforestation, construction of dams and flood control measures, and air and water pollution.

Planning and Management

The long time needed by trees and some animals to mature makes management objectives for regeneration and harvesting very difficult. Once degraded and altered mangrove swamps cannot be restored to their original state. A typical example is acid sulphate soils in many mangrove and tidal forest areas, which becomes difficult to reclaim. Some of the socio-economic management issues are risk of conflict arising from tenure and property rights of the forest resources by the state or state agencies, jurisdiction of different institutions, and lack of awareness of benefits and economic values of forest resources.

3.2.3 Waste Disposal and Pollution in Coastal Areas

1. Sources of Pollution

The principal sources of water pollution, in general, according to nature of its origin, are: (1) point sources at outfall sewers or drainage channels, (2) non-point or diffuse sources conveying pollutants from land-based activities, and (3) atmospheric deposition entering aquatic environment. Usually the sources of pollution are coastal and upstream points, the latter lying mostly within the wider drainage basins and beyond. The main point sources are municipal and industrial wastewater discharges as treatment plant effluents, and development sites such as land clearance and excavation. A diffuse source is a broad unconfined area from which pollutants enter a water body. Agricultural land drainage/ or surface run-off of crop treatment residues and animal wastes, and urban storm run-off enter the water body all along a stream as they flow through the area.

The distinction between point and diffuse sources of pollution is illustrated in Figure 3.11. Various point and diffuse sources of pollution resulting from multiple uses of coastal resources are given in Table 3.14.

Table 3.14: Various Point and Diffuse Sources of Water Pollution in Coastal Area

Point Sources	*Diffuse Sources*
Domestic wastes from urban settlement, ports and beach resorts	Pesticides and nutrients from agriculture
Heavy metals, chemicals from industries	Nutrients and organic matter load from aquaculture farms
Oil slicks from marine and land-based activities	Drainage from livestock pastures
Nutrients and organic matter load from feedlots	Urban storm run-off
Cooling water discharge	Sand and gravel mining
Sediment discharge from development sites	

Because of its natural salinity and tremendous amount of dilution the ocean apparently provides, there is a natural tendency to consider the ocean as a convenient 'sink' or receptacle for wastes of all kinds. In coastal areas, all the sewage and industrial effluents discharged into streams and rivers ultimately make its way to the ocean. Treated effluent is discharged directly into the ocean. Estuaries being adjacent to the land are the first marine areas to receive wastes carried by the river flow.

Atmospheric deposition occurs from two categories of sources: short residence type materials such as metals released from nearby smelting activities, and long residence time components, such as organic pollutants, distributed widely on regional or even global scales. In fact, many contaminants discharged to estuaries and rivers reach the open sea regions in a relatively small fraction of the material. Atmospheric input for most contaminants is much more significant than riverine input (GESAMP, 2001). GESAMP (1990) report shows that the marine pollution is derived mainly from land-based sources and the atmosphere (Figure 3.12), and river runoff and land-based discharges directly affect the coastal waters.

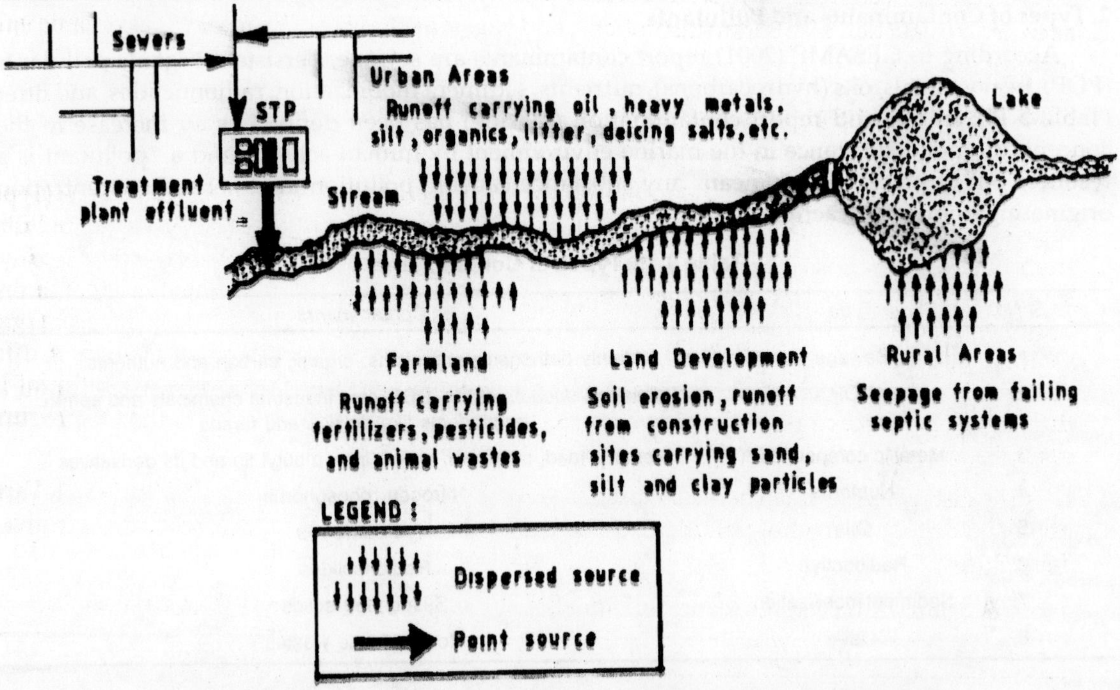

Figure 3.11: Schematic View of Dispersed and Point Source Pollutants

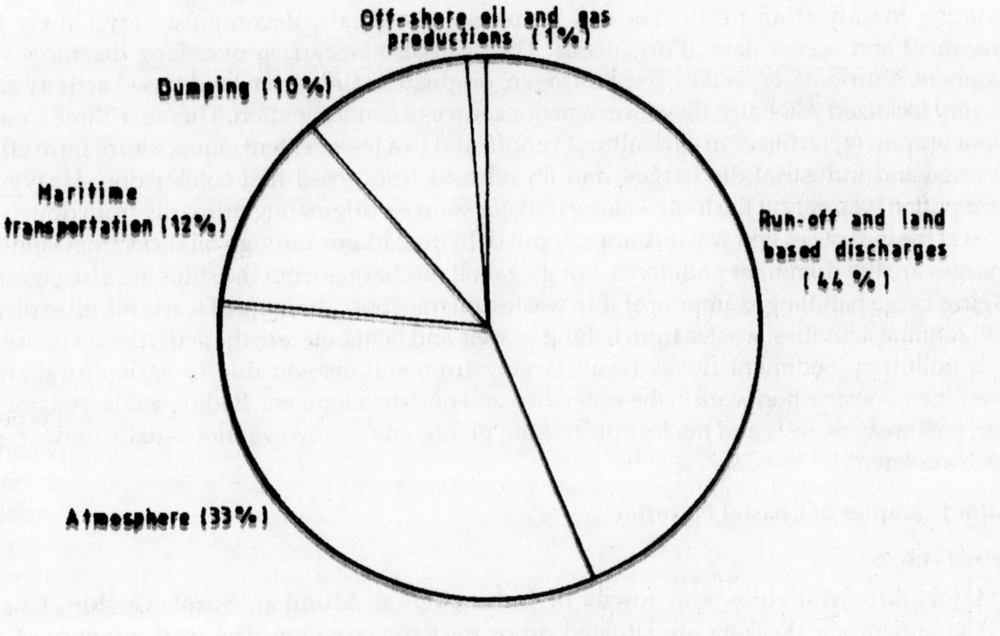

Figure 3.12: Major Sources of Marine Pollution (After GESAMP, 1990)

2. Types of Contaminants and Pollutants

According to GESAMP (2001) report contaminants are sewage, persistent organic pollutants (POP), heavy metals, oils (hydrocarbons), nutrients, sediment mobilization, radionuclides, and litter (Table 3.15). In the said report contamination (marine) has been defined as an increase in the concentration of a substance in the marine environment by human activity, and a "pollutant is a resource out of place" which means any substance causing pollution in abnormal concentration originating from human activities.

Table 3.15: Types of Contaminants

Sl.No.	Type	Contaminants
1.	Sewage	Mainly pathogenic organisms, organic carbon and nutrients
2.	POP	Pesticides such as aldrin, DDT, etc.; industrial chemicals and some by-products like dioxins and furans
3.	Metallic compounds	Mercury, lead, cadmium, nickel, iron, tributyl tin and its derivatives
4.	Nutrients	Nitrogen, phosphorus
5.	Oils	Hydrocarbons
6.	Radioactive	Radionuclides
7.	Sediment mobilization	Suspended solids
8.	Litter	Mostly plastic wastes

Sewage, as only domestic or combined with industrial discharges, is a major source of contamination in marine and coastal waters in the global scale. Persistent organic pollutants (POP), originating mainly from pesticides and industrial chemicals, decompose very slowly in the environment and accumulate in organisms. They can also be carried over long distances via the atmosphere. Nutrients, especially fixed nitrogen, originate mainly from land-based activity and are often very localized. Globally, they form a major source of contamination. There are three sources of nutrient output: (*a*) fertilizer in agricultural runoff, and to a lesser extent, aquaculture farm effluent; (*b*) sewage and industrial discharges, and (*c*) releases from fossil fuel combustion. Heavy metal contamination is largely at the local scale near major sources, originating primarily from downstream sources at the industries and waste dumps. Input of hydrocarbons through oil slicks from land-based discharges are the dominant pollutants, but illegal oil discharges from the ships are also significant. Dredging, cargo handling, dumping of ship wastes, oil transport, draining of waste oil, oil exploration and oil refining activities, wastes from fishing vessels and boats, etc. are the activities responsible for coastal pollution. Sediment fluxes result largely from soil erosion due to agricultural activity, deforestation, construction work in the watershed and port development. Radionuclide contamination is from past weapon tests and nuclear processing plants, and is however not usually under concern of risk assessment.

3. Major Examples of Coastal Pollution

Industrial Wastes

Major industrial cities and towns of India such as Mumbai, Surat, Cochin, Chennai, Vishakhapatnam and Kolkata are situated on or near the coastline. The total quantity of waste discharged by the industries has been estimated to be approximately 700 Mm^3 per year (DOD, 1996).

Pollution potential in industrial effluents in coastal areas of the maritime states in India is 3.73 Mm^3/d (CPCB, 1996). Pollutants present in industrial effluents show a wide diversity in composition. They range from toxic heavy metals such as cadmium, lead, mercury and arsenic discharged from electroplating, chemical and metallurgical industries to organic substances from sugar mills, refineries, distilleries and tanneries. The Hooghly river estuary shows an estimated concentration of heavy metals about 1.45 percent of the total suspended solids (FAO/NACA, 1995). The river Kalu, north of Mumbai, has a mercury concentration exceeding 100 mg/l (Sa de Elrich, 1999).

In coastal waters total atmospheric input of fixed nitrogen mainly from burning of fuels by industry and vehicles in European and North American countries has increased by 50–200 per cent during the past 50 years. This input falls not only directly on the water surface but also are falling on watersheds and entering coastal waters via rivers and streams. The input of atmospheric fixed nitrogen to the open sea from the energy sources and fertilizers will rise significantly particularly from the developing countries. The estimated increase in the formation of nitrogen oxides (NO_x) from energy use between 1990 and 2020 has been shown to be 40 per cent against 87 per cent increase in fertilizer nitrogen production in Asia, whereas it is 10 per cent against 2 per cent in North America (GESAMP, 2001).

The major international route for oil tankers originating from the Gulf transports huge quantities of crude oil to India. Any major accident in the area spills few thousand tons of oil into the sea. In addition to offshore oil exploration and production activities any transfer operations of oil cause spillage of oil. In the global scale, the overall total influx of oil to the sea from sea-based activities is of the order of 850,000 tons per year, with an additional 350,000 tons estimated from coastal refineries, storage and transportation facilities, oil seeps and other unknown sources (GESAMP, 2001).

Domestic Sewage

Domestic sewage is discharged mostly in semi-treated or untreated conditions into streams and rivers, lakes and estuaries in all but a few cities in India. In general the condition is alarming in developing countries, where an estimated 90 percent of waste water is discharged directly to rivers and streams without any waste processing treatment (WMO, 1997). The total volumes of discharge from Kolkata and Mumbai are 350 and 365 million tonnes per year, respectively. The Mahim bay in Mumbai receives 64 million tonnes of domestic sewage and 0.9 million tonnes of industrial effluents every year (FAO/NACA, 1995).

Agricultural Wastes

Agricultural nutrients and pollutants enter river, ground water and marine systems from both diffuse sources (cultivated plot) and point sources (dairies, feedlots, processing plants). The nutrient and pollutant losses from diffuse sources depend on the types of fertilizers, manures and pesticides, and their present application rates previous records. The losses from point sources depend on location, type and number of animals raised in a feedlot; and waste treatment, transport and disposal. Location, type and quality of agricultural produce processed, and the waste treatment, transport and disposal will affect the nutrient and pollutant losses from agro-industrial processing units. India, being one of the major agricultural countries, generates wastes annually to the extent of 5 million tonnes of fertilizer residues, 55,000 tonnes of pesticide residues and 1,25,000 tonnes of synthetic detergents. It has been estimated that nearly 25 percent of all these residues find their way through the rivers into the coastal environment.

Solid Wastes

Disposal of solid wastes such as plastic materials on land and from ships results in littering of beaches. An international coastal cleanup in 75 countries in 1997 showed valuable data on sources and amount of coastal debris. 2.8 million kg of dangerous and unsightly trash cleaned from 14,000 km of beach comprises the majority as plastic materials (62 per cent), followed by metal (12 per cent), glass (10 per cent), paper (10 per cent) and wood (3 per cent). Plastics have increased from 54 per cent in 1993 to 62 per cent in 1997 (Center for Marine Conservation, 1998).

4

Impact of Coastal Resource Use on the Environment

Coastal areas–the niche of highly diverse and significant natural resources of the earth–have come under conditions of severe degradation and loss of resources in the recent decades. Marine, estuary and coastal wetland areas, constituting mainly the coastal environment, provide the unique biological ecosystems. The ecosystems are strongly influenced by social and economic systems, and population pressure and increasing demands for space and resources undermine the sustainable use of coastal areas and of their resources. Globally, environmental degradation has continued and even intensified in many cases. However, degradation is much more severe in the coastal areas than in the open ocean.

GESAMP (1991) defines marine pollution as "introduction by man, directly or indirectly, of substances or energy into the marine environment (including estuaries) resulting in such deleterious effects as harm to living resources, hazards to human health, hindrance to marine activities including fishing, impairment of quality for use of sea water and reduction of amenities". GESAMP (1991) further advocates a concept of environmental capacity defined as a property of the environment and described as the "ability of a receiving system or ecosystem to cope with certain concentrations or levels of waste discharges without suffering any significant deleterious effects" (Barg, 1997). These definitions and concepts discriminate between the terms "contamination" meaning increased concentration of substances with no significant adverse effects, and "pollution" signifying the occurrence of adverse effects, as has been mentioned earlier.

The nature and severity of the environmental impacts is usually judged on local, regional and global scales. On a global basis, problems have been categorized as (*i*) actual damage or compromise to marine resources, and (*ii*) potential threats of damage. The most serious problems affecting the quality and uses of the marine and coastal environment to the extent of actual damage or potential threats of damage are (GESAMP, 2001):

1. Alteration and destruction of habitats,
2. Effects of sewage on human health,
3. Hypernutrification and eutrophication,
4. Decline of fish stocks and other renewable resources, and
5. Changes in sediment flows due to hydrological changes.

Impacts created in the coastal areas due to these problems have been judged with respect to the four categories of problems: food security, public health, ecosystem health and economic health. Although agriculture and fisheries (including aquaculture) contribute to food production and income generation in many countries, some of their activities can degrade coastal ecosystems, severely damage the renewable resources and cause adverse effects, directly or indirectly, on food security. The human health concerns due to impacts on coastal environments are infections during bathing and boating in coastal waters through direct contact and by consumption of seafood through indirect contact. A number of suitable ecosystem health indicators exist ranging from reductions in populations of many organisms due to contamination to eutrophication due to nutrient discharge. Habitat loss such as of coral reefs and mangrove forest is of increasing concern for ecosystem health and biodiversity. Degradation of habitats adversely affecting fish abundance, quality of seafood due to toxicity problem, and fish kills are examples of economic and social benefit loss.

4.1 Impacts on Coastal Environment

4.1.1 Alterations and Destruction of Habitats

Habitat destruction and ecosystem alterations are the most widespread human impact on coastal environment. These impacts are caused by physical (*e.g.* landfills, sedimentation), chemical (*e.g.* water pollution) or biological means (*e.g.* introduction of non-indigenous species).

Developmental activities in the hinterland areas of the coast lead to changes in water, sediment and nutrient fluxes through river drainage basins to the sea. Most of the world's major river deltas suffer from decreased transport of sediments. Developmental activities in the coastal foreshore areas have an impact on the environment in terms of flow modification, of turbidity generation, and of effects on biological communities in beach, littoral and sub-littoral zones. Thus the natural coastline considerably changes and the areas previously covered by dunes, wetlands and mangroves significantly reduce. These ecosystems suffer all over the world, and as a result fisheries and aquaculture are affected due to degradation of fish spawning nursery and feeding grounds.

Pollution from sewage and various chemical compounds as outlined in the previous section significantly affect the ecosystem, even to the extent of destruction of the whole ecosystem. Many contaminants, if assimilated by marine organisms, can affect their viability and/or reproductive capability. Nutrients promote blooms of phytoplankton and benthic algae, which increase water turbidity, reduce light penetration and thus adversely affect the pelagic and benthic communities such as coral reefs and sea grass beds. Accidental or deliberate introduction of alien species changes the ecological balance of the region into which they are introduced. The adverse effects can be reduction or even extinction of indigenous species, damage to fisheries, and wholesale changes to ecosystem.

4.1.2 Effects of Marine Pollution on Human Health

Coastal waters contaminated by sewage and other wastewater discharges give rise two primary concerns of human health: direct contact through bathing, swimming and boating activities, and indirect contact through the consumption of seafood. Marine coastal beaches contaminated with pathogens from sewage pollution give rise to risk of enteric and respiratory infections and diseases among bathers even from marine waters with a mean concentration of 50 faecal streptococci/100 ml. Outbreaks of cholera and typhoid fevers, viral diseases and other illnesses are frequently traced to pathogen-contaminated seafood and bathing waters. These health risks are particularly high in regions where carriers of pathogens are common among the people and sewage treatment and disposal systems are not adequate.

Consumption of contaminated seafood can lead to three categories of threats to human health: microbial, chemical and radiobiological. Several diseases, as mentioned before, transmitted through consumption of seafood such as shellfish, often raw or lightly steamed, fall within the microbial threat to human health. In the marine environment pathogenic bacteria can remain viable for several weeks and viruses can survive for months. Filter-feeding shellfish breed in areas often placed near the sources of pollution such as sewage outfall in the sea or estuary, and are prone to accumulating high levels of pathogen.

Microbiological, natural poisonous organic agents and human-induced contaminants fall within the chemical category and are called natural poisons, such as shellfish poisons. Many poisoning diseases such as of amnesic and paralytic shellfish, ciguatera, neurolytic and diarrheic shellfish are apparently associated with fish and seafood that feed on toxic marine algae and toxic algal blooms. Risks to human health resulting from seafood contamination of mercury, cadmium and copper remain a major concern predominantly on the local scale. However, apart from a few extreme cases of mercury and cadmium poisoning, contamination from inorganic chemicals such as transition metals are doubtful cases of most serious marine pollution from public health perspectives. Impacts of persistent organic pollutants (POP) on human health are also of doubtful significance. But there are concerns of possibility of immuno-suppression in mammals from both acute and chronic low-dose exposures.

Impact of radionucleotides on human health and environment at global and regional levels are usually of minor significance, as all practices involving production, use and disposal of radioactive materials and contamination of significant quantities of radionucleotides are under the Basic Safety Standards of International Atomic Energy Agency (IAEA). IAEA (1993) conducted an evaluation of comparative risks associated with chemical carcinogens and radionucleotides through seafood consumption. Naturally occurring radionucleotides can pose a greater risk than that from disposed radioactive materials in the sea. On the other hand, some chemicals of the POP group carry risks either similar or larger than those arising from sea disposal of radioactive wastes.

4.1.3 Hypernutrification and Eutrophication

Eutrophication in coastal waters has increased and become widespread globally due to the input of nutrients, particularly nitrogen and phosphorus compounds. Substantial and measurable increase in the concentrations of dissolved nutrients is termed as hypernutrification, whereas the consequent significant increase in phytoplankton growth and productivity is termed as eutrophication. The phenomena of *eutrophication* are used in limnology to describe the normal processes that occur in lakes and lead to their extinction by gradual filling and aging. In freshwater environment dissolved inorganic phosphorus is the most important growth limiting substrate, whereas dissolved inorganic nitrogen is the most important in the marine environment. The term *eutrophication* is used synonymously with pollution in reference to lakes. A more accurate characterization to the problem would be the term *cultural eutrophication*, meaning the acceleration and hastening of the natural aging process because of human activity.

The process eutrophication is directly related to the aquatic food chain and is controlled by the limitations of plant nutrients. Abundant nutrients unbalance the normal succession in the food chain and promote blooms of blue green algae that are not easily utilized as food by zooplankton. Thus the water becomes turbid and appears as "pea soup" under extreme conditions. Floating masses of algae are sometimes windblown to the shore, and are decomposed there. Eutrophication creates various impacts: (i) Decaying algae settle to the bottom of the coastal and marine waters and cause DO depletion which may be periodic or permanent, leading to mass mortality of fish and other organisms; (ii) Harmful algal blooms involving toxin-producing species frequently cause very serious human health problems through consumption of contaminated seafood as described before; and (iii) Eutrophication can turn marine areas into wastelands, with severe loss of productivity including the abundance of fish.

4.1.4 Decline of Fish Stocks and Other Renewable Resources

Impact on fish and other organisms relate directly to the four other problems in the marine and coastal regions listed above. As has been pointed out above, damage or destruction of ecosystem alters the habitat as a result of pollution from sewage, industrial and other wastes, and ultimately lead to reduction or even extinction of many species and damage to fisheries. Hypernutrification and eutrophication from nutrient pollution can turn rich and productive marine areas into wastelands, with severe loss of productivity including that of fisheries.

4.1.5 Changes in Sediment Flows

There are two problems on changes in sediment flows–increasing mobilization of sediments from developmental activities, and decreasing natural sediment supply in river runoff. The rate of deforestation due to hinterland and foreshore development activities is a major cause of increased sediment mobilization. Figure 4.1 shows area deforested and rate of deforestation between 1980-1990 in developing countries (GESAMP, 2001). The other causes of increased sedimentation are erosion and diversion of water courses. Material eroded from the coastline or brought to the sea from other sources by river and its tributaries is usually transported in the inshore areas. Beaches, estuaries and tidal flats are such areas where accumulation of suspended sediment is likely to be high (Boaden and Seed, 1985). In estuaries sediment settles down in specific areas and forms sediment bars, sometimes in the neighborhood of tidal creeks. In temperate areas, such increased sedimentation gives rise to blanketing of benthic community. In tropical areas, coral reefs are damaged significantly. Promoting vegetation including mangrove forests decrease the wave action and reduce sedimentation. A dense permanent grass and weed vegetation prevents fresh deposited sediment being eroded. In the Sunderban areas of the Bay of Bengal coast a luxurious mangrove population enhance mud deposition especially in the upper intertidal zones and provide resistance to bank erosion by their dense root system (Bose *et al.*, 1991).

Decrease in natural sediment supply to coastal waters is caused by the reduced flow of rivers and thereby leads to accelerated coastal erosion. Water flow being the most effective transport mechanism of river bank erosion, any fluctuating river flows cause problem in coastal ecosystem (Bose *et al.*, 1991). In the world, very few rivers run entirely free of man-made obstructions such as dams, which significantly reduce particle discharges from the rivers. Excavation work, forestry and agriculture are also responsible for degradation of water courses. Reduced sediment supply gives rise to decrease in

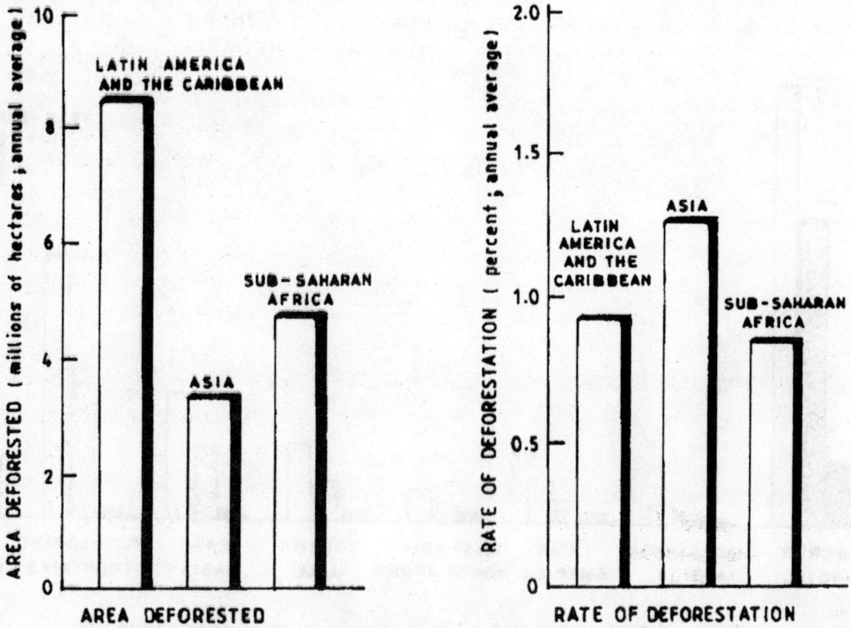

Figure 4.1: Loss of Tropical Forest in Developing Regions, 1980–1990 (After GESAMP, 2001)

natural inflow of fresh water, certain chemicals, nutrients and suspended particles and to under-nourishment of beaches and fine shelf sediments.

4.2 Potential and Specific Cases of Impacts

4.2.1 Agricultural Activities

Table 4.1 shows the major activities of coastal agriculture that can adversely change coastal environment and create impacts on coastal ecosystems. These impacts results from the four categories of problems as described before, and range from decreased fish yields through human health hazard to coastal erosion. It is an overview of potential impacts. The specific impacts of agricultural activities can be best illustrated with agricultural intensification by increasing use of fertilizers and pesticides in the global scale (Figures 4.2 and 4.3). The amounts of nitrogenous compounds and persistent organic pollutants (POPs) entering the marine environment have continued to increase and create impacts in most regions of the world (GESAMP, 2001).

4.2.2 Capture Fisheries and Coastal Aquaculture Activities

Potential impacts on the coastal ecosystems and on social sectors corresponding to the changes on the environment as a result of capture fisheries and aquaculture in the coastal waters are shown in Table 4.2. The table gives an overview of the impacts of coastal aquaculture, besides that of capture fisheries. The main adverse effects of various culture systems and of species cultured with respect to concerns of key areas of ecosystem have been described with specific examples in Tables 4.3–4.5.

Type and degree of actual damage by aquaculture development will depend on the method of aquaculture, level of production, and the biological, chemical and physical characteristics of the coastal area. Some impacts have been studied in detail, and some are either poorly understood or

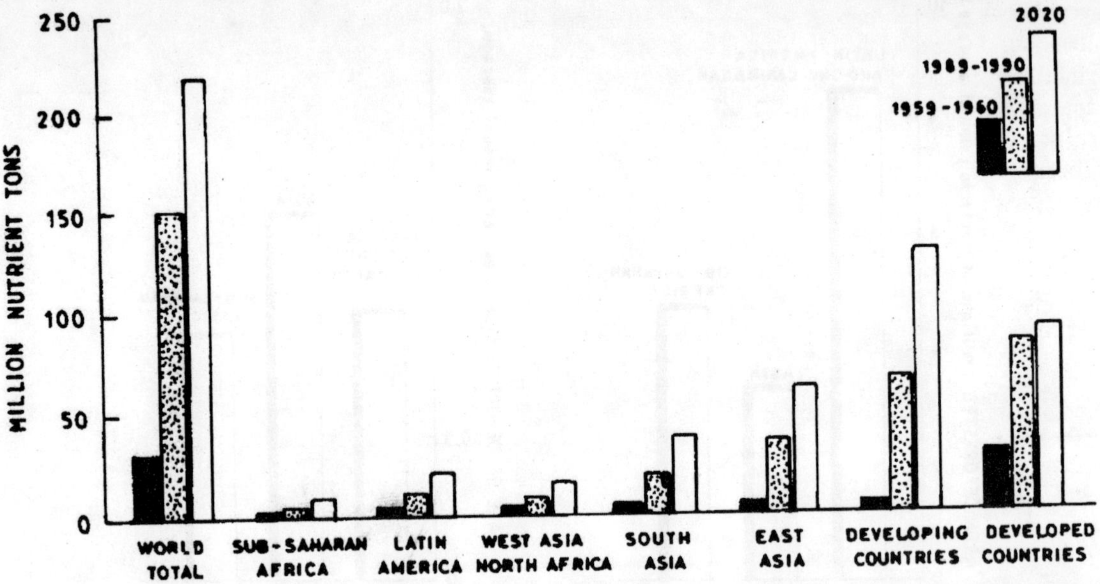

Figure 4.2: Estimated Growth in Fertilizer Use, 1960–2020 (From WRI, 1998)

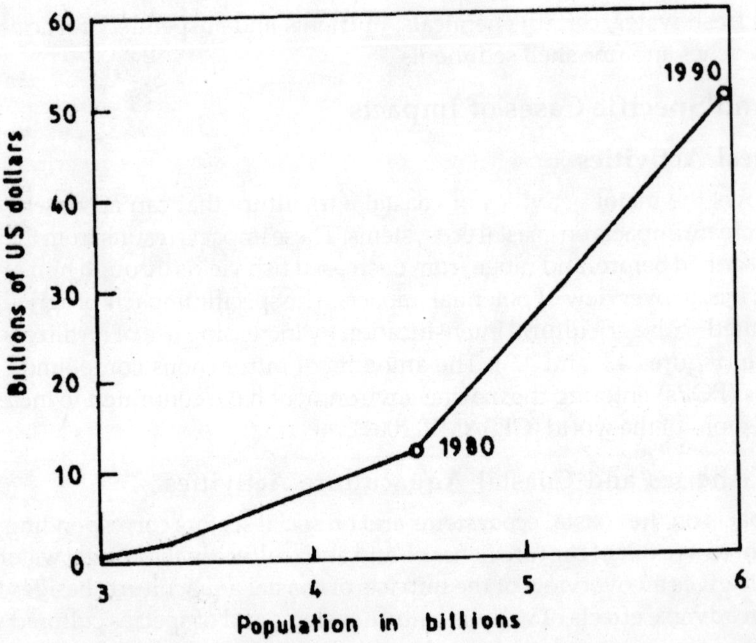

Figure 4.3: World Sales of Pesticides in Relation to Global Population, 1970–1990 (From WRI, 1998)

perceived as potential impacts. There are few examples of eutrophication from coastal waters due to nutrient and organic enrichment. In recent past there have been many instances of mass mortality of farmed fish caused by harmful algae, which, however, could not be evidenced against the release of waste compounds from the fish farms (GESAMP, 1991). There is some water quality deterioration in local scale such as in Kandaleru creek in Andhra Pradesh, India, where the effluent water shows far higher concentrations of all water quality parameters than the influent water during the harvest time (FAO/NACA, 1995). In general there is lack of field data specific to the amount and quality of effluent loadings from shrimp ponds and related ecological problems (Barg, 1997). A 2001 survey of 60 shrimp farms in Thailand indicated that effluents loads have sharply declined in recent years due to dramatic reduction in water exchange rate in order to avoid disease. Water exchange rate declined to less than 2 per cent per day, and water usage to 1.5 m^3/kg in intensive ponds (Chamberlain, 2002).

Table 4.1: Potential Environmental Changes and Impacts of Agriculture on Coastal Ecosystem: An Overview

Activity	*Environmental Change*	*Impact on Ecosystem*
Estuary flood control and alterations to the coastal river flows	Increased estuary salinity, decreased circulation, sediment trapping, shoreline erosion	Decreased crop and fish yield, increased water-borne diseases of fish
Agricultural pesticides	Toxic pollution of estuaries, inshore waters and corals	Fish kills, reduced fish yields, toxic fish consumption
Fertilizer use	Increased amount of nutrients, eutrophication, and pollution of estuaries and corals	Fish kills, reduced fish yield
Irrigation from coastal aquifers	Depletion of coastal aquifers, salt water intrusion	Ground water contamination for drinking purpose
Overcropping or grazing in coastal watersheds	Watershed erosion, increased sedimentation, turbidity and deposition in floodplain	Decreased fish yield, obstruction of navigation channels, corals and beaches covered with sediments
Overgrazing of coastal dunes	Trampling and/or overgrazing of beach and dune stabilizing vegetation	Initiation or increased dune migration on to agricultural/ urban areas, and on infrastructure
Agro-industries	Organic and toxic pollution, eutrophication	Fish kill, reduced fish yield, reduction in tourism and recreation

Source: Extracted and modified from Barg, 1997 and Scialabba, 1998

Table 4.2: Potential Environmental Changes and Impacts of Capture Fisheries and Coastal Aquaculture on Coastal Ecosystem: An Overview

Activity	*Environmental Change*	*Impact on Ecosystem*
	Capture fisheries	
Intensive localized fishing effort competition between onshore and offshore fishing for same stocks	Harvesting targeted species at a rate greater than sustainable yield	Decreased fish yield
Fishing with explosives Fishing with trawling and other moving gears	Perturbation of the benthos Coral reef and other critical habitats degraded and destroyed	Decreased fish yields, decreased tourism and recreation value
Fishing of fish meal species	Overexploitation	Decreased fish yields

Contd...

Table 4.2–Contd...

Activity	Environmental Change	Impact on Ecosystem
	Coastal aquaculture	
Reclamation of mangrove	Destruction of mangrove, filling and canalization, increased estuarine salinity	Decreased fish yields, reduced filtration capability, increased coastal erosion, flooding and storm damage
Commercial shrimp farm development	Acquisition of land used by peasants farmers/fishers	Social impact such as multi-user conflict of coastal resources
Intensive fish and shellfish farming	Nutrient enrichment and eutrophication of coastal waters, particularly semi-enclosed embayments	Decreased fish yields, possibility of mass mortality of farmed fish and shellfish
	Enrichment of benthic ecosystem, formation of anoxic sediments	Long-term viability of a farm site damaged
Collection of wild seeds	Overexploitation	Decreased fish yields
Overstocking beyond the capacity of coastal embayments, *e.g.* for oyster culture		
Withdrawal of ground water at a rate greater than normal recharge	Salt water intrusion of aquifer leading to increased salinity, ground water contamination	Water scarcity for human use

Source: Extracted and modified from GESAMP, 1991; Barg, 1997 and Scialabba, 1998.

Table 4.3: Some Specific Instances of Fisheries and Aquaculture Impacts on Coastal Ecosystems

Key Areas of Ecosystems	*Activities*	*Specific instance*	*Reference*
Shrimp culture: Virgin mangrove swamps	Shrimp pond construction	About 200 000 ha of mangrove destroyed in the Philippines and 25 per cent of mangrove resource lost in Thailand.	GESAMP, 1991
		In Andhra Pradesh (A.P.), India, shrimp pond construction caused about 80 per cent conversion of mangroves.	Hein, 2001
		Affected rural communities dependent on mangrove resources for their livelihood in Asian and Latin American countries.	Scialabba, 1998
Use of brackish water and ground water	Construction of channels for water supply, drainage and pumping	Affected quality of drinking water wells in 7 and salinization of paddy lands in 19 out of 26 villages in Nellore district in A. P., India	Hein, 2001
		In Kandaleru creek of A. P., India accumulation of flood water in the hinterland villages due to blockage of natural drain by dyking	FAO/NACA, 1995
		Salt loading estimated to the extent of 5.4 MT per ha annually in low-salinity shrimp farming central region of Thailand	Szuster and Flaherty, 2001
		Land subsidence in Taiwan due to excessive pumping of ground water	GESAMP, 1991
Nutrient and organic enrichment	Fertilizers, manures and feeds	Distinctly black color of pond sediment and steady decrease of average DO level of the supply water to the ponds over the years at A.P. coastal region of India	FAO/NACA, 1995
	Discharge of farm effluent	Considerable water requirement to remove excess nutrients and organic matter from ponds to the extent of 11000–21430 m^3/t production for semi-intensive and 29000–43000 m^3/t for intensive culture in Taiwan	Barg, 1997

Contd...

Table 4.3–Contd...

Key Areas of Ecosystems	*Activities*	*Specific instance*	*Reference*
Use of chemicals (bioactive compounds)	As therapeutants, disinfectants, biocides	Retarded growth and mortality due to bacterial infection, and high values of COD during the peak discharge period in shrimp ponds in Nellore district of A.P., India.	FAO/NACA, 1995
		Abuse of antibiotic in Taiwan hatcheries resulted in poor resistance and survival of shrimp.	Barg, 1997
Collection of wild seeds	Supplementary stocking of extensive and semi-intensive ponds	In coastal areas of eastern states of India shrimp fry collection and large scale destruction of by-catch fry and juveniles create ecological imbalance. For every *Penaeous monodon* fry total number of other species caught is estimated to be as high as 191 in one collection area. A total of 49 species of finfish and 11 species of crustaceans formed the bycatch.	FAO/NACA, 1995

Table 4.4: Some Specific Instances of Finfish Culture Impacts on Coastal Ecosystems

Activity	*Specific Instances*
Mangrove Areas	
Milkfish and mullet culture in brackish water ponds	Extensive land degradation with potential problems related to acid sulphate soils
Nutrient and Organic Loading	
Intensive salmonids culture in ponds and tanks	Yearly oxygen demand and nutrient release per ton production of about 300 kg BOD_5, and 10 kg Total-P and 81 kg Total-N in trot ponds in a Danish farm.
Marine cage farms	Effluent loading expressed as g/kg fish/24 h: 0.5–1.4 g suspended dry matter, 0.01-0.05 g Total-P, 0.15-0.30 g Total-N and 0.1-0.2 g Total NH_3-N in Atlantic salmon farm.
	Annual loading of 2 ton of P, 18 ton of N and 100 ton of BOD_7 from a typical Norwegian net pen farm.
	Release of up to 76 per cent of the carbon and 76 per cent of N fed to cage-farmed salmon into marine environment as solid and soluble wastes.
	Effects on the micro-benthos underneath and around cage culture of salmon in Norway, Scotland and Ireland, and yellowtail in Japan.
Use of Chemicals	
Therapeutic treatment	Approximately 70-80 per cent antibiotics reaches the environment, being deposited in sediments and capable of causing antibacterial effects for 12 weeks after cessation of treatment.
	Detectable levels of oxytetracycline in blue mussels located 80 m from a fish farm using this antibiotic.
	Potential toxicity of organo-P compounds to non-target species.
Introduction of alien species	Competition of Pacific salmon with Atlantic salmon in spawning and nursery areas, and importation of diseases in Norway.
Salmon culture	Tilapia mossambica created a "real nightmare" in Philippines.
Brackish water milkfish culture	

Source: Extracted from GESAMP, 1991 and Barg, 1997.

Table 4.5: Some Specific Instances of Bivalve Culture Impacts on Coastal Ecosystems

Activity	Specific Instances
Reduction in primary productivity in high density culture over a large area	
Oyster culture	Culture of 50,000-60,000 oysters reduced the amount of seston by 76-95 per cent in Japan.
Mussel culture	Culture of green mussels removed 60 per cent of available food in New Zealand.
Physico-chemical changes of substrate in the culture site and benthic communities beneath the suspended culture	
Oyster culture	Biodeposition in the form of faeces and pseudo-faeces to the extent of 16 ton over a 9-month growing season of 420,000 oysters in Hiroshima Bay.
	A 1500 m² oyster farm deposited about 10 ton of dry matter accumulating about 10 cm per farming season in Sweden.
	30 per cent of oyster and mussel farms in France are periodically abandoned or relocated.
Physical effect due to culture structures	Modify current velocity and direction of water movements affecting patterns of erosion and sedimentation of particulate matter.
Introduction and transfer of species	
	Parasites and diseases introduced when European flat oyster was transferred from North America to Europe and Pacific oyster from Japan to North America.

Source: Extracted from GESAMP, 1991 and Barg, 1997.

Deposition of organic fish farm and bivalve waste has deteriorated the benthic ecosystems in the vicinity of the culture site. Interaction of large-scale bivalve cultivation with the marine food web created impact in coastal embayments. High BOD of waste and deposits can result in oxygen depletion in the bottom water, but in general large-scale depletion of oxygen in coastal waters is unlikely (GESAMP. 1991).

Impacts of some farms of aquaculture on natural habitat such as mangrove are better documented. Causes for mangrove degradation are many, and only in last decades an important factor has been the construction of shrimp ponds. In India mangrove conversion has been undertaken by all types of farms. Although in India only 5 per cent of the ponds are constructed in mangroves, it is substantial in some areas such as Godavari delta in A. P. Remote sensing studies indicated about 80 per cent of the conversion of mangroves due to pond construction in A. P., India (Hein, 2001). In general, clearing of mangroves for shrimp pond construction has been exaggerated, and it is difficult to assess in terms of quantity and quality of the mangrove area impacted. In Asia many of the shrimp farms had been built in intertidal areas previously cleared of mangroves decades or even centuries earlier. In fact the overall loss on the global scale due to shrimp farm development is less than 3 per cent of the historic mangrove loss (Chamberlain, 2002).

Studies on use of chemicals (including pesticides and antibiotics) has shown that there are cases of longevity of antibiotics in animal tissues for as much as 77 days in rainbow trout after cessation of treatment, and cases of degradation of environment by antibiotics such as antibacterial effects remaining for long periods. The problem with pesticides is, however, incompletely understood, but evidence has shown that some of these chemicals have adverse environmental effects. Development and transfer of antibiotic resistance in microbial communities such as in shrimp pathogens and native aquatic bacteria has been reported (GESAMP, 1991).

The mass mortality among black tiger shrimp (*Penaeous monodon*) caused by viral diseases and subsequent collapse of the intensive shrimp aquaculture industry in Taiwan during 1987–89 has been cited as a model disaster for farmers everywhere. This disaster spread in several other countries in Southeast Asia. These diseases were a direct result of ecosystem mismanagement resulting from poor farming practices such as pond ageing due to accumulation of organic matter, abuse of chemicals, and water pollution in the coastal areas (Lin, 1989).

4.2.3 Multiple Activities

Table 4.6 shows potential environmental changes and impacts of multiple activities such as sewage and industrial waste disposal; tourism, urban and port development; coastal oil development; and mining and dredging. Threats to human health due to marine pollution and contamination problems through sewage discharge have been discussed in the previous section. An example of contamination of cultured shellfish and threat to human health is that of Manila Bay, Philippines. High microbial loading to the extent of coliform count of 1000 per 100 ml in the major oyster and mussel culture sites at the Manila Bay gave rise to an outbreak of gastro-enteritis in Singapore due to consumption of oyster meat imported from the Philippines (Barg, 1992).

Table 4.6: Potential Environmental Changes and Impacts of Uses or Activities other than Agriculture, Fisheries and Aquaculture on Coastal Ecosystems

Activity	*Environmental Change*	*Impact on Ecosystem*
Domestic and tourism sewage, and industrial waste disposal	Estuary, beach pollution	Decreased fish yields, contamination of fish, shellfish and beach water
Tourism, urban, port and harbor development	Filling of wetlands (including mangroves)	Decreased fish yields, decreased aquaculture areas
	Withdrawal of groundwater at a rate greater than natural recharge, saltwater intrusion of aquifer	Contamination of groundwater
	Coral reef pollution	Decreased tourism and recreation attraction
Coastal oil development, offshore oil transportation, chronic oil release or accidental oil spills	Oil pollution of estuarine and inshore waters	Decreased fish yields, decreased recreation or tourism quality
Coastal mining, dredging for construction materials	Increased sedimentation and turbidity, change in composition of bottom sediments, pollution of coral reefs	Decreased fish yields, decreased recreation or tourism attraction

The most striking example of contaminants affecting human health is the Minamata Bay disease, which caused poisoning of both animals and humans through seafood contamination by mercury. Itai itai disease was another example of high cadmium consumption through seafood. High heavy metal contents exceeding the WHO standards were found in the analysis of fish and shellfish samples from Jakarta Bay of Indonesia. The sources of the heavy metals were from industrial effluents (Barg, 1997).

Most of the environmental impacts of catastrophic oil spills are relatively short-lived, in spite of the potential disruptions of flora and fauna of the marine populations. Expanding coastal populations and tourism have impacts of increasing human pressures and represent one of the greatest threats for marine coastal ecosystems. Critical coastal resources, such as mangroves and coral reefs are being plundered in the name of development. Human activities have already destroyed half of the world's mangroves and either damaged or destroyed coral reefs in at least 93 of the 110 countries in which they occur. Thus gradual removal of the coastal habitats on which almost 80 per cent of the world fisheries are located is jeopardizing human food security (GESAMP, 2001).

4.3 Integrated Ecosystem Approach of Resource Use

Presently all countries with a coastline are concerned with sustainable management of coastal resource systems with an objective to use the many benefits provided by coastal ecosystems for human well-being. In general, the benefits to people, called "services" or "goods and services"–the terms commonly used in economics, are categorized as provisioning, regulating and cultural, which directly affect people, and as supporting, which are necessary to maintain all other services. Provisioning services provide food, water and several others to humans. Regulating services mean benefits obtained from regulation of ecosystem processes such as regulation of floods, drought, land degradation and disease. Cultural services are non-material benefits people derive from ecosystems through recreation and aesthetic experiences. Figure 4.3 shows the various ecosystem services linked to human well-being as outlined by WRI (2003 b).

The humanity has always depended on the services provided by the ecosystems. In recent decades human well-being is affected by the gaps between the ecosystem service supply and ever-growing demands on the degraded ecosystems. The subsequent diminishing prospects of sustainable development affect individuals, communities and nations through impacts on security, health and social relations. The conceptual framework for the Millennium Assessment (WRI, 2003 b) assumes a dynamic interaction between people and ecosystems with particular attention of the linkages between ecosystem services and human well-being. The assessment uses the full range of ecosystems–from undisturbed natural forests to landscapes with mixed pattern of human use and ecosystems managed and modified by humans, such as agricultural land and urban areas. Further, biodiversity and ecosystems are considered as closely related concepts, as living organisms from terrestrial, marine and other aquatic ecosystems are part of the ecological complex.

For assessment, the ecosystems have been categorized into ten systems (Table 4.7). Each category comprises a number of ecosystems and is not mutually exclusive, with each overlapping the other. For example, a wetland ecosystem may be considered both in the "coastal" as well as in the "inland water" systems. For convenience of assessment the various factors that affect ecosystems directly either through natural processes or through human interventions have been considered:

1. Changes in local land use and land cover
2. Modification of river flow
3. Species introduction and removals
4. Harvest of crops, fish and wildlife
5. Discharge of pollutants

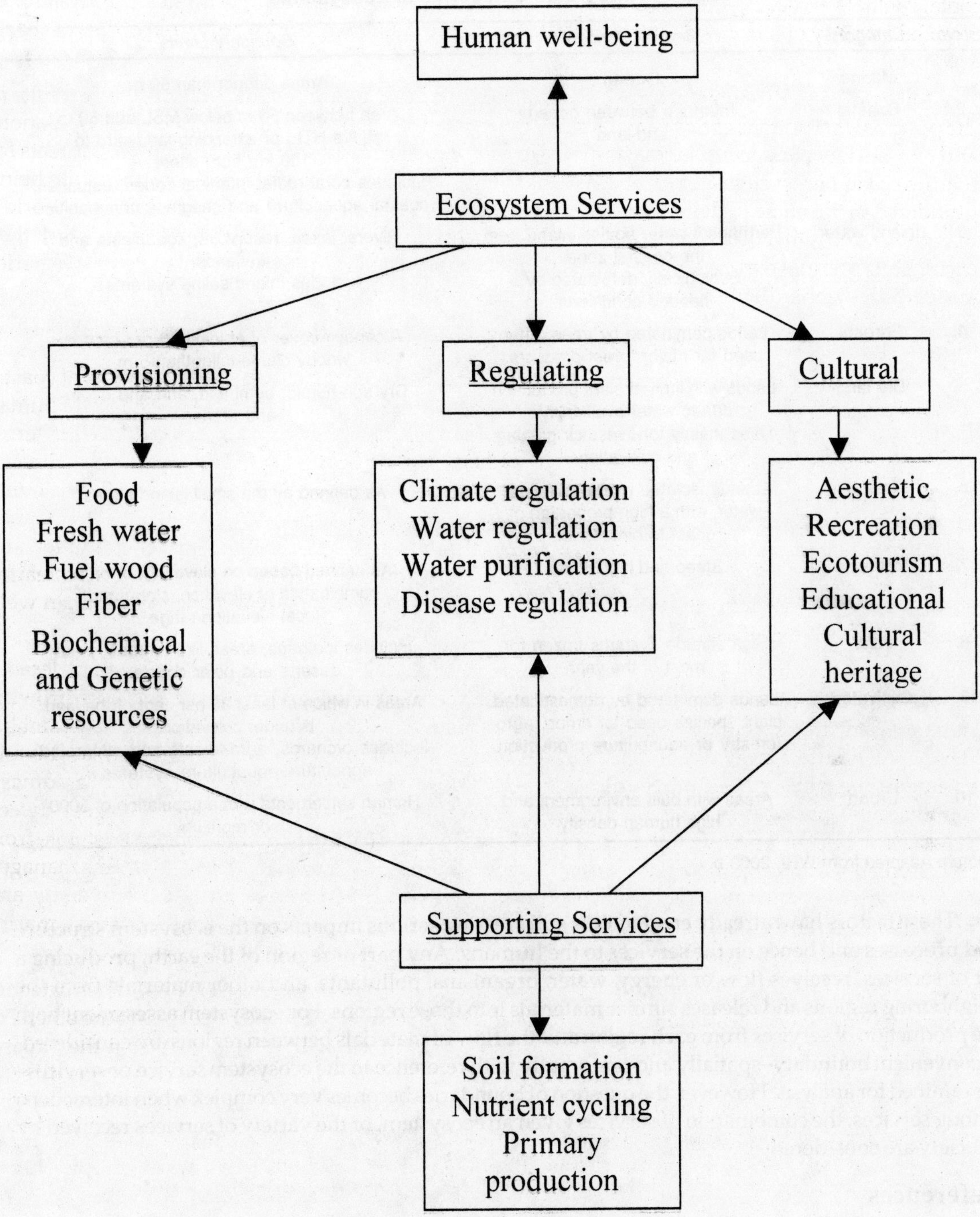

Figure 4.4: Ecosystem Services Linked to Human Well-being (Adapted from WRI, 2003 b)

Table 4.7: Integrated Categories of Ecosystems

Sl.No.	Category	Central Concept	Boundary Limits
1.	Marine	Ocean	Areas deeper than 50 m
2.	Coastal	Interface between ocean and land	Area between 50 m below MSL and 50 m above HTL, or extending landward to 100 km from shore. Includes coral reefs, intertidal zones, estuaries, coastal aquaculture and seagrass communities.
3.	Inland water	Permanent water bodies inland from the coastal zone. Ecologically dominated by flooded conditions	Rivers, lakes, reservoirs, floodplains and wetlands. Includes inland saline systems.
4.	Forests	Lands dominated by trees, often used for major forest products.	A canopy cover of at least 40 per cent by woody plants taller than 5 m.
5.	Dry land	Lands with limited plant production due to water availability. Uses mainly for livestock grazing and cultivation.	Dry sub-humid, semi-arid, arid and hyper–arid areas.
6.	Island	Lands isolated by surrounding water, with a high proportion of coast to hinterland.	As defined by the small island states.
7.	Mountain	Steep and high lands	As defined based on elevation, or on a combination of elevation, slope and local elevation range
8.	Polar	High altitude systems frozen for most of the year	Includes ice caps, areas like tundras, polar deserts and polar coasts.
9.	Cultivated	Lands dominated by domesticated plant species used for crops, agro-forestry or aquaculture production.	Areas in which at least 30 per cent of the land is under cultivation. Includes orchards, agro-forests and integrated agriculture-aquaculture systems.
10.	Urban	Areas with built environment and high human density.	Human settlements with a population of 5000 or more.

Source: Adapted from WRI, 2003 b.

These factors have already created or is still creating serious impacts on the ecosystem structure and processes and hence on the services to the humans. Any part or region of the earth, producing a set of services, receives flow of energy, water, organisms, pollutants, and other materials from the neighboring regions and releases similar materials into those regions. For ecosystem assessment both the production of services from each region and the flow of materials between regions are considered. A convenient boundary–spatially and temporally with reference to the ecosystem service or services–is examined for analysis. However, the question of boundaries becomes very complex when interactions among services, the combination of services given an ecosystem, or the variety of services received by a society are considered.

References

Agarwala, V. P. 1985. *Forests in India–Environmental and Production Frontiers.* Oxford and IBH Pub. Co., New Delhi.

Alcantara, L. 1982. Variation of fish pond layouts for different types of brackish water management. *Seminar on Fish Pond Engineering*. Surabaya, Indonesia.

Bardach, J. E., Ryther, J. H. and McLarney, W. O. 1972. *Aquaculture–The Farming Husbandry of Freshwater and Marine Organisms*. John Wiley and Sons.

Barg, U. C. 1997. Guidelines for the promotion of environmental management of coastal aquaculture development. *FAO Fisheries Technical Paper. No. 328*. Rome, FAO. 1992. Daya Pub. House, New Delhi.

Barnes, R. S. K. 1984. *Estuarine Biology*. Edward Arnold, London.

Beveridge, M. C. M. 1984. Cage and pen fish farming: Carrying capacity models and environmental impact. *FAO Fisheries Technical Paper No. 255*. Rome, FAO.

Biswas, K. P. 1996. *A Text Book of Fish, Fisheries and Technology*. Narendra Pub. House, Delhi.

Boaden, P. J. S and Seed, R. 1985. *An Introduction to Coastal Ecology*. Blackie, Glasgow and London.

Bose, A. N., Ghosh, S. N., Yang, C. T. and Mitra, A. 1991. *Coastal Aquaculture Engineering*. Oxford and IBH Pub. Co., New Delhi.

Boyd, C. E. 2003. Guidelines for aquaculture effluent management at the farm-level. *Aquaculture* 226, 101-112.

Boyd, C. E. and Gautier, D. 2000. Effluent composition and water quality standards. *Global Aquaculture Advocate* 3 (5), 61-66.

Briggs, M. R. P. and Funge-Smith, S. J. 1994. A nutrient budget of some intensive marine shrimp ponds in Thailand. *Aquaculture and Fisheries Management* 25, 789-811.

Burke, L., Kura, Y., Kassem, K., Revenga, C., Spalding, M. and McAllest, D. 2000. *Pilot Analysis of Global Ecosystems (PAGE): Coastal Ecosystems*. World Resources Institute (WRI), Washington D. C.

Center for Marine Conservation. 1998. 1997 International Coastal Cleanup: Results, Washington D. C. Center for Marine Conservation.

Central Pollution Control Board (CPCB) Report. 1996. *Pollution Potential of Industries in Coastal Areas in India*. CPCB, New Delhi.

Chamberlain, G. 2002. Sustainable shrimp farming: issues and non-issues. *INFOFISH International* 2, 26-33.

Chen, T. P. 1976. *Aquaculture Practices in Taiwan*. Fishing News Books, Surrey.

Clark, J. R. 1977. *Coastal Ecosystem Management–A Technical Manual for the Conservation of Coastal Zone Resources*. Wiley-Interscience, New York.

Csavas, I. 1994. Important factors in shrimp farming. *World Aquaculture* 25, 34-56.

Das Gupta, A. 1998. Groundwater and the environment. In *Water Resources Environment Planning, Management and Development*, A. K. Biswas, ed, Ch. 5. Tata McGraw-Hill, New Delhi.

DOD (Department of Ocean Development). 1996. *Report of the Expert Committee on Ocean Part of the Coastal Zone*. Government of India, New Delhi.

FAO. 1994. Mangrove Forests Management Guidelines. *FAO Forestry Paper No.117*.Rome, EAO.

FAO.1995. Review of the state of world fishery resources: inland capture fisheries. Inland Water Resources and Aquaculture Service, Fishery Resources Division. *FAO Fisheries Circular* No. 885. Rome, FAO. 63 p.

FAO.1997. Review of the state of world aquaculture. Inland Water Resources and Aquaculture Service, Fishery Resources Division. *FAO Fisheries Circular* No.886 Rev. 1. Rome, FAO. 163 p.

FAO.1999. Review of the state of world fishery resources: inland fisheries. Inland Water Resources and Aquaculture Service, Fishery Resources Division. *FAO Fisheries Circular* No.942. Rome, FAO. 53 p.

FAO. 1999. *Background Paper No. 6. FAO/Netherlands Conference on Multifunctional Character of Agriculture and Land*, Rome, FAO.

FAO. 2000. *FAO Web Site Data Base*: www.fao.org, Rome.

FAO. 2001.Aquaculture Production 1999. *FAO Yearbook–Fisheries Statistics vol. 88/2.*

FAO. 2003. *FAO Web Site Data Base*: www.fao.org, Rome.

FAO. 2004. The State of World Fisheries and Aquaculture (SOFIA), 2004. FAO Fisheries Department, FAO, Rome.

FAO/NACA. 1995. Report on a Regional Study and Workshop on the Environmental Assessment and Management of Aquaculture Development (TCP/RAS/2253) *NACA Environment and Aquaculture Development Series No. 1. Annex II-5 Country Reports*: India. Network of Aquaculture Centers in Asia-Pacific, Bangkok, Thailand.

Ganapati, P. N. and Sharma, D. V. R. 1965. Mixing and circulation in Gautami-Godavari estuary, *Curr. Sci.* 34 (22), 631-32. Cited in Jhingran, 1991.

GESAMP (IMO/FAO/Unesco-IOC/WMO/WHO/IAEA/UN/UNEP Joint Group of Experts on the Scientific Aspects of Marine Environmental Protection, 1990). *The State of the Marine Environment. GESAMP Reports and Studies* No. 39. 111 p.

GESAMP (IMO/FAO/Unesco /WMO/WHO/IAEA/UN/UNEP Joint Group of Experts on the Scientific Aspects of Marine Pollution, 1991). *Global Strategies for Marine Environment Protection. Rep. Stud. GESAMP* 45, 36 p.

GESAMP (IMO/FAO/Unesco-IOC/WMO/WHO/IAEA/UN/UNEP Joint Group of Experts on the Scientific Aspects of Marine Environmental Protection) and Advisory Committee on Protection of the Sea (2001). *Protecting the Oceans from Land-based Activities.* Report Studies GESAMP No. 71, 162 p.

Hein, L. 2001. Towards improved environmental and social management of shrimp farming. *Discussion paper. FAO Investment Center*, Rome, FAO.

Hu, S. H. 1986. Introduction to the design of coastal aquaculture systems in Taiwan. *Aquacultural Eng.* 5, 235-252.

IAEA. 1993. Risk comparison relevant to sea disposal of low level radioactive waste. *IAEA TEC-DOC-725.* International Atomic Agency, Vienna. Cited in GESAMP, 2001.

IIM. 1981. *Marine Fish Marketing in India.* Vol. I: Summary and Conclusions. Center for Management in Agriculture, Indian Institute of Management Ahmedabad.

IIM. 1983. *Inland Fish Marketing in India.* Vol. I: Overview, Summary and Conclusions. Center for Management in Agriculture, Indian Institute of Management Ahmedabad.

Jhingran, V. G. 1991. *Fish and Fisheries of India.* Hindustan Pub. Corp, Delhi.

Iversen, E. S. 1976. *Farming the Edge of the Sea.* Fishing News Books, Surrey.

Joseph, K. O., Gupta, B. P., Arasu, A. R. T., Muralidhar, M. and Krishnani, K. K. 2003. Effect of shrimp farming on nitrogen levels in the waters of Kandaleru creek, Andhra Pradesh. *Indian J. Fish.* 50 (3), 291–296.

Klemetson, S. L. and Rogers, G. L. 1985. Aquaculture pond temperature modeling. *Aquacultural Eng.* 4, 191-208.

Lawson, T. B. 1997. *Fundamentals of Aquacultural Engineering*. Indian Ed., CBS Publishers and Distributors, New Delhi.

Liao, I. C. 1986. General introduction to the prawn pond system in Taiwan. *Aquacultural Eng.* 5, 219-233.

Liao, I. C. 1992. Marine prawn culture industry of Taiwan. In *Marine Shrimp Culture: Principles and Culture,* W. Fast and L. J. Lester, eds., pp. 653-675. Elsevier, Amsterdam.

Lin, C. K. 1989. Prawn culture in Taiwan: What went wrong? *World aquaculture* 20(2), 19-21.

LOICZ (Land-Ocean Interaction in the Coastal Zone), 1998. *Coastal Typology Dataset*. Texel, The Netherlands: LOICZ International Project Office.

Maity, D. K. 1990. *Project Planning, Design and Implementation of Shrimp Farms at Digha, West Bengal.* Unpublished Report.

Mal, B. C. 1995. *Soil and Water Conservation Engineering*. Kalyani Publishers, Ludhiana.

Milne, P. H. 1972. *Fish and Shellfish Farming in Coastal Waters.* Fishing News Books, Surrey.

Ministry of Environment and Forests. 1994. *Annual Report*. Government of India, New Delhi.

Mukherji, D. K. 1978. *Rice in West Bengal. In Rice in West Bengal*, Vol. I. Directorate of Agriculture, Government of West Bengal, Kolkata.

Naik, M. G., Marakala, C., Rajesh, K. M. and Mridula, R.K. 2005. Ecology and biodiversity of microfauna in a mangrove fringed lagoon, south west coast of India. Abstract of the Seventh Indian Fisheries Forum, 8-12 November, 2005, Bangalore. Asian Fisheries Society, Indian Branch, Mangalore, India.

Odum, E. P. 1963. *Ecology*. Amerind Pub. Co., New Delhi.

Paliwal, K. V. 1972. *Irrigation with Saline Water*. Water Technology Center, Indian Agricultural Research Institute (JARI), New Delhi.

Primavera, J. H. 2002. Environmentally responsible grow-out systems in aquaculture: An introduction to thematic session one. In *Proc. of the ASEM Workshop on Aquaculture Challenges in Asia,* Beijing, 27-30 April, 2002.

Pritchard, D. W. (1967). What is an estuary: Physical view point: 36. In *Estuaries*, G. H. Lauff, ed., American Association for the Advancement of Science, Washington D. C. Cited in Jhingran, 1991.

Raghavaiah, C. V., Jana, M. K. and Ghosh, B. C. 1995. *Rain-fed lowland and deepwater rice management*, 45-56. In *Rice Management and Biotechnology*. S. Kannaiyan, ed. Associated Pub. Co., New Delhi.

Ramachandran, S. 2003. Coastal zone management in India. An overview, Ch. 1. In *Coastal Urban Environment*. R. Ramesh and S. Ramachandran, eds. Capital Pub. Co., New Delhi.

Rosenthal, H. 2002. Interactions between coastal resource users: aquaculture, shipping. In *Proc. of the ASEM Workshop on Aquaculture Challenges in Asia*, Beijing, 27-30 April, 2002.

Sa de Elrich. 1999. Present state of the coastal environment in India. *Technical Session of the Meeting of Experts on Coastal Development Issues in India and the South Asian Region*. Goa, India.

Sampath, V. 2003. Coastal fisheries and urbanization, Ch. 2. In *Coastal Urban Environment*. R. Ramesh and S. Ramachandran, eds. Capital Pub. Co., New Delhi.

Scialabba, Nadia, ed. 1998. *Integrated Coastal Area Management and Agriculture, Forestry and Fisheries.* FAO Guidelines. Environment and natural Resources Service, FAO, Rome.

Space Applications Center (SAC). 1992. Coastal Environment. *SAC Report. RSAM/SAC/COM/SN/11/92.* Department of Space, Government of India, New Delhi. Cited in Ramachandran, 2003.

Srivastava, H. C. 1991. Industrialization of aquaculture towards blue revolution. In *Aquaculture Productivity*, V.R.P. Sinha and H.C. Srivastava, eds. Oxford and IBH Pub. Co., New Delhi.

Szuster, B. W. and Flaherty, M. 2000. Inland low-salinity shrimp farming in the central plains region of Thailand. In *The Chao Phraya Delta: Historical Development, Dynamics and Challenges of Thailand's Rice Bowl.* F. Molle and T. Srijantr, eds. Kasetsart University, Bangkok, 159-170

Tait, R. V. 1981. *Elements of Marine Ecology. An Introductory Course.* Butterworths.

Upadhyay, A. S. 1994. *Handbook on Design, Construction and Equipments in Coastal Aquaculture (Shrimp Farming).* Allied Publishers, Bombay.

UN. 1994. *World Urbanization Prospects: The 1994 Revision*, UN Department for Economic and Social Information and Policy Analysis, Population Division, New York.

Wang, Q., Yang, C. and Yu, Y. 1995. The shrimp farming industry in China: past development, present status, and perspectives on the future. In *Swimming through Troubled Water. Proceedings of the Special; Session on Shrimp Farming Aquaculture '95*, C. L. Browdy and J. S. Hopkins, eds., pp. 1-12. World Aquaculture Society, Baton Rouge, Louisiana, USA.

Wheaton, F. W. 1977. *Aquacultural Engineering*. Wiley-Interscience, New York.

Wood, S., Kate, S. and Scherr, S. J. 2000. *Pilot Analysis for the Global Ecosystem (PAGE): Agroecosystems* A joint study by the International Food Policy Research Institute (IFPRI) and World Resources Institute. World Resources Institute. Washington D. C.

WRI. 1998. *World Resources 1998-1999.* World Resources Institute, Oxford University Press, New York, p. 369

WRI. 2003 a. *Annual Report*. World Resources Institute, Washington D. C., Island Press.

WRI. 2003 b. Ecosystem and human well-being: *A Report of the Conceptual Framework Working Group of the Millennium Ecosystem Assessment*. World Resources Institute, Washington D. C., Island Press

Part II:
WATER QUALITY

5

Water Quality Characteristics

5.1 Water Quality Parameters

Broadly water quality characteristics are determined by various parameters that affect the environmental conditions of growth and survival of the species. The environmental standards of culture are maintained for various systems and species and the critical conditions of growth and survival are directly associated with the biological processes such as respiration, metabolism and ionic regulation, and physical processes such as transfer of gases and ions across gill surfaces (Muir, 1982). Deterioration of environmental conditions affects these processes and the tolerance limits are set by judging in terms of growth rate, food consumption, disease resistance and product quality. The most common water quality requirements are oxygen, ammonia, nitrite, nitrate, pH, carbon dioxide, alkalinity, hardness, suspended solids, and dissolved organic substances. Extensive reviews and summaries of these requirements have been published in the literature (Alabaster, 1974; Parkins, 1976; Wickins, 1981; Muir, 1982; Poxton and Allouse, 1982). The water quality standards mostly have been determined with an objective to sustain the fisheries and the natural environment.

5.1.1 Classification of Water Quality Parameters

According to Stickney (1979), the water quality parameters can be categorized under two headings:

1. Parameters that are altered by biological activity in the water body, called "non-conservative" parameters:
   (*a*) Dissolved oxygen (DO)
   (*b*) Primary productivity
   (*c*) Plant nutrients

2. Parameters not affected by biological activity, called "conservative" parameters–
   (*a*) Temperature
   (*b*) Salinity
   (*c*) Suspended solids
   (*d*) Alkalinity

The other type of classification is as follows

1. *Physical parameters*: Temperature, Suspended solids
2. *Chemical/biochemical parameters:* DO, pH, Dissolved gasses, Nutrients, Salinity, Hardness, Organic and Inorganic pollutants
3. *Biological/Microbiological*: Predators, Faecal matters/Bacterial contaminants

1. Dissolved Oxygen

Nearly all aquatic organisms must derive their oxygen from water available as dissolved in the water. The level of DO present in pond water is the most important parameter of water quality. Maintenance of DO in pond water is, therefore, very critical to survival and growth of aquatic organisms.

The DO in pond water has two different sources–atmospheric air, and aquatic plants through photosynthesis. Oxygen comprises 20.95 percent by volume of air. Dissolution of oxygen takes place by diffusion through surface of contact with pond water and is too slow to be of any practical significance. The aquatic plants produce oxygen as a by-product of photosynthesis, which gets dissolved in water and provides a major source of oxygen.

Oxygen is sparingly soluble in water. According to Henry's law, at a constant temperature in a state of equilibrium the concentration of DO in water is in direct proportion to the partial pressure of oxygen in the atmosphere. At atmospheric pressure at standard conditions (0°C and sea level), the partial pressure of oxygen in dry air is

$$p_{O2} = 760 \times 0.2095 = 159.2 \text{ mm Hg.}$$

In contact with water oxygen will enter water from air until the pressure (tension) of oxygen in the water is equal to the partial pressure of oxygen in the air. For practical purposes, the solubility of oxygen is usually given as the concentration (mg/l) at equilibrium (saturation) for water in contact with dry air at 760 mm Hg and at specific temperatures and salinity (Table A 2).

The saturation concentration of oxygen is dependent on temperature, salinity and atmospheric pressure. Dependence of temperature and salinity is shown in Figure 5.1. At constant temperature the saturation DO concentration is linearly related to salinity. As the temperature increases, the solubility of oxygen in water decreases. The solubility correction for pressure can be done from the formula:

$$DO_p = DO_{po} \times (P_o - p_v / 760 - p_v) \quad (5.1)$$

where, DO_p and DO_{po} are solubility of oxygen at a pressure P and at 760 mm pressure, respectively, P_o = observed atmospheric pressure, and p_v = vapor pressure of water at the temperature of DO measurement. For practical purposes, the vapor pressure values can be neglected, and the correction can be approximated as

$$DO_p = DO_{po} \times P_o/760 \quad (5.2)$$

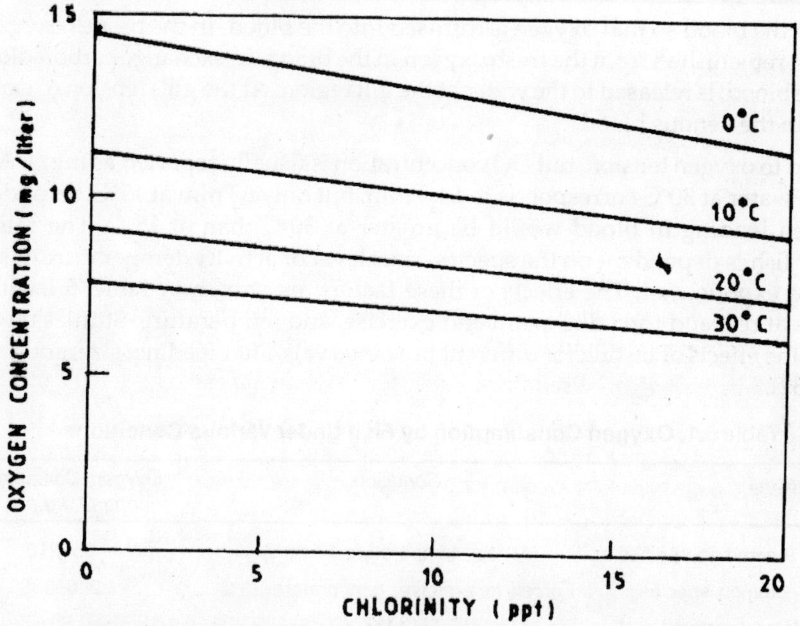

Figure 5.1: Dependence of Saturation Concentration of Oxygen in Water on Water Temperatures and Chlorinities (Data from APHA, 1989)

Oxygen solubility may also be expressed as oxygen tension and percent saturation. Oxygen tension represents the partial pressure of oxygen in the atmosphere required to hold a certain oxygen concentration in the water. DO concentration of 5 mg/l in water at 25°C will be in equilibrium with oxygen tension of (5/8.24) x 159.2 = 96.6 mm Hg and per cent DO saturation of water will be 5/8.24 = 60.7.

Example 5.1

Calculate the oxygen tension and per cent saturation of DO in a pond water sample with a temperature of 28° C and a DO concentration of 5.3 mg/l when the atmospheric pressure is 740 mm Hg. Assume the sample salinity is less than 100 mg/l.

Solution

At 28° C saturation concentration of DO = 7.81 mg/l.

Oxygen partial pressure or oxygen tension in the air at 760 mm pressure

= 159.2 mm

Oxygen tension at DO concentration of 5.3 mg/l = (5.3/7.81) x159.2 = 108.04 mm

∴ At 740 mm corrected oxygen tension = 108.04x 740/760 = 105.2 mm

and DO concentration = 5.3x 740/760 = 5.16 mg/l

% DO saturation = 5.16/7.81 = 66.08.

Oxygen requirement by fish–Fish require oxygen for respiration. During respiration, gaseous exchange takes place by simple diffusion across the gills. At the gills the oxygen tension is higher in the water than in the blood so that oxygen is diffused into the blood. In the tissues oxygen is quickly used up and then replenished from the fresh oxygen in the blood. In exchange carbon dioxide, which is loaded into the blood, is released to the water at the gill region. At the gills tension of carbon dioxide is less than that in the venous blood.

Fish respond to oxygen tension, but DO concentration is usually reported as mg/l. A 4-mg/l DO concentration in water at 30°C corresponds to 84.5 mm, but only 63 mm at 15°C. So under saturated condition oxygen loading in blood would be greater at 30°C than at 15°C. The rate of oxygen consumption by fish is dependent on the species, size, level of activity, temperature of surrounding water, and stress experienced. The effects of these factors are shown in Table. 5.1 with respect to different species at rest and forced to swim and exercise, and temperature. Studies on catfish have shown to reflect the effects of fasting for different time intervals, after feeding, size, and DO content of water (Boyd, 1982).

Table 5.1: Oxygen Consumption by Fish Under Various Conditions

Species	*Condition*	*Oxygen Consumption (mg O_2/kg per h)*
Freshwater fish, 9 common species	At rest, temperature 17-20°C	65–210
Freshwater, 5 common species	Forced to exercise, temperature 20°C	266–888
Catfish (*Ictalarus catus*)	At 11°C	60
	At 25°C	276
	At 28°C, fasted overnight	380
	At 28°C, fasted 3 days	290
	At 28°C, fasted 9 days	290
	Immediately after feeding	520
	One hour after feeding	680
	At 25-28°C, DO 1 mg/l	90
	At 25-28°C, 2 mg/l	190
	At 25-28°C, 4 mg/l	390
	At 25-28°C, 7 mg/l	690
	At 7 mg/l, size 10 g	1050
	At 7 mg/l, size 50 g	750
	At 7 mg/l, size 100 g	625
	At 7 mg/l, size 500 g	480
	At 7 mg/l, size 1000 g	340
Tilapia nilotica	At 25°C, forced to swim at 30 cm/s	220
	At 25°C, forced to swim at 60 cm/s	458

Source: Extracted from Boyd, 1982.

Limiting DO concentrations for any fish are dependent on several factors as discussed above. In general, warm water fish cannot tolerate DO concentrations below 3 mg/l. DO less than 1 mg/l will be fatal if exposure is more than a few hours. Between 1 and 5 mg/l fish survive but with poor reproduction and slow growth if exposed continuously. The duration should not exceed 16 h in a day, if the DO is less than 5 mg/l. At no time, it should be less than 3 mg/l. Above 5 mg/l fish can survive indefinitely and grow and reproduce normally (Boyd, 1982).

2. Primary Productivity and Nutrients

The basic aquatic food chain or ecosystem operating within a pond shows that plant growth of certain kinds, such as phytoplankton, is desirable in ponds. Green plants, and in particular phytoplankton are the principal autotrophs or primary producers. Certain non-photosynthetic anaerobic bacteria and blue green algae are also autotrophic in that they are able to synthesize organic matter (*i.e.* new cell biomass) from inorganic carbon by using chemical energy derived from the cellular oxidation of inorganic substrates such as hydrogen sulphide, sulphur, nitrogen, divalent iron and hydrogen. They are collectively known as chemosynthetic autotrophs.

Natural productivity is very important in ponds where natural food plays a part. In these cases the improvement of natural productivity is the first goal to reach in order to increase productivity. All pond food organisms, including autotrophs and heterotrophs, consist mainly of carbon, nitrogen and phosphorus. Phytoplankton grown on a nutrient rich medium is composed of about 45-50 percent carbon, 8-10 percent nitrogen and 1 percent phosphorus on a dry basis, and so they are dependent on biological supply of these primary nutrients for their growth.

The nutrients are basically classified as major or primary nutrients, minor or secondary nutrients, and micronutrients or trace elements. Nitrogen and phosphorus, primarily as nitrate and phosphate, and potassium are particularly important as the major nutrients. They are added in the form of fertilizers. Calcium, magnesium, sulphate, and silica are added as minor nutrients in the form of salts. Micronutrients are chlorides, manganese, iron, copper, zinc, cobalt, vanadium, and some others are trace elements.

3. Temperature

Temperature is one of the most important among the external factors influencing the growth and metabolism of fish. Temperature affects directly the rate of various biochemical reactions, and indirectly the solubility of oxygen in water. Under otherwise favorable conditions the optimum temperatures of many "cold-water" and "warm water" fishes are 14–18°C and 24–30°C, respectively. Within supportable limits high temperatures are best for the production of cultivated fish. Higher production in tropical region is due to the higher temperature than that in the temperate region.

Fish, being cold-blooded animals, change their body temperatures with the outside temperature. In general three thermal characteristics are considered in fish growth: thermal death point, acclimation or acclimatization temperature, and temperature shock. Thermal death occurs outside a zone of resistance, where the fish can tolerate for limited periods only but not indefinitely. Fish are under stress at temperatures somewhat below the upper zone, or above the lower zone of resistance. A zone of tolerance is defined as combination of acclimation and environment temperature in which fish can live indefinitely. Acclimation should be slow enough for adaptation to the changing temperature (Wheaton, 1977).

4. Salinity

The term 'salinity' is used to define the inorganic solute substances or all dissolved ions in water. This definition is necessary to describe the inorganic solute concentration in the saltwater systems such as seawater and brackish water. Salinity is the total amount of solid materials in grams contained in one kilogram of seawater, when all the carbonate has been converted to oxide, the bromine and iodine replaced by chlorine, and all organic matter completely oxidized.

Salinity is determined in terms of only one chemical, *i.e.* the chloride content, which is the major chemical in seawater and accounts for about 55 percent of the salinity. This is based on the law of constancy of composition. The law states that the proportions of each major constituent of seawater are virtually constant irrespective of the absolute concentration of the total solids. In fact, seawater has a relatively constant ionic composition, but is more concentrated in ions than freshwaters.

Salinity is calculated from Knudsen's formula:

$$S\ ‰ = 0.30 + 1.805 \times \text{chlorinity, or}$$

$$S\ (mg/l) = 30 + 1.805 \times \text{chlorinity,} \tag{5.3}$$

where, chlorinity is expressed as g/kg (or parts per thousand, ppt) of a sample of seawater which is identical with 107.87 g silver just necessary to precipitate the halides in 0.3285 kg seawater. Salinity is often expressed as mg/l, which becomes 'ppt' on dividing by 1000. Chlorinity is determined by titration of a given volume of saltwater with standard solution, when the color change occurs after the last halide has been precipitated. For practical purposes chloride concentration (in g/kg solution) is nearly equal to chlorinity in seawater. The empirical relationship is redefined as

$$\text{Salinity} = 1.8066 \times \text{Chlorinity} \tag{5.4}$$

based on the electrical conductivity of seawater relative to a specified solution of potassium chloride in water. The scale is dimensionless and the traditional dimension 'ppt' no longer applies (APHA, 1989).

Salinity of natural water ranges between essentially zero and more than 40 ppt. In general, seawater varies from 33 to 37 ppt, and freshwater less than 0.5 ppt. In brackish water region the salinity gradually increases and different zones of brackish water develop. In seawater the breadth of fluctuation extends between zero and 41 percent at the surface depending on the amount of river inflow, the extent of evaporation and quantity of rain. The salinity ranges in intermediate and deep waters are less than those at the surface. The values for deep waters usually lie between 34.5 and 35 ppt (Poxton and Allouse, 1982).

Surface waters are usually dilute solutions of bicarbonates and carbonates of alkaline earth metals, calcium and magnesium being the dominant ones. However, ionic concentration of surface water changes from region to region. In arid regions surface waters have large proportions of sodium, sulphate and chloride. Coastal tropical waters are often low in alkaline earth bicarbonates but rich in sodium and chloride. Coastal ponds filled with well waters may have large proportions of sodium and bicarbonate (Boyd, 1982).

Salinity is considered as a water quality factor for aquaculture on the basis of permissible limit of survival and growth of various species. As salinity is related to osmotic pressure of water, an osmotic pressure of approximately 6 atmospheres in blood of freshwater fish, equivalent to 7000 mg/l of sodium chloride, can withstand the fish life. In general, freshwater fish and aquatic organisms can live well in water with dissolved solids up to 2000 mg/l. As a rule, other water quality factors remaining the same, salinity does not have great effects on pond productivity and hence on management (Boyd, 1982). However, some species of fresh and brackish water are sensitive to sudden changes in salinity. Both food intake and conversion efficiency were found to be increased with salinity (Poxton and Allouse, 1982).

5. Suspended Solids

Suspended solids make the water turbid, and turbidity of water is caused by bits of particulate matter found in the water column. They are made up of three types of materials: (*i*) sediment particles;

(*ii*) organic materials–detritus composed of plant and animal remains, waste food particles, and faecal material; and (*iii*) phytoplankton cells and other living microorganisms. Sediment particles, commonly known as mud, are composed largely of silt and clay, and are suspended in the water column due to mixing by wind or water current. The particles may be washed into the water from the land during periods of runoff.

The size of solid particles suspended in water range from 0.01mm to 1 cm. The suspended particles smaller than about 1mm and larger than individual molecules (about 0.001 mm) exhibit colloidal properties. The colloidal particles contribute little to the normal turbidity of water, but they are largely responsible for the color of the natural waters.

In most cases some turbidity is not objectionable, since most culture animals can tolerate extremely high levels of suspended solids, such as 100,000 mg/l (Boyd, 1982). Turbidity due to plankton is usually desirable, because plankton blooms favor growth of fish food organisms, and also prevent underwater growth of unwanted weeds. There are certain undesirable effects of suspended solids:

1. High turbidity restricts light penetration and reduces primary productivity due to shading.
2. Inorganic clay particles lead to clogging of the gills or irritation of gill filaments and other membranes. Settling of many particles may lead to the burial of eggs, larvae, or fry of benthic organisms, or of some species.
3. Increase in potential for disease may occur due to particulate matter providing large surface area for growth of fungi and bacteria.
4. High levels of suspended matter may lead to filling of ponds.

6. pH

Like any other liquid, water can be acid, neutral or alkaline. The water used in aquaculture is not chemically pure and contains, in solution, different substances, which give it an acid, neutral or alkaline reaction. The intensity of these characteristics is measured by determining the exact quantity of hydrogen ions, or hydrogen ion concentration, as expressed in pH.

In case of fish culture, the stability or variability of culture water pH counts most. Most of the species are adapted to an average pH value and do not easily withstand sudden and strong variations of pH. Biological conditions are better when the pH of water is quite constant rather than in water undergoing considerable variations. Neutral or slightly alkaline pH between 7.0 and 8.0 is the best quality water suited to aquaculture. Conditions of pH more acid than 6.5, and more alkaline than 9–9.5, for long periods, diminish reproduction and growth of fish. pH 4 and 11 are acid and alkaline death points, respectively (Figure 5.2).

7. Alkalinity and Hardness

The alkalinity of water is a measure of its capacity to neutralize acids. The term total alkalinity refers to the total concentration of bases present in water, expressed as mg calcium carbonate per liter of water. Since alkalinity of waters is due principally to salts of weak acids and strong bases, and such substances act as buffers, alkalinity is thus a measure of buffer capacity.

The equilibrium ionization reaction resulting from dissolution of carbon dioxide in water show that certain amount of carbon dioxide must be present to maintain a given amount of bicarbonate in solution:

$$CaCO_3 + CO_2 + H_2O = Ca\,(HCO_3)_2 \tag{5.5}$$

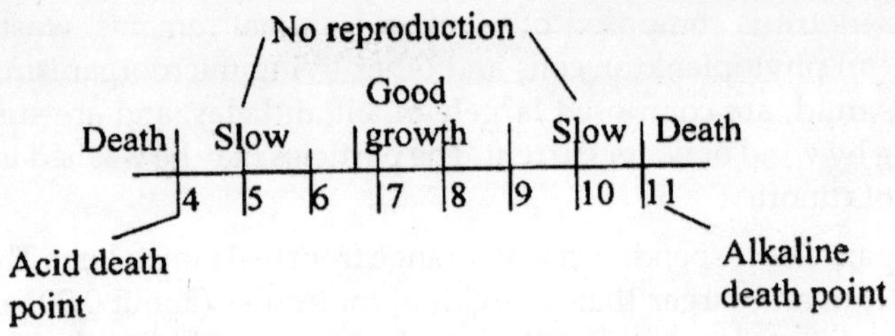

Figure 5.2: Range of pH for Various Reactions to Fish Life

This reaction shows that increase in carbon dioxide content dissolves more calcium carbonate, while its decrease precipitates calcium carbonate. Dilute bicarbonate solutions are weakly alkaline, as shown below:

$$HCO_3^- + H^+ \rightarrow H_2O + CO_2 \qquad (5.6)$$

$$HCO_3^- \rightarrow H^+ \rightarrow CO_3^{2-} \qquad (5.7)$$

Hydrogen ion produced in the last reaction is used up in the second reaction. It can be shown that concentration of carbon dioxide and carbonate (CO_3^{2-}) being the same, pH will be approximately 8.3.

Natural waters that are considered productive should contain 40 mg/l or more alkalinity. However, the greater productivity of water results from increase in phosphorus and other nutrients increase along with total alkalinity. Alkalinity of fertilized pond ranges from 20 to 120 mg/l. Ponds with total alkalinity of 200–300 mg/l in lakes tend to give low productivity due to shortage of carbon dioxide.

Hardness, or total hardness, is defined as the total concentration of calcium and magnesium ions, expressed as mg/l of $CaCO_3$. The term, originally used with fresh water, was developed as a measure of the ability of water to precipitate soaps. Usually bivalent calcium and magnesium ions precipitate soaps and contribute entirely to hardness. But some polyvalent ions like aluminium, iron, manganese, strontium and zinc, and hydrogen ion also precipitate soaps.

Total hardness is usually related to total alkalinity, because the former is due to the cations and the latter due to anions in a solution of carbonate, and cations of hardness are normally derived from the solution of carbonate minerals. For example, total hardness of waters in arid regions may be considerably greater than total alkalinity because of the concentrations of the ions due to evaporation. Ground waters in coastal regions sometimes have high alkalinity and low hardness (Table 5.2).

In aquaculture total hardness is not so important as total alkalinity. Water should contain a small concentration of calcium and magnesium, in fish culture, which is approximately present if total harness exceeds 20 mg/l.

8. Dissolved Gases

There are two categories of dissolved gases present in water–(i) the gases that do not react with water to an appreciable extent: nitrogen, oxygen, hydrogen, methane, and (ii) gases that ionize in water or react with water to produce ions: carbon dioxide, hydrogen sulphide, sulphur dioxide, ammonia.

Table 5.2: Total Hardness and Total Alkalinity of Water from Some Different Ponds

Type of Water	*Total Alkalinity (mg/l)*	*Total Hardness (mg/l)*
Pond on sandy soil	13.2	12.9
Pond on acidic, clay soil	11.6	12.3
Pond on calcareous soil	51.1	55.5
Pond filled with soft but alkaline well water	93.0	15.1
Pond in arid region	346	708

Source: Boyd, 1982.

Nitrogen

Nitrogen is the most common inert gas dissolved in freshwater and saltwater systems. Absorption of nitrogen from the atmosphere containing 78 percent nitrogen is more than that of oxygen in the water. It depends on temperature, salinity, partial pressure of nitrogen in water, and motion of the air-water interface. An increase in temperature or salinity decreases the nitrogen saturation concentration in water.

Nitrogen is the gas most commonly associated with "gas bubble" disease, the condition that results when fish are subjected to waters supersaturated with nitrogen. Common causes of gas supersaturation are air injection under high pressures, hydroelectric projects and use of power plant cooling waters for aquaculture.

Hydrogen

The presence of hydrogen in natural waters is due primarily to its production by bacteria in anaerobic decomposition of organic matter.

Methane

It is commonly called 'marsh gas' evolving in natural water as a by-product of anaerobic decomposition. Anaerobic conditions are found in lakes below the thermocline during summer stratification. The gas is not directly toxic in low concentrations. However, since the gas is produced only under anaerobic conditions, it indicates lack of oxygen. In culture systems carrying a heavy biological load, small anaerobic areas, if develop, may be fatal to fish, resulting in collapse of the system. Therefore the presence of methane must receive immediate attention.

Carbon Dioxide

Carbon dioxide plays significant roles in many ways to make the aquatic environment fit for life. First, carbon dioxide participates in biological processes such as photosynthesis and respiration. Second, it maintains a natural buffer system of relatively constant pH in the aquatic environment. Third, it provides carbon skeleton in life system and forms numerous carbon compounds essential for life by carbon fixation.

Carbon dioxide in natural water comes from various sources:

1. Respiration of plants and animals.
2. Bacterial decomposition of organic matter under both aerobic and anaerobic conditions.

3. Chemical reactions between acids and various carbonate compounds.
4. Ground water seepage through soils where decomposition is occurring, and
5. Atmosphere containing 0.03 percent carbon dioxide.

Carbon dioxide is highly soluble in water, solubility depending on the temperature and/or salinity (Figure 5.3). In comparison with carbon dioxide, oxygenconcentration in air is about 700 times greater than the carbon dioxide concentration. But the water in equilibrium with atmosphere contains oxygen only 1.09 times that of carbon dioxide, because of its much lower solubility than that of carbon dioxide.

When carbon dioxide goes into solution in water a series of chemical equilibrium equations sets in:

$$CO_2 + H_2O \leftrightarrow H_2CO_3 \leftrightarrow H^+ + HCO_3^- \leftrightarrow H^+ + CO_3^{2-} \tag{5.8}$$

It shows that carbon dioxide is found in water in four forms: free gas (CO_2), carbonic acid (H_2CO_3), bicarbonate radical (HCO_3^-) and carbonate radical (CO_3^{2-}). The reactions show that pH increases as the reactions shift toward the right and decreases as they shift toward left. However, the system is highly buffered and strongly resists any change in pH.

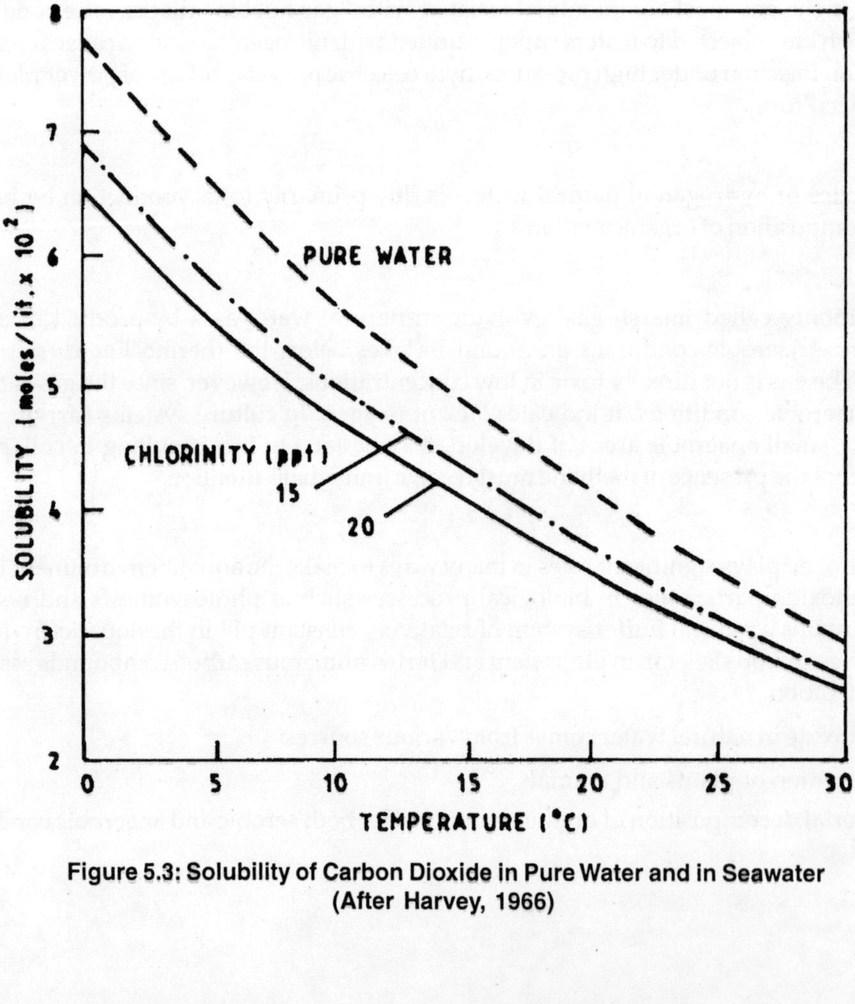

Figure 5.3: Solubility of Carbon Dioxide in Pure Water and in Seawater (After Harvey, 1966)

Hydrogen Sulphide

Hydrogen sulphide (H_2S) is produced naturally by aerobic bacteria, and also by the anaerobic decomposition of organic matter. Sulphates and other oxidized sulphur compounds can be reduced by the anaerobic bacteria as shown below:

$$SO_4^{2-} + 8\,H^+ \rightarrow S^{2-} + 4\,H_2O \tag{5.9}$$

Weather conditions, water level fluctuations, water temperature and pH affect the amount of hydrogen sulphide produced. Hydrogen sulphide ionizes in water as follows:

$$H_2S \leftrightarrow H^+ + HS^- \tag{5.10}$$

$$HS^- \leftrightarrow H^+ + S^{2-} \tag{5.11}$$

Un-ionized hydrogen sulphide is highly toxic to fish, but the ions produced in water are not appreciably toxic. Increase in pH decreases the proportion of un-ionized hydrogen sulphide. Hydrogen sulphide can occur naturally at levels, which are detrimental to fishery production and survival.

Dissolved hydrogen sulphide gas is not found in water containing even a trace of DO. Oxygen oxidizes the hydrogen sulphide to water and free sulphur or to sulphate as follows:

$$2\,H_2S + O_2\,(aq) \leftrightarrow 2\,H_2O + 2\,S \tag{5.12}$$

$$H_2S\,(aq) + 2\,O_2\,(aq) \leftrightarrow 2\,H^+ + SO_4^{2-} \tag{5.13}$$

By aeration hydrogen sulphide can be removed readily from the water, accomplishing oxygen dissolution at the same time. As the hydrogen sulphide is oxidized, HS^- ions react with hydrogen ions of the water to form more hydrogen sulphide gas. This in turn will be oxidized by DO to the extent that DO is present in the water.

Sulphur Dioxide

Sulphur dioxide (SO_2) reacts with water as follows:

$$SO_2\,(aq) + H_2O \leftrightarrow H^+ + HSO_3^- \tag{5.14}$$

The gas is completely ionized at all pH above 4. Both sulphur dioxide and bisulphite are strong reducing agents, and bisulphite reacts with DO as follows:

$$HSO_3^- + \tfrac{1}{2}\,O_2\,(aq) \leftrightarrow H^+ + SO_4^{2-} \tag{5.15}$$

Sulphur dioxide enters natural waters from combustion fumes in the air and as a partially oxidized product of bacterial decomposition of organic matter.

Ammonia

Sources of ammonia are fertilizers, feed, fish excrement and microbial decomposition of nitrogenous compounds. Ammonia is the main product of both aerobic and anaerobic decomposition of nitrogenous organic matter, but smaller amounts of ammonium hydroxide (NH_4OH) and ammonium carbonate ($(NH_4)_2CO_3$) are also produced. Bacterial deamination (*i.e.* removal of an amino group from an amino acid) is another source of ammonia in aquatic system.

Ammonia reacts with water to ionize as follows:

$$NH_3 + H_2O \leftrightarrow NH_4^+ + OH^- \tag{5.16}$$

Equilibrium concentrations of ammonia and ammonium ion are dependent on pH and temperature, the values of which are obtained from Figure 5.4 or in the tabular form. The effect of pH on un-ionized ammonia concentration is greater than that of temperature. For example, water at pH 8 and 30°C contains 2 mg/l of total NH_3-N, *i.e.* 0.16 mg/l un-ionized NH_3. When pH is increased to 9.2 at the same temperature, the same water contains 1.0 mg/l. un-ionized NH_3. The sum of un-ionized ammonia and ammonium ion is called total ammonia nitrogen (NH_3-N).

Ammonia is highly toxic to aquatic organisms. Ammonium ion form appears to be tolerated in much higher concentrations by aquatic organisms. At a pH below 7, very little ammonia is present, but above 7, significant concentrations of ammonia are present and could cause lethal or sub-lethal toxic effects. Salt water systems, which are usually maintained between 7 and 8.5, tend to have greater ammonia toxicity risk. With increasing ammonia level in water, ammonia excretion by fish decreases

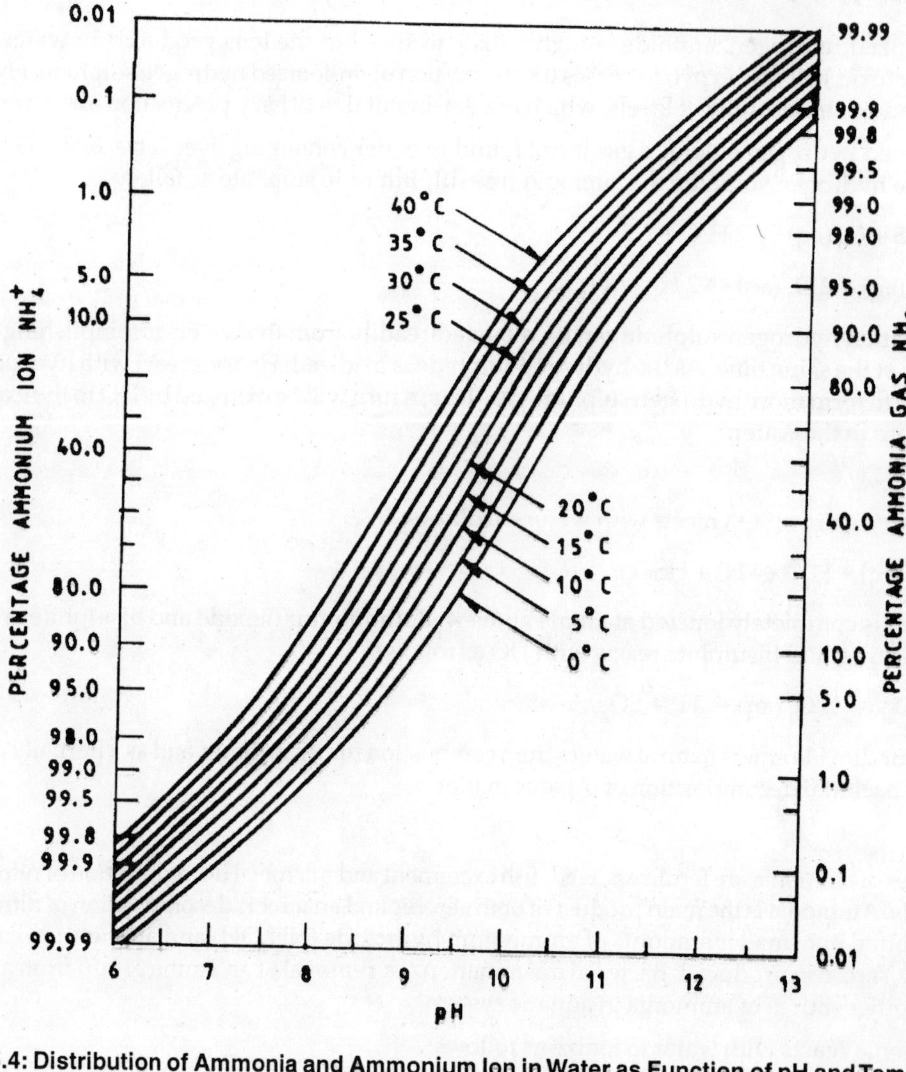

Figure 5.4: Distribution of Ammonia and Ammonium Ion in Water as Function of pH and Temperature (After Wheaton, 1977)

and hence blood ammonia and pH level increases. Ammonia also increases oxygen consumption by tissues, damages gills, reduces internal ion concentration and reduces oxygen transport capacity of blood. Ammonia is more toxic when DO concentration is low (Boyd, 1982).

9. Biological Parameters

The biological parameters significant to water quality characteristics are mainly related to (a) microbial food webs that have direct impact on productivity, and (b) environmental pollution of the water body that can adversely affect the physiological states of the aquatic organisms.

Microbial population in ponds constitutes mainly bacteria, microalgae and protozoa. Bacterial population consists of free-flowing planktonic bacteria, bacteria attached to particles, and benthic or sediment bacteria. Most of the bacteria in a pond are saprophytic heterotrophs, which obtain their organic matter for growth and energy by decomposing detritus. There are also autotrophic bacteria, which involve in nutrient cycling. Examples of these two types of bacteria are cellulose decomposers, amylolytic bacteria, protein decomposers, methane producers, phosphorus solubilizers, nitrogen fixers, nitrifiers, sulphate reducers, etc. These reactions play a vital role in maintenance of pond ecosystem, without which all the nutrients would be tied up in dead matters and no aquatic organisms would live.

The microalgae and blue green algae, *i.e.* the cyanobacteria, are the principal primary producers in ponds. Microalgae are extensively used as feeds for shrimp and fish larvae. Besides being used as feeds, they help in improving DO level and water quality in ponds. The algae that die enter the pool of detritus. The detritus is grazed upon principally by protozoa, which in turn are eaten by the zooplanktons.

The major potential biological pollutants causing water quality problems are infectious and toxic agents. These are (*a*) parasites such as those of public health significance, *e.g.* trematodes, nematodes, etc, (*b*) pathogenic bacteria, *e.g. salmonella, shigella, E. coli, Vibrio cholerae, Vibrio parahaemolyticus, Clostridium botulinum*, etc., and (*c*) biological toxins such as Scrombotoxin. A large number of fish species, both marine and freshwater, may serve as a source of medically important parasites. Some of these parasites are highly pathogenic, causing human infection through consumption of raw or inadequately cooked fish.

Diseases are among the greatest problems in maintaining the water quality and the natural environment for growth of aquatic species. Water pollutants, said to be potential for causing water quality problems, are mostly pathogenic bacteria and viruses, and are introduced into fishponds and watercourses by intentional or unintentional addition of human and animal wastes. Any existing or potential sources of pollution near the aquaculture farm and the nature of pollutants that are likely to affect the water supply to the farm should, therefore, be the major concern of the farming practice.

5.2 Fundamental Principles

Understanding of the fundamentals of chemical and biological phenomena causing changes in the quality of water have become critical in pursuing the various management strategies in aquaculture practice. Over the years, intensification of old problems and the introduction of new ones have led to basic changes in the philosophy of aquacultural engineering practice. Solution of these problems and development of processes to minimize them requires a basic knowledge of chemical and biological phenomena.

Equilibria between water and water vapor, air and dissolved oxygen in water, and two or more compounds dissolved and dissociated into ions in water are some examples of equilibrium processes involved in aquatic systems. Since the quality of water depends to a considerable extent upon its content of dissolved substances, an understanding of various equilibria and reaction rates involved in aqueous systems is an essential preliminary to a rational attack on the water quality management problems.

Water quality management and treatment processes involve a number of chemical and biological reactions. In general two aspects are pertinent to these reactions: one, regarding the equilibrium of the reaction and the other, the rate at which the equilibrium is attained. Reactions, which proceed very fast, reach an equilibrium state quickly. Many reactions proceed slowly and require to be expressed in term of rate of reaction so that the reactions can be dealt with on a practical basis. This is true for biodegradation or bio-transformation reactions, microbial growth and decay, aeration, disinfection, chemical hydrolysis, and oxidation-reduction. Reaction kinetics is concerned with rate or velocity of reactions, which, among other functions, decide on the design of the process system or the equipment in which the reaction occurs or is carried out.

This chapter outlines the fundamental principles of equilibrium relationships, process kinetics and biological reaction systems. The reader is referred elsewhere for more detailed discussions. Several good references relevant to these topics are Glasstone (1942), Walas (1959), Levenspiel (1972), Smith and Van Ness (1975), Bailey and Ollis (1977). For data and constants, Handbook of Chemistry and Physics (1984) may be consulted.

5.2.1 Equilibrium Relationships

1. Some Thermodynamic Concepts of Equilibria

Chemical Potential and Fugacity

Thermodynamically any equilibrium process is defined by Gibbs (1875–78) in terms of a quantity called "chemical potential" μ, which must have the same value for a given component, say in each phase at equilibrium. The quantity potential is a driving force in the transfer of material from one phase to another. Gibbs defined the chemical potential of a pure component in the following manner:

$$\mu = F = H - TS \tag{5.17}$$

The free energy F for one mole (molar free energy) is equivalent to the chemical potential. The enthalpy (H) and entropy (S) refer to a single mole of material.

For the equilibrium process of evaporating a mole of liquid water,

$$\Delta\mu = 0 \text{ or } \mu_L = \mu_G \tag{5.18}$$

It shows that for the equilibrium process there is no change in the chemical potential of water as a result of its transfer from liquid to gas.

For a pure component in a constant temperature process

$$\left(\frac{\partial \mu}{\partial p}\right)_T = v \quad \text{or} \quad \left(\frac{\partial \mu}{\partial p}\right)_T = \frac{RT}{p} \tag{5.19}$$

$$\text{or } \mu = RT \ln p + \theta \tag{5.20}$$

At very low pressures certain generalizations can be made regarding the behavior of all substances. For this reason, a new variable, related to μ and called "fugacity" ("escaping tendency") f, is introduced. Lewis (1923) defined fugacity for a pure component by relating it to chemical potential as follows:

$$\mu = RT \ln f + \theta \tag{5.21}$$

where, θ is a function of temperature only and μ is based on one mole of material.

It is stated that the fugacity is equal to the pressure near zero or at any pressure where the ideal gas law is valid, *i.e.* $f^* = p^*$, or in a liquid vapor mixture,

$$\bar{f}_{gas} = \bar{f}_{liquid}$$

where, the superscript * symbolizes a pressure approaching zero, and the superscript –, in solution or mixture. For multi-component single phase systems eqs (5.20) and (5.21) become

$$\mu_k = RT \ln \bar{f}_k + \theta \tag{5.22}$$

where, μ_k = chemical potential of component k in the mixture. At very low pressure the mixture will exhibit behavior closely approaching that of an ideal gas. Therefore, the partial pressure of k in the mixture ($\bar{p}^*$) will be given by the Dalton's law:

$$\bar{p}^* = p^* y_k \tag{5.23}$$

where, y_k = mole fraction of k in the gaseous mixture, and p^* is total pressure.

For ideal solution, a simple relationship for the effect of composition on fugacity of a gas is

$$\bar{f}_k = f_k\, y_k \tag{5.24}$$

If eq. (5.24) is applied to a vapor phase, which follows ideal gas behavior, the fugacities may be replaced by pressure. The result is Dalton's law, which states that the partial pressure of a component in a gas is equal to the total pressure multiplied by the mole fraction. Eq. (5.23) is an application of this equation at pressures approaching zero.

Applying the concept to liquid solutions,

$$\bar{f}_k = f_k x_k, \tag{5.25}$$

where, $\bar{f}_k$ is equal to f_k in the gas phase in equilibrium with the liquid solution. Under ideal condition $\bar{f}_k$ is equivalent to $\bar{p}_k$ in the gas, and f_k = vapor pressure p'_k of the ideal gas over the liquid. Therefore, eq. (5.25) becomes

$$\bar{f}_k = \bar{p}_k = p_k' x_k \tag{5.26}$$

This expression is known as Raoult's law.

Activity and Activity Coefficient

In many cases it will be more convenient to deal simply with the ratio of the fugacity in the given state to that in some arbitrary standard state at the same temperature. Lewis calls the ratio the "activity"

$$a = f/f^{o}, \tag{5.27}$$

where, f^{o} represents fugacity of the substance in the state of an ideal gas, *i.e.*

$$f^{o} = f^{*} = p^{*}$$

From eq (5.21), $\mu - \mu^{o} = RT \ln f/f^{o}$,

$$\text{or } \mu = \mu^{o} + RT \ln a \tag{5.28}$$

The activity may be defined directly on the basis of this equation, instead of referring it to fugacity. Fugacity is described as being to gases what activity to ions in solution.

Actually activity function is used only in treating solutions. At higher dilutions, the ions in solutions are further apart than at more concentrated solutions and the forces of attraction become less. If the solution under consideration is not ideal, the activity coefficient factor γ is introduced, which is one for an ideal solution. The difference between unity and the actual value of γ in a given solution is a measure of the departure from ideal behavior in that solution. When the solution behaves ideally (infinite dilution), the activity coefficients in term of mole fraction can be equated to that in term of concentration, *i.e.*

$$a = \gamma c \tag{5.29}$$

If actual concentrations are used in place of activities, the error in numerical calculations from the equilibrium relationships is not great for dilute aqueous solutions.

Equilibrium Constant and Standard Free Energy Change

Chemical reactions occur due to the activity of molecules or ions, and the number of active molecules or ions determine the rate of the reaction. The reactants and products are both being active, the reactions can proceed in both directions at the same time, and it is the net result that is observed at equilibrium.

For the generalized chemical reaction:

$$aA + bB + \ldots \leftrightarrow lL + mM + \ldots,$$

where, a, b, ...l, m,... are the number of moles of the substances A, B,...L, M,...taking part in the reaction. The equilibrium constant K becomes

$$K = \frac{k_r}{k_b} = \frac{a_L^{\,l} a_M^{\,m}}{a_A^{\,a} a_B^{\,b}} \tag{5.30}$$

where, k is the velocity constant for a given temperature, and r, b refer to forward and backward reactions, respectively. Eq. (5.30) is the exact form of law of mass action, applicable to any system, ideal or not.

In phase equilibria the transfer of material from one phase to another involves no change of free energy. Similarly, under chemical reaction equilibrium conditions there is no free energy change, *i.e.*

$\Delta F = 0$ at constant T and p.

For any component A,

$$\bar{F}_A = RT \ln \bar{f}_A + \theta$$

At the same temperature but at another condition, *i.e.* pressure p′ (standard state),

$$F_A^{o} = RT \ln f_A^{o} + \theta$$

So,

$$\bar{F}_A = F_A^{o} + RT \ln (\bar{f}/f^{o})_A \tag{5.31}$$

Applying the free energy concept into the generalized chemical reaction, we get, in terms of activity

$$\Delta F^{o} = -RT \ln \frac{a_L^{\,l}\, a_M^{\,m}}{a_A^{\,a}\, a_B^{\,b}} \tag{5.32}$$

where, ΔF^o represents the change in free energy for the reaction when each of the reactants and products is in its standard state. Eq. (5.32) is the basic relationship between the standard free energy change and the equilibrium constant becomes

$$\Delta F^{o} = -RT \ln K \text{ or } K = e^{-(\Delta F^{o}/RT)} \tag{5.33}$$

If ΔF^o is negative, K> 1, *i.e.* the equilibrium yield is high, and if positive, K< 1, *i.e.* the equilibrium yield is low.

The standard state for pure component is at the specifications of temperature (T) and pressure (P), and for solution at T, P, and composition. For gases, it is for pure component at temperature of reaction (25°C), and unit fugacity. For solids, liquid and solution present in the reaction mixture, the standard state of $f^o = 1$ is not convenient. So eq. (5.33) is not used. Eq. (5.32) must be used for equilibrium calculation.

Example 5.2

Deduce a relationship between the standard free energy change and activity of an ionic solution, where the standard state is referred to at unit molal activity.

Solution

For gases and volatile or non-ionical solutes, $\Delta F^o = RT \ln p/p^o$, which with $p^o = 1$ atm, becomes

$\Delta F^o = RT \ln p$

For ionic solutions, the standard state is referred to at unit molal activity.

$$\text{Now } p = \frac{24.45}{V} \cdot \frac{T}{298} = \frac{24.45T}{298}[c]$$

where, pV = RT = 24.45 litre-atm at 25°C, 24.45T/298 is the volume in litres occupied by a mole of the constituent at a partial pressure of 1 atm and temperature T, and [c] is the molal concentration in moles/liter, of the constituent. Since for a gas and non-ionized solutes, activity coefficient $\gamma = 1$ at low pressures and low concentrations, respectively, [c] = a.

$\therefore p = 24.45$ a, assuming $T/298 \cong 1$

Using this value of p and Gibb's definition of partial molal free energy, it can be shown that $\Delta F^o = RT \ln a$

Example 5.3

Calculation of vapour pressure of water using eq. (5.33).

Solution

Reaction: $H_2O\,(l) \leftrightarrow H_2O\,(g)$

F°(1 atm) =–56690–54635 cal/mole

Δ F° =–2055 cal

So for gases at 1 atm pressure, $K = p_w = \exp(-2055/1.985 \times 298) = 0.031$ atm.

Example 5.4

Determination of ionization constant for the first ionization of carbonic acid at 25° C.

Solution

Reaction: $H_2CO_3\,(aq) \leftrightarrow H^+\,(aq) + HCO_3^-(aq)$

ΔF° =–623.4 0–587.1

For the reaction $\Delta F^o = 36.3$ kJ, and K = exp (-36,300/8.314x298) =–14.65

Therefore, $K = 4.34 \times 10^{-7}$

2. Ionic Equilibrium in Water

Activity and Activity Coefficient

Reactions related to water quality characteristics are mostly ionic in nature. The inorganic constituents of water, after dissolution, yield ions in solution to a greater or lesser extent depending on the nature of the constituent. Even if ionization is complete, as is the case with simple salts in aqueous solution, there may still be a tendency for ion-pairs to form in relatively concentrated solution, so that dissociation is not necessarily complete. A general equilibrium reaction of ionization can be set up as follows:

$$MA \leftrightarrow M^+ + A^-$$

where M^+ and A^- represent free ions and MA is the undissolved portion of the electrolyte which includes both un-ionized molecules and ion pairs.

K, often called "dissociation constant" or "ionization constant", becomes

$$K = \frac{a_{M^+} a_{A^-}}{a^{MA}} \quad (5.34)$$

or in terms of activity coefficient and ionic concentration

$$K = \frac{C_{M^+} C_{A^-}}{C_{MA}} \times \frac{\gamma_{M^+} \gamma_{A^-}}{\gamma_{MA}} \tag{5.35}$$

However, in dealing with ionic equilibrium, with certain aspects of kinetics in solution, substitution of concentration for activity is frequently not possible. For solid solutes such as electrolytes in a very dilute solution, the mole fraction of a solute is proportional to its concentration (c), moles/liter and molality (m), moles/1000g. At an infinite dilution, when a solution behaves ideally, the activity coefficients based on x, c, and m are all unity, *i.e.* $\gamma = 1$ in the relations $a = \gamma x$, $a = \gamma c$, $a = \gamma m$. Therefore the standard states for ions are chosen as a hypothetical ideal solution of unit activity and activity of an ion becomes equal to its concentration at infinite dilution.

For the undissociated part of the electrolyte, a definition of mean activity and mean activity coefficient is introduced. The electrolyte $M_{\nu+}A_{\nu-}$ ionizes as

$$M_{\nu+}A_{\nu-} \leftrightarrow \nu_+ M^+ + \nu_- A^-$$

to yield ν_+ number of M^+ ions and ν_- number of A^- ions. The mean activity $a_\pm$ of the electrolyte is defined by

$$a_\pm = (a_+^{\nu+}\, a_-^{\nu-})^{1/\nu} \tag{5.36}$$

where, $\nu = \nu_+ + \nu_-$ is the total number of ions.

The mean activity coefficient of the electrolyte is defined by

$$\gamma_\pm = (\gamma_+^{\nu+}\, \gamma_-^{\nu-})^{1/\nu} \tag{5.37}$$

which in terms of concentration c

$$\gamma_\pm = a_\pm / c_\pm \tag{5.38}$$

where, $c_\pm$ is the mean concentration of the electrolyte. These relations indicate that for conversion of c to activity the mean activity coefficient for various concentrations of an electrolyte must be known.

Ionic Strength

The activity of an ion or molecule can be found by multiplying its molar concentration by an activity coefficient, as indicated by eq. (5.29). For many practical purposes, rough calculations obtained by use of molar concentration rather than activities are sufficient. However, in certain cases with deviation from ideality it may be practical and desirable to obtain more precise results. Lewis and Randall recognized that the activity coefficients for ions in an electrolyte were related to the concentration of charged particles in the solution. They introduced the concept of *ionic strength*, which represents the variation of activity coefficient with concentration, and is a measure of the intensity of the electrical field due to the ions in a solution. It is defined as half of the terms obtained by multiplying the molality (gram ionic weights/1000 g) or concentration, of each ion (c_i, gram ionic weight/liter), present in the solution by the square of its valence (z_i) as:

$$I = \tfrac{1}{2} \sum c_i z_i^2 \tag{5.39}$$

where, I = ionic strength.

It is important to keep in mind that the ionic strength is a general property of the solution and not a property of any particular ion in the solution.

In dilute solutions the activity coefficient of a given strong electrolyte is approximately the same in all solutions of a given strength. As the concentration is increased the specific influence of the added electrolyte becomes manifest. Variation of the activity coefficient of an ion with the ionic strength of the medium is expressed in the form of Debye-Huckel equation (1923) as follows:

$$\log_{10} \gamma_i = -\frac{Az_i^2\sqrt{I}}{1 + a_i' B\sqrt{I}} \tag{5.40}$$

where, A and B are constants for the solvent at a specified temperature and dielectric constant, and a_i' is the average effective diameter of the ions. This theory is applicable only to dilute solutions with ionic strength less than about 0.1. Excepting seawater, most waters are more dilute than this. The Debye-Huckel constants of water and the values of a_i' are given in Table 5.3 and Table 5.4, respectively.

Table 5.3: Debye-Huckel Constants of Water

Temperature (°C)	*A*	*B*
0	0.488	0.325
15	0.500	0.328
25	0.509	0.330
30	0.514	0.331

Table 5.4 : Values of a_i'

Ions	a_i'
OH^-, HS^-, K^+, Cl^-, NO_2^-, NO_3^-, NH_4^+	3
Na^+, PO_4^{3-}, SO_4^{2-}, HPO_4^{2-}, HCO_3^-, $H_2PO_4^-$	4
CO_3^{2-}	5
Ca^{2+}, Cu^{2+}, Zn^{2+}, Mn^{2+}, Fe^{2+}	6
Mg^{2+}	8
Al^{3+}, Fe^{3+}, H^+	9

In terms of mean activity coefficient of the electrolyte, eq. (5.40) is transformed as follows

$$\log_{10} \gamma = -\frac{Az_+z_-\sqrt{I}}{1 + Ba_i'\sqrt{I}} \tag{5.41}$$

The mean ionic diameter a_i' must be assumed to fit the data. However, for very dilute solutions $Ba_i'\sqrt{I}$ is small compared with unity and may be neglected. Then the limiting equation of Debye-Huckel is

$$\log_{10}\gamma = -Az_+z_-\sqrt{I} \tag{5.42}$$

Calculation of activities of ion in pond water can be done from Debye-Huckel equations. The activity coefficient of ions would be one at infinite dilution and c = a. At higher concentrations, electrostatic interactions increase and thereby effective concentrations of ions are reduced.

Ion-pairs

Although many salts are probably regarded as completely or almost completely ionized at all reasonable concentrations, the ions are not necessarily free to move independently. As a result of electrostatic attraction, cations and anions are strongly attracted to each other and form a certain proportion of ion-pairs. So at any instant a number of ions are made unavailable for transport of current. The electrolyte may be completely ionized, but not necessary completely dissociated. At high dilution the solute is both ionized and dissociated completely.

When ion-pairs exist, the equilibrium constant becomes

$$K = \frac{a_{M^+} a_{A^-}}{a_{ion-pair}} \tag{5.43}$$

or

$$K = \frac{\gamma_1(1-\alpha)c \times \gamma_2(1-\alpha)}{\gamma_2 \alpha c} \tag{5.44}$$

where, α is the fraction of association, which depends on dielectric constant and mean diameter of ions. For very dilute solutions, *i.e.* when c is small, the activity coefficients are almost unity, and α is negligible in comparison with unity. Then eq. (5.44) becomes

$$K = c/\alpha \tag{5.45}$$

Free ions and ion-pairs cannot be distinguished by analytical methods. Thus, sulphate in solution, which may be distributed among SO_4^{2-}, $CaSO_4^{\circ}$, $MgSO_4^{\circ}$, KSO_4^- and $NaSO_4^-$, can only be calculated as SO_4^{2-} by measuring the total SO_4^{2-} concentration and the actual ionic concentration, and the actual ionic concentration is less than the measured ionic concentration in solution. Further, ion-pairing has little effect on ionic activity calculations in weakly mineralized water (I<0.002 M) or when total dissolved ion is less than 500 mg/l. The equilibrium constants for ion pairs formed by major ions in natural water are given in Table 5.5.

Table 5.5: Equilibrium Constants for Ion Pairs at 25°C and Zero Ionic Strength in Natural Waters

Ion Pair	*K*
$CaSO_4^{\circ}$	5.25×10^{-3}
$CaCO_3^{\circ}$	6.30×10^{-4}
$CaHCO_3^+$	5.5×10^{-2}
$NaSO_4^-$	2.4×10^{-1}
$NaCO_3^-$	5.35×10^{-2}
$NaHCO^{\circ}$	1.78
KSO_4^-	1.1×10^{-1}

Ionic Concentration and Ionic Balance in Water

Natural water contains various inorganic solutes dissolved in it, which exist in the form of cations and anions. Most of the ions that are present in appreciable quantities and affect the water quality are listed in Table 5.6. Surface waters are usually dilute solutions of alkaline earth carbonate and bicarbonate, such as of calcium and magnesium. Concentrations of these ions depend on the type of soil, watershed, ground water and climate of the region (Boyd,1982).

Seawater contains much higher concentrations of total ions in comparison to surface waters, *e.g.* 35,000 mg/l corresponding to 35 ppt salinity. It has a relatively constant ionic composition and has much more ionic concentration than fresh waters. In a particular water the concentrations of cations and anions in milliequivalents per liter nearly balance because of the principle of electrical neutrality. The ionic balance principle is useful in water analysis for aquaculture. If the cation and anion concentrations do not agree reasonably well, the analysis would be in error.

Table 5.6: Cations and Anions of Water

Cations	*Anions*
Ca^{2+}, Mg^{2-}, Na^+, K^+, Fe^{2+}, Mn^{2+}, Zn^{2+}, H^+, NH_4^+	HCO_3^-, CO_3^{2-}, SO_4^{2-}, HPO_4^{2-}, $H_2PO_4^-$, HS^-, $HsiO_3^-$, $H_2BO_3^-$, Cl^-, F^-, NO_3^-, OH^-

Example 5.5

Calculate the activity of HCO_3^- ion in pond water, which contains the ions given below. What would be its activity in a solution, when the ionic strength changes to 0.01 M? Water temperature is 30°C (*Source* of data: Boyd, 1982).

Ions	*Molecular Weight*	*Concentration*	
		mg/l	*moles/l × 10³*
HCO_3^-	61	40	0.66
SO_4^{2-}	96.06	10	0.10
Cl^-	35.45	12	0.34
Ca^{2+}	40.08	60	1.50
Mg^{2+}	24.31	10	0.40
K^+	39.10	80	2.05
Na^+	22.99	12	0.52

Solution

From eq. (5.39), substituting the values of ionic concentration and valence, we get

I = 0.0058 M

At 30° C, activity coefficient of HCO_3^- (γ_{HCO3}^-) is calculated from eq. (5.40):

$-\log \gamma_{HCO3}^- = -0.0355$, and $\gamma_{HCO3}^- = 0.922$

$\therefore$ a = 0.922 × 0.66 × 10^{-3} = 0.00061 M

At I = 0.01 M, γ = 0.901, and a = 0.00059 M

Example 5.6

Anion-cation balance: Water samples from two fishponds contain the ion concentrations given below. Find out the anion cation balance (Data adapted from Boyd, 1982).

Solution

Ions	*Milliequivalent/Litre*	
	Pond A	*Pond B*
HCO_3^-	1.00	0.23
CO_3^{2-}	0	0
SO_4^{2-}	0.09	0.12
Cl^-	0.19	0.34
Ca^{2+}	0.98	0.27
Mg^{2+}	0.12	0.14
K^+	0.04	0.19
Na^+	0.19	0.03
Total cation:	1.33	0.69
Total anion:	1.28	0.63

There is a little disparity between the concentrations of anion and cation, which shows the major cations and anions are nearly in balance.

3. Ionization of Acids and Bases

The pH Concept

The concept of pH was developed from the ionization reaction of water:

$$2H_2O\ (l) = H_3O^+ + OH^-$$

The hydrogen ion is the hydrated proton, commonly represented as H_3O^+. The equilibrium constant for this reaction is very large, *i.e.* 10^{-14}. Since the ionization of water is very slight, the activity of the water in any aqueous solution will be constant and the equilibrium constant relationship of water becomes:

$$K_w = K_{a_{H_2O}} = a_{H^+} a_{OH^-} \quad (5.46)$$

where, K_w is called the "ion product" of water. In pure water, $a_{H+} = a_{OH-}$, or in terms of concentration $c_{H+} = c_{OH-} = \sqrt{K_w} = 10^{-7}$ at 25° C.

A convenient means of expressing the concentration of hydrogen ions without involving negative exponents is the pH scale in which pH is defined as:

$$pH = -\log_{10} a_{H^+} \quad (5.47)$$

Thus for a solution of pH 4, $a_H^+ = 10^{-4}$. Similarly a_{OH-} can also be expressed on a pOH scale by definition.

Table 5.7: Ion Product of Water at Various Temperatures

Temperature (C)	$K_w \times 10^{14}$	*Temperature (C)*	$K_w \times 10^{14}$
0	0.114	30	1.470
10	0.292	40	2.919
25	1.008	60	9.614

Equilibrium Relations of Carbon Dioxide in Water

Natural waters mostly have pH between 6.5 and 9. But there are many factors, which affect the pH of water. The most dominating factor is carbon dioxide. Carbon dioxide gas reacts with water to ionize as follows:

$$CO_2 + H_2O \leftrightarrow H^+ + HCO_3^-$$

The equilibrium relation at 25° C is

$$K_1 = \frac{\gamma(H^+)\gamma(HCO_3^-)}{\gamma(CO_2)_{aq}} = 4.31 \times 10^{-7} = 10^{-6.37} \quad (5.48)$$

The equilibrium constant can be utilized to calculate pH of a particular solution of water.

The dissociation constant of H_2CO_3 into CO_3^{2-} via HCO_3^- is

$$K_2 = \frac{\gamma(H^+)\gamma(CO_3^{2-})}{\gamma(HCO_3^-)} = 4.69\times10^{-11} = 10^{-10.33} \quad (5.49)$$

The small value of K_2 indicates that carbonate concentrations are negligible in water containing dissolved carbon dioxide.

At a particular value of pH, carbon dioxide concentration decreases to a very low value and carbonate concentration reaches also to a very low but detectable value, which can be calculated as shown in Example 5.9.

The relationship of pH and the concentrations of various forms of carbon dioxide in solution in fresh water are shown in Figure 5.5. At low pH the presence of free carbon dioxide will be more than that at about pH 8, when nearly 100 percent of the carbon dioxide is converted to bicarbonate ion. Above pH 8 carbonate ions increase in larger proportions. Above pH 12 almost all carbon dioxide is in the carbonate form. Natural waters may have pH anywhere between 2 and 12, but most fresh water systems fall between pH 6 and 9, with dominant concentration of bicarbonate ions.

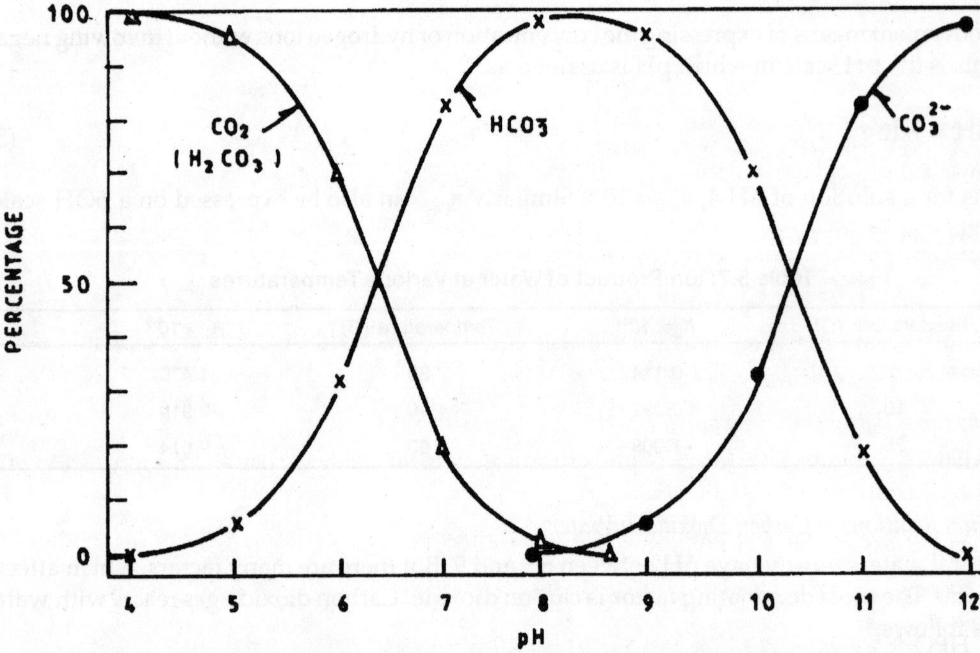

Figure 5.5: Relationship Between Concentrations (per cent) of CO_2, HCO_3^- and CO_3^{2-} Ions in Solution in Fresh Water and pH (From Boyd, 1982)

The alkalinity of a bicarbonate solution at equilibrium is due to the fact that the solution contains both an acid, carbonic and a base, carbonate in equal concentrations. As an acid, carbonic acid releases hydrogen and carbonate ions as shown in eq. (5.48), and as a base carbonate releases hydroxyl ion in water:

$$CO_3^{2-} + H_2O \leftrightarrow HCO_3^- + OH^-, K_3 = 2.14 \times 10^{-4} = 10^{-3.67} \quad (5.50)$$

Since hydroxyl ion concentration increases relative to hydrogen ion concentration, pH rises. In fact, carbonate is a stronger base than carbon dioxide is an acid ($K_1 = 10^{-6.37}$).

Combining eq. (5.48) and (5.49), we get

$$2HCO_3^- \leftrightarrow CO_2 + CO_3^{2-} + H_2O, K_4 = 1.05 \times 10^{-4} = 10^{-3.98} \quad (5.51)$$

To shift the reaction to the right, carbon dioxide is to be removed so that more carbon dioxide and carbonate are formed. In natural pond waters carbon dioxide is consumed during photosynthesis. Carbonate gets hydrolyzed and one bicarbonate ion is formed, according to eq. (5.50), replacing one bicarbonate ion out of two ions required for the reaction (5.48). However, equilibrium of the reaction (5.49) is maintained, since carbon dioxide is removed. Therefore, as bicarbonate and hydrogen ions decrease, carbonate and hydroxyl ions increase, causing the pH to rise.

Carbon dioxide is removed from the pond water system containing calcium by precipitation of calcium carbonate as follows:

$$Ca^{2+} + CO_3^{2-} \leftrightarrow CaCO_3, K_5 = 5.01 \times 10^{-3} = 10^{-8.3} \quad (5.52)$$

Since precipitation will limit the accumulation of carbonate, the rise in pH will be moderated.

Equilibrium Relations of Ammonia

The ionization reaction of ammonia in water:

$$NH_3 + H_2O \leftrightarrow NH_4^+ + OH^-$$

has the equilibrium constant as

$$\frac{\gamma(NH_4^+)\gamma(OH^-)}{\gamma(NH_3)} = 1.81 \times 10^{-5} = 10^{-4.74} \quad (5.53)$$

at 25°C. This relation shows that nearly all the ammonia gas in dilute solution is ionized below pH 6, but is not ionized above pH 11 (Figure 5.4).

Equilibrium Relations of Phosphates

All the three inorganic forms of phosphorus in natural water are usually the ionization products of ortho-phosphoric acid (H_3PO_4 aq):

$$H_3PO_4 \text{ aq} \leftrightarrow H^+ + H_2PO_4^-, K_1 = 7.5 \times 10^{-3} = 10^{-2.13}$$

$$H_2PO_4^- \leftrightarrow H^+ + HPO_4^{2-}, K_2 = 6.2 \times 10^{-8} = 10^{-7.21}$$

$$HPO_4^{2-} \leftrightarrow H^+ + PO_4^{3-}, K_3 = 4.8 \times 10^{-13} = 10^{-12..32}$$

all at 25°C. These relations and Figure 5.6 indicate the percentage of the ions present, in dilute solutions, are less than 10 percent of the monohydrogen phosphate at pH less than 5, less than 10 percent of dihydrogen phosphate at pH less than 1, and more than 99 percent of the orthophosphate at pH less than 2. It is evident that phosphate alkalinity in natural water consists of both di–and mono-hydrogen phosphates

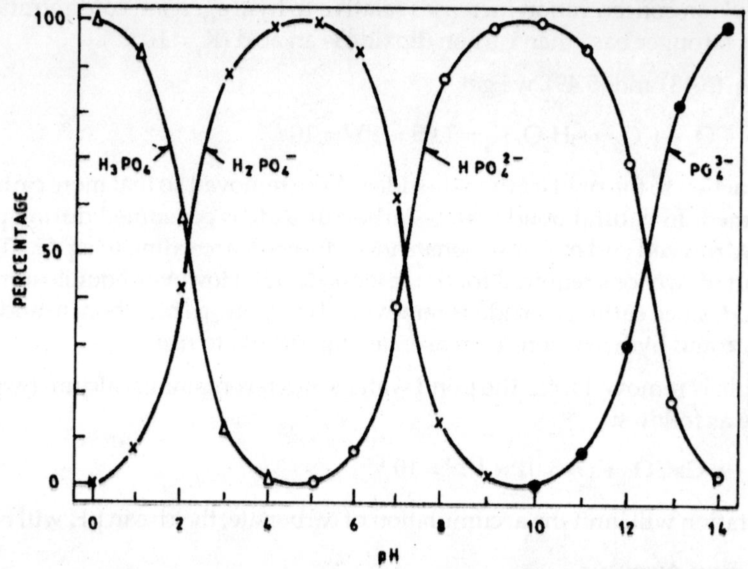

Figure 5.6: Relationship Between Concentrations (per cent) of Orthophosphate Ions and pH in Water (From Boyd, 1982)

4. Solubility Relationship

A fundamental concept is that all solids, no matter how insoluble, are soluble to some degree. In contact with water they do dissolve, slightly, and form equilibrium. In fact the equilibrium that exists between crystals in the solid state and its ions in solution is amenable to consideration under the equilibrium relationship. For example, tricalcium phosphate ionize as follows:

$$\underline{Ca_3(PO_4)_2} \Leftrightarrow 3Ca^{2+} + 2PO_4^{\ 3-},$$

where, $\underline{Ca(PO_4)_2}$ represents the solid phosphate, whose concentration can be regarded as a constant in the equilibrium relationship. Thus

$$\frac{(Ca^{2+})^3(PO_4^{\ 3-})^2}{(Ca_3(PO_4)_2)} = K \text{ or } (Ca^{2+})^3(PO_4^{\ 3-})^2 = KK_s = K_{sp} \quad (5.54)$$

The constant K_{sp} is called the solubility-product constant. The K_{sp} value of eq. (5.54) at 25° C is $1x10^{-27}$. The solubility-product principle is basic to explaining the phenomena of precipitation and solution of precipitates. In an unsaturated solution, the product of the molar concentration of ions is less than the K_{sp} value, whereas it is the greater than the K_{sp} value in supersaturated solution. The pH of water affects the equilibrium relationship, which in turn can affect solubility.

Example 5.7

Calculate the ion product constant of water at 25° C and 1 atm pressure.

H_2O (l) $\leftrightarrow$ $H^+ + OH^-$

F° (u.a.) =–56690 0–37585 cal/mole

Solution

$-\Delta F^{\circ} = -19105$ cal/mole

$K_w = \gamma^2 [H^+][OH^-] = \exp(-19105/1.985 \times 298) = 1\text{x}10^{-14}$

Example 5.8

Calculate the pH of pure water in contact with atmosphere containing 0.033 percent carbon dioxide or 0.48 mg/l carbon dioxide concentration. What would be the pH if the concentration of carbon dioxide rises to 30 mg/l?

Solution

Total concentration of $CO_2 = 0.48$ mg/l $= 1.091 \times 10^{-5}$ mole/l or $10^{-4.96}$ M

From eq. (5.46), assuming $[H^+] = [HCO_3^-]$, we get $[H^+]^2 = 10^{-11.33}$

or $H^+ = 5.67$

If CO_2 concentration becomes 30 mg/l, pH = 4.8

Example 5.9

Calculate pH of water, when the concentrations of both carbon dioxide and carbonate are very low.

Solution

Assuming $[CO_2] = [CO_3^{2-}]$, we get from eq. (5.48) and (5.47)

$[H^+]^2 [CO_3^{2-}] / [CO_2] = 10^{-16.70}$

or $[H^+] = 10^{-8.35}$, *i.e.* pH = 8.35

5.2.2 Process Kinetics

1. Rate of a Chemical Reaction

Reaction kinetics is concerned with rate or velocity of reactions. Reaction kinetics, among other functions, decides on the design of the process system or the equipment in which the reaction is carried out or occurs. In a chemical process the reacting materials, which constitute different chemical species form into new and different materials. When this occurs by rearrangement or redistribution of the constituent atoms to from new molecules, we call it reaction. Rate of a reaction may be defined in a number of ways. Quantitatively the reaction rate is expressed as the number of units of mass of some component i, formed or transformed per unit time t per unit volume of the system. When the system volume remains constant, the rate of reaction can be expressed as follows:

$$r = \frac{dn_i}{Vdt} = \frac{moles\ i\ present}{(volume\ of\ system)(time)} = \frac{dC_i}{dt} \tag{5.55}$$

where, $C_i = n_i/V$ is called the concentration of the component i.

In terms of amount of i that has been transformed in this time, $x = n_o - n$, the rate is

$$r = \frac{dx}{Vdt}, \text{ where } n_o = \text{number of moles at zero time.}$$

The functional relation between rate and concentration is called a rate equation or rate expression. For a general stoichiometric chemical reaction

$$aA + bB + \ldots.. \rightarrow lL + mM + \ldots..$$

the rate of disappearance of A is expressed by the rate equation

$$r_A = \frac{dn_A}{Vdt} = k\left(\frac{n_A}{V}\right)^p \left(\frac{n_B}{V}\right)^q \ldots = k\left(\frac{n_{A_o} - x}{V}\right)\left(\frac{n_{B_o} - \frac{bx}{a}}{V}\right) \quad (5.56)$$

$$= k(C_A)^p (C_B)^q$$

where, a, b,....l, m,.... are the number of moles of the components, k is called rate constant, and p, q,...are the powers to which the concentrations are raised, called the order of the reaction. Thus the reaction is p-th order with respect to A, q-th order with respect to B, ... and the sum of the exponents p +q +....= n-th order for the whole reaction. Order is strictly an empirical concept applicable to rate equation of the from of eq. (5.56). The dimensions of the rate constant k for n-th order reaction rate are

$(\text{time})^{-1}$ $(\text{concentration})^{1-n}$,

which for a first order reaction become simply $(\text{time})^{-1}$. Commonly used concentration units are mol / liter, mol / cm^3, or pressure in atm, and time units is 's'.

The determination of reaction rate reduces inevitably to experimental determination of concentration as a function of time, under controlled conditions, often isothermal. The problem is to fit the rate equation to these data, and find a correlation. Generally such an equation is found by trial. Most often the stoichiometry of the reaction suggests the form of rate equation to try first. If this proves incorrect, then for each equation a unique rate equation is to be fitted to the data.

The kinetic data can be analyzed by two procedures: the integral and differential methods. In the integral method of analysis, a particular from of the rate equation is assumed and the data of concentration vs. time should fit the equation in straight line form. In the differential method of analysis data fitting of the rate equation is done directly, and without any integration. The integral method is easy to use and is applied for relatively simple rate equations. The differential method is useful in more complicated situation.

2. Kinetic Models of Homogeneous Reactions

A homogeneous reaction takes place in one phase only, whereas a heterogeneous reaction requires the presence of two phases to proceed. Most gas phase and liquid phase reactions are homogeneous in nature. Homogeneous reactions are single and multiple types.

(1) In a single reaction: A + B → P, the rate equation is

$$r_A = -k\,C_A\,C_B \quad (5.57)$$

In multiple reactions, such as in series:

$$A \xrightarrow{k_1} B \xrightarrow{k_2} C$$

and in parallel:

$$A \xrightarrow{k_1} B,\ A \xrightarrow{k_2} C, and\ A \xrightarrow{k_1} P,\ B \xrightarrow{k_2} Q$$

the rate equations are: for series reactions

$$r_A = -k_1 C_A,\ r_B = k_1 C_A - k_2 C_B,\ r_C = k_2 C_B \tag{5.58}$$

and for parallel reactions

$$r_A = -(k_1 + k_2) C_A,\ r_B = k_1 C_A,\ r_C = k_2 C_A \tag{5.59}$$

In reversible reactions, where the products recombine and form the original reactants,

$$A \xrightarrow{k_f} B + C \xrightarrow{k_b} A$$

$$-r_A = k_f C_A - k_b C_B C_A \tag{5.60}$$

For an irreversible unimolecular first order reaction, $A \rightarrow$ products,

$$r_A = -\frac{dC_A}{dt} = kC_A \text{ or integrating } -\ln\frac{C_A}{C_{A_o}} = kt \tag{5.61}$$

where, C_{Ao} = initial concentration at $t = 0$.

In terms of fractional conversion x_A of the given reactant A,

$$\frac{dx_A}{dt} = k(1 - x_A) \tag{5.62}$$

where, $x_A = (n_{Ao} - n_A)/n_{Ao}$ or $n_A = n_{Ao}(1 - x_A)$.

Eq. (5.62) on rearrangement and integration becomes

$$\int_o^{x_A} \frac{dx_A}{1 - x_A} = k\int_o^t dt \text{ or } -\ln(1 - x_A) = kt \tag{5.63}$$

which is equivalent to eq.(5.61). A plot of (C_A/C_{Ao}) vs t gives a straight line through the origin.

(2) For a bimolecular second order reaction with rate equation given in eq. (5.57), the rate equation is

$$r_A = -k C_A C_B$$

Assuming that amounts reacted at any time t are equal, *i.e.* $C_{Ao}x_A = C_{Bo}x_B$, and writing

$$-r_A = C_{A_o}\frac{dx_A}{dt} = k(C_{A_o} - C_{A_o}x_A)(C_{B_o} - C_{A_o}x_A),$$

we get on integration

$$\ln\frac{1-x_A}{1-x_B}=\left(C_{B_o}-C_{A_o}\right)kt \tag{5.64}$$

(3) In irreversible reactions in series

$$A\xrightarrow{k_1}R\xrightarrow{k_2}S$$

the integrated forms of the rate equations, shown in (5.58), are:

$$C_A=C_{A_o}e^{-k_1t} \tag{5.65}$$

$$\frac{dC_R}{dt}+k_2C_R=k_1C_{A_o}e^{-k_1t},$$

which on integration becomes

$$C_R=C_{A_o}k_1\left(\frac{e^{-k_2t}}{k_2-k_1}+\frac{e^{-k_1t}}{k_1-k_2}\right) \tag{5.66}$$

From eqs. (5.65) and (5.66), with total number of moles: $C_{Ao}=C_A+C_R+C_S$,

$$C_S=C_{A_o}\left(1+\frac{k_2}{k_1-k_2}e^{-k_1t}+\frac{k_1}{k_2-k_1}e^{-k_2t}\right) \tag{5.67}$$

Eq. (5.67) shows that the concentrations of A, R and S vary with time. Figure 5.7 shows the general characteristics of the concentration-time curves for three components.

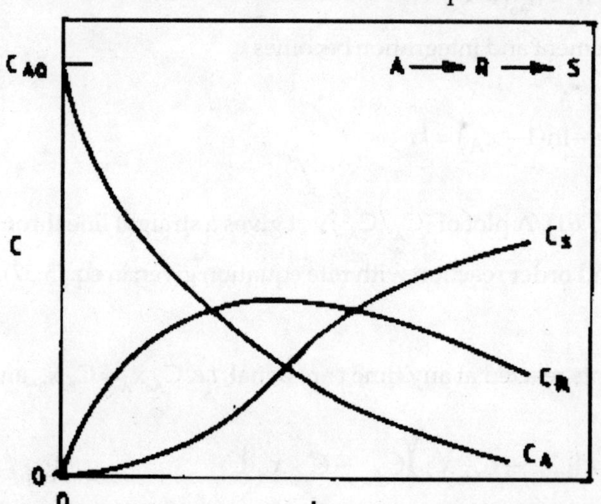

Figure 5.7: Typical Concentration-Time Curves for Reactions A → R → S (After Levenspiel, 1972)

A decreases exponentially, R rises to maximum and then falls, and S rises continuously, the greatest rate of increase of S occurring where R is a maximum.

When $k_2 >> k_1$, eq. (5.67) reduces to

$$C_S = C_{A_o}\left(1 - e^{-k_1 t}\right)$$

i.e. the rate is determined by k_1. When $k_1 >> k_2$, eq. (5.67) becomes

$$C_S = C_{A_o}\left(1 - e^{-k_2 t}\right),$$

i.e. the rate is determined by k_2. Thus for reactions in series it is the slowest step that has the greatest influence on the over-all reaction rate.

To find the maximum concentration of R,

$$\frac{dC_R}{dt} = 0,\ or\ t_{\max} = \frac{\ln(k_2 / k_1)}{k_2 - k_1} = \frac{1}{k_{\log mean}} \quad (5.68)$$

$$\text{or, } \frac{C_{R,\max}}{C_{A_o}} = \left(\frac{k_1}{k_2}\right)^{k_2/(k_2 - k_1)} \quad (5.69)$$

In irreversible reactions in parallel

$$A \xrightarrow{k_1} R\ A \xrightarrow{k_2} S,$$

the k values can be found by integrating all three differential rate equations shown in eq. (5.59). The first integrated equation is

$$-\ln\frac{C_A}{C_{A_o}} = (k_1 + k_2)t \quad (5.70)$$

The second equation is obtained by dividing the second and the third differential equations and integrating

$$\left(\frac{C_R - C_{R_o}}{C_S - C_{S_o}}\right) = \frac{k_1}{k_2} \quad (5.71)$$

Figure 5.8 shows the plots for evaluation of k_1 and k_2.

(4) Zero-order reactions

Fundamentally no single reaction proceeding in a forward direction only can have a true order less than unity, but in practice, many biological reactions exhibit zero order characteristics, *i.e.* within certain limits, the reaction proceeds steadily independent of the amount of reactant present. This

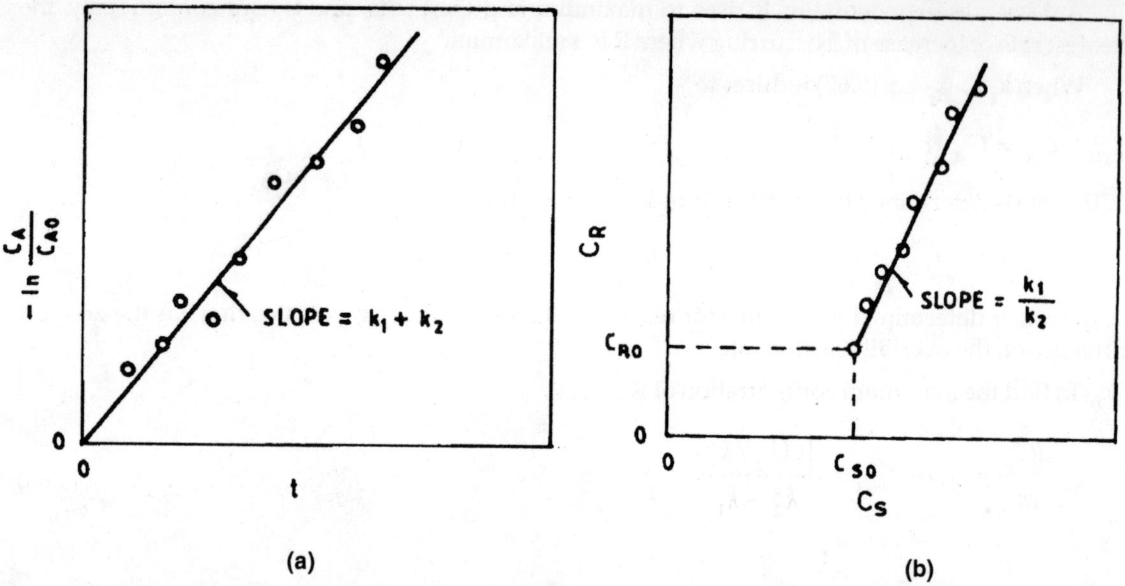

Figure 5.8: Plots for Evaluation of Rate Constants for Reaction of A → R and A → S (a) From eq. 5.70 and (b) From eq. 5.71 (After Levenspiel, 1972)

means that the rate of reaction is not proportional to the amount of reactant present, so that it can be written as:

$$-r_A = \frac{dC_A}{dt} = k \tag{5.72}$$

When a reaction is basically monomolecular but is operated in presence of excess of one reactant, the condition depresses the order by one. The most common example is that a simple enzyme catalyzed reaction behaves as a monomolecular system when the proportion of the enzyme catalyst is high but changes to zero–order in presence of excess reactant. Typical examples are non-enzymatic browning, inhibited rancidity and many quality losses in frozen foods.

From eq. (5.72), rearranging and integrating between C_{Ao} at t = 0 to t = t, we get

$$C_{Ao} - C_A = kt \tag{5.73}$$

For a zero order reaction, a plot of the C_A vs. time yields a straight line with slope equal to the rate constant k in units of concentration/unit time.

Example 5.10. Modeling of Biological Filtration Kinetics

An empirically based model has been proposed by Weatherley (1984) for the kinetics of oxidation of ammonia and nitrite in a marine biological filter. At low concentration of the metabolites encountered in aquacultural waters and for the case of steady biomass conditions, the rate of oxidation of two primary metabolites, ammonia and nitrite ion, has been approximated by first order kinetics. Thus for the reactions

$$NH_4^+ \xrightarrow{k_1} NO_2^- \xrightarrow{k_2} NO_3^-$$

the rate of oxidation of ammonia by *Nitrosomonas* and the rate of oxidation of nitrite in the presence of ammonia, *Nitrosomonas* and *Nitrobacter* are given by

$$\frac{dC_1}{dt} = -k_1C_1 \quad and \quad \frac{dC_2}{dt} = k_1C_1 - k_2C_2$$

where, $C_1 = NH_3$ concentration, mg/l, and $C_2 = NO_2^-$ concentration, mg/l.

k_1 can be calculated from: $\ln C_1 = k_1t$ + constant, or $C_1 = C_{1o} \exp(-k_1t)$, where C_{1o} = initial NH_3 concentration. For calculation of k_2

$$\frac{dC_2}{dt} = k_1C_{1_o} \exp(-k_1t) - k_2C_2$$

If $k_2 < k_1$, NO_2^- will give a maximum value during the course of reaction. At maximum value of C_2, $dC_2/dt = 0$, $C_2 = C_{2max} = 0$, $t = t_{max}$. So

$$k_1C_{1o} \exp(-k_1t_{max}) - k_2C_{2max} = 0,$$

$$\text{or, } k_2 = \frac{k_1C_{1_0} \exp(-k_1t_{max})}{C_{2\,max}}$$

Since k_1 and k_2 are functions of pH and temperature, they can be expressed as

$\ln k_1 = 0.0413\,T - \varphi_1$ and

$\ln k_2 = 0.0255\,T + \varphi_2$,

where, φ_1 and φ_2 are functions of pH alone, as expressed below:

$\varphi_1 = 104.65 - 37.886\,pH + 4.6789\,pH^2 - 0.1944\,pH^3$

$\varphi_2 = 2.281 - 0.3983\,pH - 0.0381\,pH^2$

For calculation of C_2, we get

$$C_2 = \frac{C_{1_o}k_1}{k_2 - k_1}\{\exp(-k_1t) - \exp(k_2t)\}$$

The experimental and predicted values of k_1 and k_2 obtained by Weatherley (1984) on the kinetic study of ammonia oxidation are shown in the following table:

Trial Number	C_{O2} (mg/l)	pH	Salinity (ppt)	Temperature (°C)	k_1		k_2	
					Expt.	Pred.	Expt.	Pred.
1	7.7	7.56	31	22.1	0.462	0.454	0.1195	0.0959
2	7.85	7.5	29.1	21.7	0.344	0.434	0.121	0.1107
3	7.59	7.59	30.5	21.8	0.485	0.454	0.0827	0.0925
4	7.74	7.74	33.5	21.8	0.428	0.483	0.0764	0.0798

3. Effect of Temperature on Reaction Rate

The effect of temperature on reaction rate is expressed through the effect of temperature on rate constant. Such temperature dependency has been in practically all cases to be well represented by Arrhenius' law. Arrhenius (1889) found that reaction velocities were a function of temperature and he proposed an empirical equation as:

$$k = Ae^{-E/RT} \text{ or } \ln k = \ln A - E/RT \tag{5.74}$$

where, A = constant called frequency factor, E = activation energy of the reaction, R = universal gas constant, and T = absolute temperature. Arrhenius also suggested that molecules will react only if they possess at least a certain minimum amount of energy known as "activation energy". In other words, this is the energy required to get the molecules into an active state.

Arrhenius' equation is widely applicable to reactions in solution and to heterogeneous reaction. The equation fits well over wide temperature ranges and is a very good approximation to the true temperature dependency. Subsequent development of the kinetic theory oı gasses and transition state theory support this conception.

The temperature dependency of reaction is determined by the activation energy and temperature level. From Arrhenius 'law, a plot of ln k vs. 1/T gives a straight line, with large slope for large E and small slope for small E (Figure 5.9).

5.2.3 Biological Reaction Systems

There are two domains of biological reaction systems to the interest of aquacultural engineers: reactions involved in aquatic eco-system and treatment of farm effluent by biological process. Bacteria, micro algae and protozoas all have major roles in aquatic eco-systems. The interactions of algae and bacteria and factors affecting their productivity in the pond water control the microbial processes in the food web. Anaerobic processes are significant in detritus decomposition in pond sediments, where bacteria are the principal consumers of oxygen. Biological treatment of aquaculture effluents is the most common water purification system in use for recirculating culture practice as well as for discharge into streams and seas. In all these cases the general outline of the biological process is that microorganisms present in the natural eco-system react with various types of reactants, called substrates, grow and reproduce. As a result, either some endproducts are produced, *e.g.* alcohol and acids, or microorganisms grow consuming the substrates.

The basic reaction scheme of the biological processes can be written as shown in Figure 5.10.

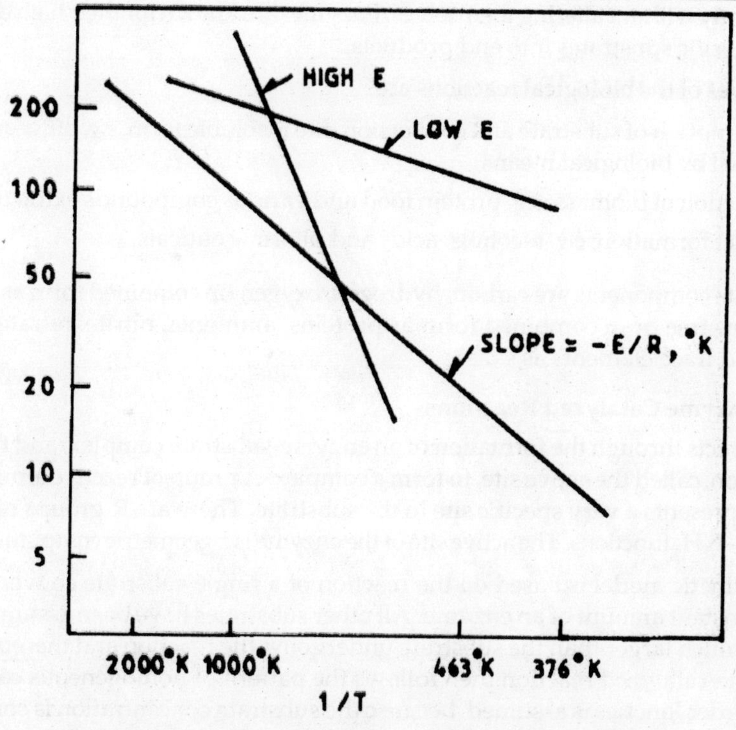

Figure 5.9: Temperature Dependency of Reaction Rate (After Levenspiel, 1972)

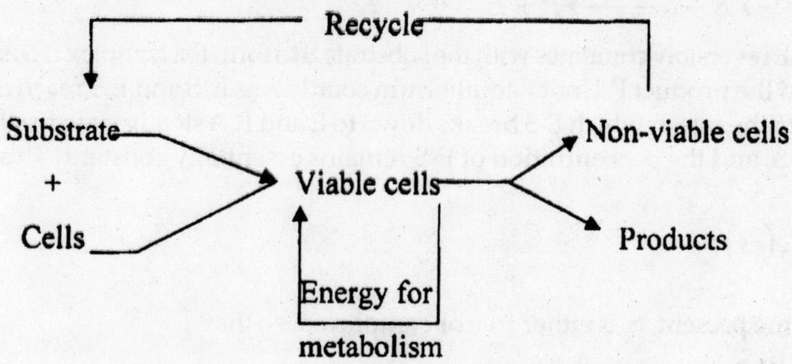

Figure 5.10: Basic Biological Reaction Scheme

A living microorganism takes in many chemical species called nutrients (substrates) from its environment, grows, reproduces, and releases metabolic products into its surroundings. Three types of reactions are carried out within the cell: degradation of nutrients, biosyntheses of small molecules, and biosyntheses of large molecules from the small molecules. Each of these reactions is carried out by an enzyme present inside the cells. Enzymes are proteins and act as catalysts increasing the speed of the reactions greatly without altering themselves. Enzymes are known for their high degree of efficiency in converting specific substrates into end products.

The objectives of the biological reactions are:

1. Consumption of substrate and conversion into insoluble form, *e.g.* effluent degradation and disposal by biological means,
2. Production of biomass, *e.g.* protein food and various compounds extracted from cells, and
3. Product formation, *e.g.* alcohols, acids and pharmaceuticals.

The substrate components are carbon, hydrogen, oxygen (in combined form as carbohydrate and fats), nitrogen (as free or in combined form as proteins, ammonia, nitrite, nitrate), phosphorus (as phosphates), and trace elements as salts.

1. Kinetics of Enzyme Catalyzed Reactions

An enzyme acts through the formation of an enzyme-substrate complex, and the substrate binds to a specific region, called the active site, to form a complex. A group of reactive amino acid side chains in the enzymes presents a very specific site to the substrate. These are R-groups of amino acids, and end–COOH and–NH_2 functions. The active site of the enzyme is a geometric compliment of the substrate.

The basic kinetic model is based on the reaction of a single substrate converting to product in presence of a constant amount of an enzyme. All other substrates have been assumed to be present in concentrations much larger than the substrate undergoing the reaction and thereby do not affect the rate. The enzyme-catalyzed reaction then follows the pattern of homogeneous catalyzed reactions. Further, a first order kinetics is assumed, because the substrate concentration is considerably greater than that of enzyme ($e_o << s$).

The following scheme:

$$E + S \xleftrightarrow{k_1, k_{-1}} E - S \xrightarrow{k_2} P + E$$

shows that enzyme E reversibly combines with the substrate S to form the complex E-S and is irreversibly regenerated to yield the product P. Under equilibrium conditions, E, S and E-S reach equilibrium very rapidly compared to the rate at which E-S breaks down to E and P. A steady state reaches immediately after mixing E and S, and the concentration of E-S remains essentially constant. Thus

$$v = \frac{dp}{dt} = k_2(es) \tag{5.75}$$

The total enzyme present, e_o is either free or complexed, so that

$$e_o = e + (es) \tag{5.76}$$

The kinetic expression is

$$\frac{d(es)}{dt} = k_1 e.s - (k_{-1} + k_2)(es), \tag{5.77}$$

where, the lower case letters indicate the concentrations.

For a good approximation the steady state expression is $\frac{d(es)}{dt} = 0$

Eliminating (es) and e from eqs. (5.75) to (5.77), we get

$$(es) = \frac{e_o s}{\left(\frac{k_{-1} + k_2}{k_1}\right) + s}$$

$$\therefore v = \frac{dp}{dt} = \frac{k_2 e_o s}{\left(\frac{k_{-1} + k_2}{k_1}\right) + s} = \frac{v_{max} s}{K_m + s} \quad (5.78)$$

where, v_{max} = maximum or limiting velocity that would be observed when all the enzyme is complexed or saturated with the substrate, and $K_m = \frac{k_{-1} + k_2}{k_1}$ = Michaelis constant.

The model, $v = \frac{v_{max} S}{K_m + s}$ is referred to as Michaelis-Menten model of enzyme kinetics. The derivation was done by Henri in 1902 and then Michaelis and Menten in 1913. The Michaelis constant, $K_m = k_{-1}/k_1 + k_2/k_1 = K_s + k_2/k_1$, shows that it is a dynamic or pseudo-equilibrium constant expressing the relationship between the actual steady state concentration rather the equilibrium concentration. If $k_2 << k_1$, $K_m = K_s = k_{-1}/k_1$, *i.e.* K_m is inversely proportional to the chemical affinity of enzyme for the substrate.

A graph of v vs. s drawn according to eq (5.78) is a hyperbola going through the origin with one asymptote parallel to the x-axis (Figure 5.11). The value of K_m compared to s has significance in predicting the order of reaction:

1. If $K_m << s$, $v = v_{max}$, and the reaction follows zero order kinetics.
2. If $K_m >> s$, then $v = (v_{max} / K_m)\, s$, or $v = k\,(s)$, and we have the first order kinetics with k as the first order rate constant.
3. In the region of mixed order kinetics v is still dependent on s.

Further, if $K_m = s$, then $v = v_{max}/2$. It signifies that K_m is equal to the substrate concentration at which half of the maximum possible rate is attained. K_m can be expressed as the same concentration unit as s.

Many enzymes obey the Michaelis-Menten model, in spite of the fact that they catalyze reactions with more than one substrate. The kinetics of these reactions fit the equation, provided concentrations of other substrates are held constant. In many situations, one of the substrates is water whose concentration is much larger than that of the other substrates, and therefore, is assumed to be constant.

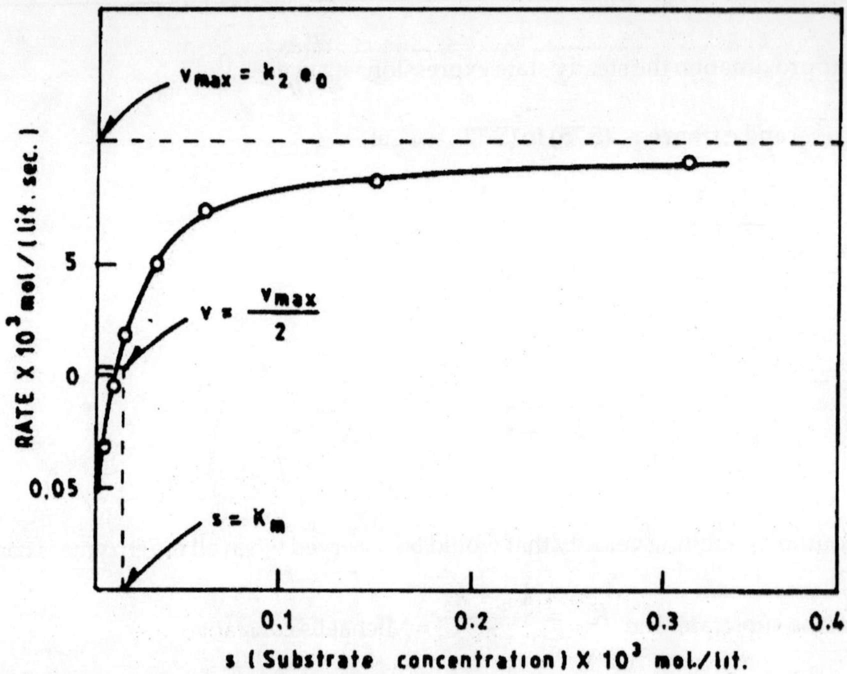

Figure 5.11: Michaelis-Menten Plot of v Vs. s

However, when two or more substrates are involved, the significance of the parameters v_{max} and K_m is different.

It is quite difficult to estimate v_{max} from Figure 5.11, since it is the limit of an asymptote of the curve. Hence $K_m = ½ v_{max}$ cannot be determined accurately. The data can be plotted in linear forms, such as in Lineweaver-Burk plot (Figure 5.12), as

$$\frac{1}{v} = \frac{K_m}{v_{max}} \cdot \frac{1}{s} + \frac{1}{v_{max}}' \tag{5.79}$$

where, the dependent and independent variables are separated.

2. Kinetics of Microbial Growth

When a small quantity of living cells (inoculum) is added (seeding) to a liquid solution of essential nutrients (medium) at a suitable temperature and pH, the cells grow(culture). The salient points of cellular growth are as follows:

1. The growth a cell can be defined as the orderly increase of all its chemical components: proteins, RNA, DNA, and intra-cellular water.
2. Bacteria and other unicellular organisms grow by increase in cell number. Higher fungi, such as molds grow by increasing the length of a vegetative structure called "mycelium".
3. "Population" growth implies an increase in number of cells as well as mass of cellular constituents. The different measures of cell mass are shown in Table 5.8.

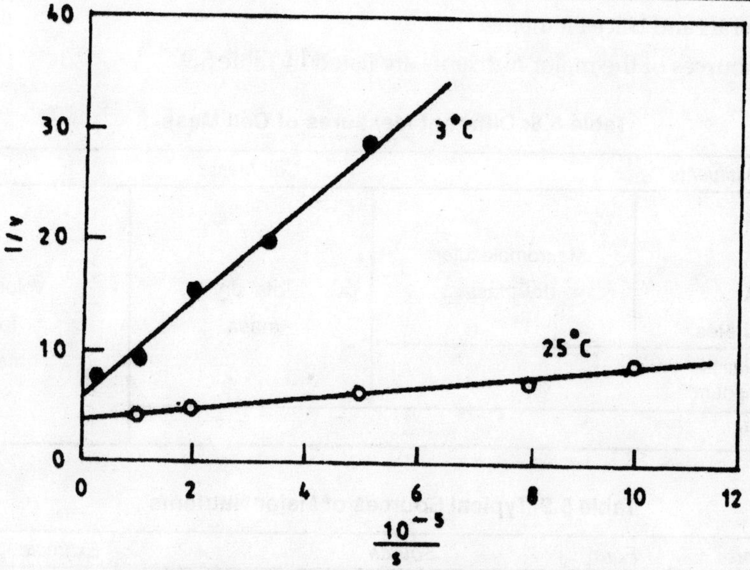

Figure 5.12: Lineweaver-Burk Plot of 1/v Vs. 1/s

4. Environmental factors of growth are temperature, pH, water and water activity, and oxygen concentration.
   (*a*) Temperature range within which organisms, in general, grow is extremely narrow:–5° to 80°C. For any particular organism it is much narrower, say 30°–40°C. Bacteria show more diversity in their response to temperature than moulds and yeast. Moulds and yeast usually do not grow well above 35–37°C, but the upper temperature limit for bacteria is around 100°C.
   (*b*) Optimum pH range is 5 to 9. Most yeast and higher fungi grow best in slightly acid pH of 5–6, but moulds and some algae grow over a wide range of pH 2–8.5. Bacteria grow readily at a pH of neutrality and mostly inhibited at a moderate acid pH 5–6.
   (*c*) Growth depends directly on water activity of the medium where the organisms grow. The growth of bacteria is impossible below water activity of 0.90. Most moulds and yeast are completely inhibited between 0.90 and 0.80.
   (*d*) From the standpoint of ability to use free oxygen, microorganisms have been classified as aerobes, anaerobes and facultative. Many bacteria and most fungi, algae and protozoa are obligate aerobes. A few bacteria and protozoa are strict anaerobes. Anaerobic environments include mud and other sediments of lakes, rivers, etc., marshes and certain sewage treatment systems. Facultative organisms include a wide variety of bacteria, protozoa, yeast and other fungi.
5. Requirements of nutrients:
   (*a*) Carbon and energy.
   (*b*) Nitrogen–inorganic and organic
   (*c*) Phosphorus

(*d*) Sulphur, and

(e) Minerals and trace elements.

Typical sources of the major nutrients are listed in Table 5.9.

Table 5.8: Different Measures of Cell Mass

Cellular Constituents		*Cell Mass*	
Protein RNA DNA Carbohydrates	Macromolecular dry mass	Total dry mass	Volume or Total mass
Lower molecular weight compounds			
Water			

Table 5.9: Typical Sources of Major Nutrients

Nutrient Elements	*Form*	*Source*	*Example*
Carbon	Combined	Carbon dioxide	
Hydrogen	Combined	Hydrocarbon	Petroleum hydrocarbons
Oxygen	Free or combined	Carbohydrate, fats	Sugars, starch, molasses, cellulose, oils and fat.
Nitrogen	Free or combined	Proteins, ammonia, nitrates	Extracts of malt, meat and yeast, oil meals, fish meal, whey, corn steep liquor
Phosphorus	Combined	Phosphates	K or NH_4 phosphates
Sulphur	Combined	Proteins, sulphates	
Minerals and trace elements	As salts		

During exponential growth of unicellular bacteria in an adequate medium the biomass is doubled by doubling of all other constituents, *e.g.* protein, and the batch growth will follow a first order autocatalytic chemical reaction:

$$\frac{dN}{dt} = \mu N \ \ or \ \ \frac{dX}{dt} = \mu X \tag{5.80}$$

where, N, X = number of cells, mass of cells per ml, and μ = growth rate constant. This is the Malthusian model of growth.

$$\therefore \frac{1}{X}\frac{dX}{dt} = \mu \tag{5.81}$$

where, m = specific growth rate, time^{-1}.

Integrating with $X=X_o$ at $t = t_{lag}$,

$$\ln\frac{X}{X_o} = \mu\left(t - t_{lag}\right) \quad or \quad \log\frac{X}{X_o} = \frac{\mu}{2.303}\left(t - t_{lag}\right) \tag{5.82}$$

From eq. (5.82), we can define another parameter, t_d, mass doubling time as

$$t_d = 0.693/\mu \tag{5.83}$$

The parameters μ and t_d characterize the batch growth and determine the effect of cell's environment on its performance.

A semi–empirical model on the concept of 'limiting substrate' postulates that of all the substrates necessary for growth all but one or two are in excess and that concentrations of these substrates control the overall rate. The growth rate changes with the concentration of 'limiting substrate' in a hyperbolic function:

$$\mu = \mu_{\max}\left(\frac{s}{K_s + s}\right) \tag{5.84}$$

where, μ_{max} = maximum growth rate when s is unlimited, *i.e.* $s >> K_s$, and K_s = saturation constant or substrate concentration at which $\mu = ½\,\mu_{max}$. From eqs. (5.83) and (5.84)

$$\frac{1}{X}\frac{dX}{dt} = \mu = \mu_{\max}\left(\frac{s}{K_s + s}\right) \quad or \quad \frac{dX}{dt} = \mu_{\max}\left(\frac{s}{K_s + s}\right)X \tag{5.85}$$

Eq. (5.85) is the Monod's expression of exponential growth of microorganisms. There are other models, but none of these has experienced the wide acceptance of that proposed by Monod.

To better describe the growth kinetics, we define some stoichiometrically related parameters:

Yield Coefficient or Growth Yield

$$Y_{X/S} = -\frac{\Delta X}{\Delta S} \tag{5.86}$$

where, ΔX is the incremental increase in cell mass which results from the utilization of the incremental amount of substrate, ΔS. Monod observed that $Y_{X/S}$ remains a constant quantity.

For organisms growing aerobically on glucose: $Y_{X/S}$ = 0.4–0. 6 g/g. Anaerobic growth is less efficient and the yield coefficient is reduced substantially. Thus

$$X - X_o = Y_{X/S}\,(s_o - s) \tag{5.87}$$

For a growth limiting substrate, $s \cong 0$, so

$$X_{max} - X_o = Y_{X/S}\, s_o, \tag{5.88}$$

where X_{max} = maximum cell concentration reached in the stationary phase of growth.

Maintenance Coefficient

There are certain reactions in cells, which consume some substrates at a small rate just to survive. Cells do not carry out metabolism until these substrates are available to it. This process is called maintenance process or endogenous metabolism.

$$\therefore \mu = \frac{\mu_{max} s}{K_s + s} - k_e \tag{5.89}$$

where, k_e = specific rate of substrate consumption for maintenance, sometimes called the "maintenance coefficient". So substrate consumption rate can be related to growth rate as

$$\frac{ds}{dt} = -\frac{1}{Y_{X/S}}\frac{dX}{dt} + k_e X \tag{5.90}$$

Example 5.11

The initial rates of an enzyme catalyzed reaction studied with various substrate concentrations are listed below:

Substrate concn. (mol/l) 10^3	0.007	0.015	0.021	0.035	0.049	0.149	0.342
Rate (mol/l.s) 10^3	0.06	0.09	0.115	0.15	0.174	0.186	0.20

Determine the values of K_m and v_{max}. Check the value of K_m from the half maximal rate.

Solution

The experimental data are plotted as shown in Figures 5.11.1 and 5.11.2 in terms of v vs. s and 1/v vs. 1/s, respectively. From the intercept of Figure 5.11.2.

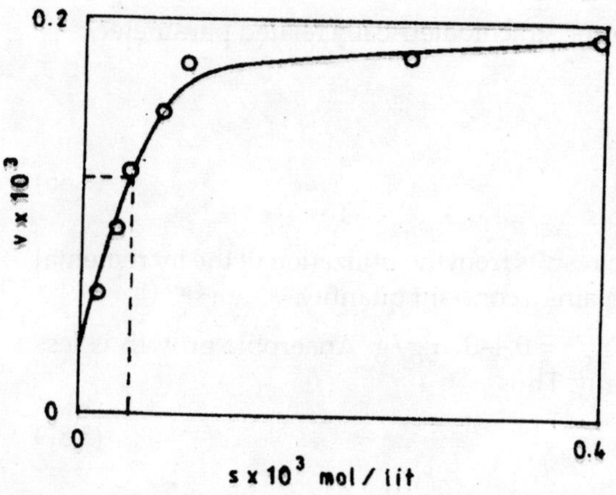

Figure 5.11.1 Plot of v vs. s of Example 5.11

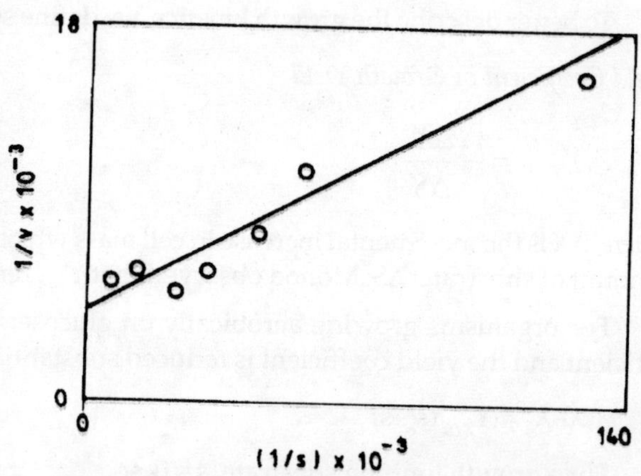

Figure 5.11.2 Plot of 1/v vs. 1/s of Example 5.11

$1/v_{max} = 4.6 \times 10^3$ and $v_{max} = 0.22 \times 10^{-3}$ mol/l.s

From the slope, $K_m = 0.019 \times 10^{-3}$ mol/l. From Figure 5.10.1, $v_{max}/2 = 0.11 \times 10^{-3}$ and $K_m = 0.02 \times 10^{-3}$ mol/l.

Example 5.12

An experiment on batch growth conducted at 20°C yielded the following data. Determine the specific growth rate and cell mass concentration at the beginning of the exponential phase.

Time (h)	1.3	2.5	4.0	5.3	7.0
Cell mass (mg/l)	3311	3631	3830	4170	4560

Solution

Plot of log X vs. time (Figure 5.12.1) yields a straight line with a slope $\mu/2.303 = 0.025$, so that $\mu = 0.06\ h^{-1}$, and from the intercept $X_o = 3041$ mg/l.

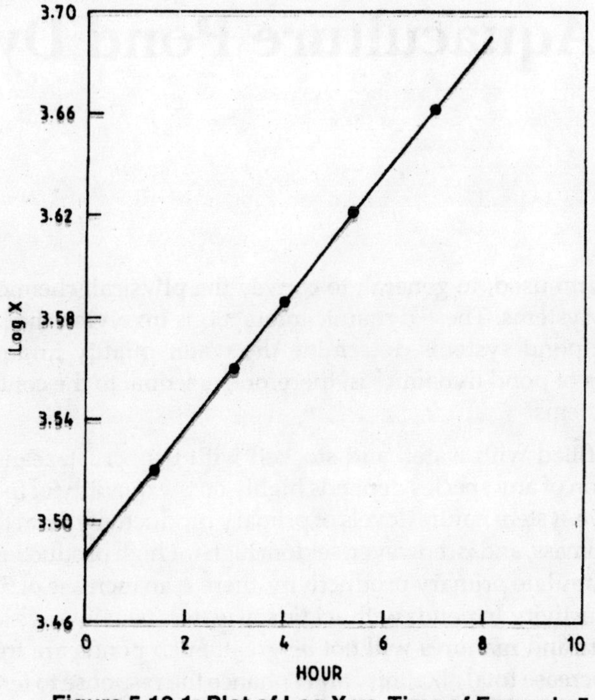

Figure 5.12.1: Plot of Log X vs. Time of Example 5.12

Example 5.13

An unstable metabolite decomposed very rapidly in aqueous solution at pH 7. At all temperatures in the range of 15° to 37°C, the decomposition proceeded with first order kinetics and the following values were obtained for its rate constant. Determine the energy of activation for the decomposition reaction.

Temperature (°C)	15	20	25	30	37
k (min⁻¹)	2.51×10^{-4}	4.57×10^{-4}	8.22×10^{-4}	1.445×10^{-3}	3.09×10^{-3}

6

Aquaculture Pond Dynamics

Pond dynamics is the term used, in general, to convey the physical, chemical, and biological factors that interact in pond systems. These dynamic interactions involving the physical, chemical and biological processes in pond systems determine the water quality in aquaculture ponds. Understanding the principles of pond dynamics is, therefore, essential to the control, manipulation and management of pond systems.

Ponds are constructed, filled with water, and stocked with fish, crustaceans or other aquatic plants and animals. Production of any species depends highly on the quantity of food available to the culture organisms. In extensive system natural levels of primary productivity form the base of the food web that culminates in fish biomass, and is, however, seldom high for high production. When fertilizers and manures are added to stimulate primary productivity, there is an increase of 5 to 10 times in fish production over natural productivity. In ponds with acidic soil waters usually have low total alkalinity, and the response to fertilizers and manures will not be great. Such ponds are treated with liming materials to reduce acidity, increase total alkalinity and enhance the response to fertilization. Further increases in production can be achieved through the use of artificial feeds to supplement the availability of natural food organisms. At this stage the culture system becomes a semi-intensive one.

With the increase of intensity of aquaculture through the application of fertilizers, manures and feed, the culture system becomes more complex with more impact on the culture environment. The overall nutrient concentration increases, plankton blooms occur, and dead organic matter settles to pond bottom. Photosynthetic oxygen production occurs during daytime. Oxygen consumption by respiration also increases with increasing biomass of the pond biota. As a result, fluctuations in dissolved oxygen occur over day and night. In pond bottom accumulation of organic matter creates anaerobic conditions, and as a result microbial metabolites such as ammonia, nitrite, and hydrogen sulphide are released into the water column. These substances are toxic to the aquatic animals. Therefore, the conditions lead to deterioration of culture environment, which, if not controlled in time by management techniques, results in disastrous consequences in the system.

6.1 Dynamics of Nutrients in Pond Ecosystem

Aquatic organisms, both plants and animals, require certain elements as nutrients in life. These are present in the aquatic body in the forms of inorganic and organic compounds and range from simple molecules to very complex ones. Carbon is the basis of life and comes next to oxygen and hydrogen as the most plentiful elements in most organisms. Nitrogen is highly essential for proteins, chlorophyll and several other nitrogenous compounds. Phosphorus is necessary for energy transfer and biochemical synthesis in the organisms. Sulphur forms part of some amino acids, vitamins and enzymes. It also plays an important role in liming of acid sulfate soils. Calcium plays a very significant role in aquatic organisms, particularly in formation of exterior shell materials of shellfish, and cartilage in some varieties of finfish.

The natural aquatic ecosystems such as fresh and seawater provide the major elements to the aquatic life. These systems are well balanced and operate in cyclic fashion through the food chain with respect to each element. However, the cycles operate quite slowly in most cases, and the productivity of plant and animal materials harvested by humans is rather low. In fact, the composition and abundance of nutrients determine the productivity of an aquatic body and the type and diversity of the aquatic life in the food chain. Hence shortage of particular forms of nutrients in the food chain decreases the productivity. There are several ways of increasing the productivity of aquatic systems. One such process is fertilization of pond by input of chemical fertilizers and organic manure, which contain some basic nutrients for plant and animal life. The input of fertilizers alters the natural nutrient balance, remove some of the nutrient deficiencies and improve the productivity. Therefore a new system of nutrient dynamics is set up, which becomes a key factor in water quality management procedure. We will discuss the nitrogen, phosphorus and carbon cycles to highlight the nutrient dynamic processes in the pond system.

6.1.1 Nitrogen Cycle

In natural waters nitrogen is available in various forms: free nitrogen gas (N_2), nitrate (NO_3^-), nitrite (NO_2^-), ammonia (NH_3), ammonium (NH_4^+), and organic forms of amino acids, proteins and particulate matter. In the nitrogen cycle nitrogen is converted from one form to another for use in the system. Conversions from one form to another occur as a result of mostly biochemical reactions. The processes involving such reactions are ammonification, mineralization, nitrification, denitrification, nitrate reduction, and nitrogen fixation (Figure 6.1).

Nitrogen derived primarily from the atmosphere, in the inert free state, is utilized only by certain nitrogen fixing bacteria and blue green algae. The quantity of nitrogen fixed depends largely upon species composition of the phytoplankton and ammonia concentration in the pond. Nitrogen fixation is a minor, but occasionally important contributor to the nitrogen budget of aquaculture ponds receiving formulated feeds.

The organic matter stage of the cycle consists of plants, animals, humus and microbial cells, where nitrogen is bound in the form of proteins and other nitrogenous compounds. When these organisms die, bacterial and chemical oxidative decomposition produces ammonia. The process is known as deammination or ammonification or mineralization. Ammonia generated by microbial decomposition and also from fertilizers and fish excrements goes into solution and exists in equilibrium with ammonium ion.

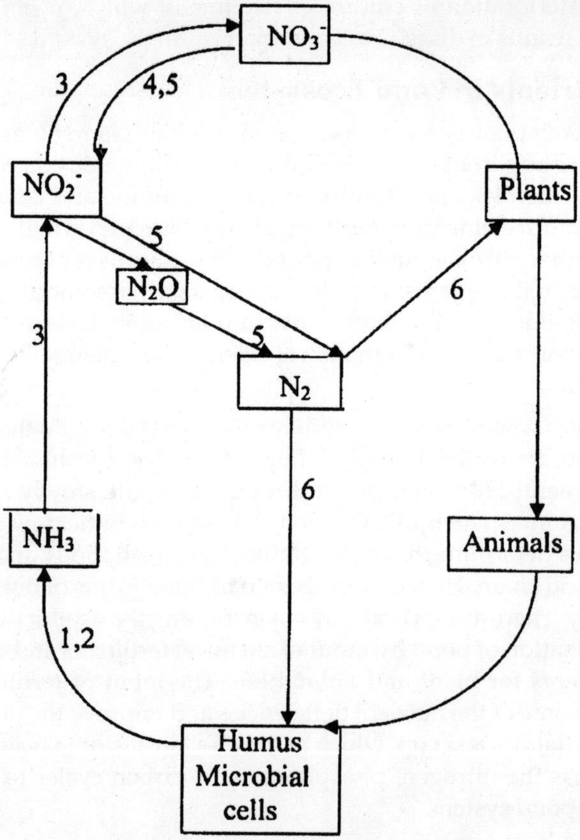

Figure 6.1: The Nitrogen Cycle (Processes: 1: Ammonification, 2: Mineralization, 3: Nitrification, 4: Nitrate reduction, 5: Denitrification, and 6: N_2 fixation)

During nitrification process nitrite and nitrate are produced by bacterial oxidation. Oxidation of nitrite to nitrate is the final step in the nitrogen cycle. Nitrate is the dissolved inorganic nitrogen form absorbed by most plants and converted to organic compound, thus completing the nitrogen cycle. Phytoplankton uptake of dissolved inorganic nitrogen is the primary pathway of nitrogen removal from pond water. Only after it has been depleted, significant quantities of nitrate will be assimilated. Phytoplankton uptake is a powerful mechanism for conversion of potentially toxic inorganic nitrogen to relatively stable organic nitrogen.

Denitrification of nitrate also can take place via nitrite and then nitrogen, or nitrous oxide (N_2O) and then nitrogen. The gaseous form of nitrogen is lost to the atmosphere. Denitrification is a bacterial process occurring in anaerobic waters and sediment. It is a process, often termed as nitrate respiration, by which bacteria use the oxidized form of nitrogen instead of oxygen.

The microbial decomposition reactions are dependent on temperature, pH, oxygen availability and nature of organic matter. The microorganisms involved in these reactions decide the optimum temperature and pH according to their optima for growth. Decomposition is favored at warm temperatures, and is generally doubled by an increase of 10°C. Bacteria grow best in neutral to alkaline environment, whereas fungi grow more in acidic one. The aerobic, anaerobic and facultative types of microorganisms present in aquatic environment oxidize the organic matter in different ways and produce different end products. Aerobic oxidation requires a steady supply of oxygen, and so takes place more rapidly in water with near saturation DO concentration. It is also more complete than under anaerobic condition. Anaerobic oxidation of organic matter is, however, less rapid, and oxidizes the carbon compounds to the extent of alcohol and acids. So in aquaculture ponds concentration of organic matter in mud increases with depth of water which is covered by anaerobic waters.

The nature of organic matter is probably the most influencing factor of microbial decomposition reactions. Decomposition of various organic molecules and the extent of decomposition by various microorganisms and their environmental conditions vary considerably. For example, cellulose and lignin are more resistant to decomposition than sugar. However, bacteria are more versatile in degrading almost any type of organic matter present in most environments than other organisms.

The most important factor is the C/N ratio of the organic compound, which decides not only the extent of ammonification and mineralization, but the need of fertilizer input in the pond also. Since the microorganisms contain more percentage of dry weight carbon than nitrogen, they absorb more carbon from the substrate and assimilate by converting the carbon to cell mass. This carbon assimilation efficiency varies with the type of microorganisms (Table 6.1). If the microorganisms have plenty of nitrogen from the substrate, *i.e.* with low C/N ratio, the necessary quantity of nitrogen will be assimilated and the excess nitrogen will be converted to ammonia by a process known as mineralization. If the organic substrate is deficient in nitrogen (*i.e.* high C/N ratio), the microorganisms do not get enough nitrogen for their growth and may remove the nitrogen from the environment for their growth thereby disbalancing the nitrogen-cycle in nature. Therefore the decomposition reactions of organic substrates with higher C/N ratio, *i.e.* less nitrogen in the environment, will be slow and incomplete. In this respect the fungi containing less nitrogen but more carbon assimilation efficiency than anaerobic bacteria are less efficient in mineralization. The examples below illustrate this aspect.

Table 6.1: Data on C and N Content and C Assimilation Efficiency of Microorganisms

	% Dry Wt. (Av.)		*Carbon Assimln. Efficiency (%)*
	C	N	
Bacteria (aerobic)	48	12.5	50–55
Bacteria (anaerobic)	48	12.5	10
Fungi	48	6	30–40

Source: Peppler, 1967; Bailey and Ollis, 1977.

Example 6.1

Calculate the quantity of nitrogen of an organic matter that will be mineralized in pond by bacterial decomposition under anaerobic condition.

Data: Organic matter: C/N = 5/1

Solution

Basis–100 kg organic matter

Substrate–C = 100 × 0.05 = 5 kg

N = 100 × 0.01 = 1 kg

Bacteria– C = 5 × 0.1 = 0.5 kg

mass = 0.5/0.48 = 1 kg

N = 1 × 0.125 = 0.13 kg

∴ N mineralized = 1–0.13 = 0.87 kg

So the substrate contains enough nitrogen for complete decomposition. This situation may arise when nitrogen is added in the form of fertilizer and organic manure in the pond.

Fertilizers usually contain nitrogen as ammonium or nitrate groups. They dissolve in water and ionize. Plants absorb and assimilate nitrogen in the form of mainly protein. In the food chain animals consume plants, and when both plants and animals die, they decompose in presence of bacteria. Thus organic residues produced during the cycle become high in nitrogen (*i.e.* low C/N) and are decomposed rapidly and excess nitrogen is mineralized. Residues low in nitrogen (*i.e.* high C/N) will not decompose completely, and the amount of organic matter will accumulate. This accumulation is enhanced more in the mud of unfertilized pond than in the fertilized pond. A simple calculation can explain the situation.

Example 6.2

Repeat the Ex. 6.1 with C/N ratio = 50/1

Solution

Substrate C = 50 kg, and N = 1 kg. Bacterial N = 1.3 kg. So the substrate N available in the organic matter will not be sufficient for bacterial decomposition. 1.3–1 = 0.3 kg N should be the minimum quantity to be supplied from fertilizer.

Example 6.3

Compare the results of Ex. 6.1 and 6.2 with that by fungal decomposition.

Solution

1. With C/N = 5/1. Fungal C = 1.75 kg, and N = 0.22 kg. So 1–0.22 = 0.78 kg N will be mineralized.
2. With C/N = 50/1, fungal C = 17.5 kg and N = 2.2 kg. So the substrate N is not sufficient for decomposition.

So more biomass is generated and less N is mineralized by fungi than by bacteria.

The importance of C/N ratio on decomposition of organic matter, although oversimplified, has been worked out in agricultural soil. Few studies show that the degradation of aquatic plant residues is regulated by C/N ratio, and with aquatic plant residues having more nitrogen content more nitrogen was mineralized than with those containing less N (Boyd, 1982).

6.1.2 Phosphorus Cycle

Like nitrogen phosphorus is recycled in natural water through a complex cycle of simple and complex compounds of phosphorus. But unlike nitrogen cycle, it is a biogeochemical cycle, since a chemical equilibrium exists between phosphate in mud and in water, which dominates the cycle. Further, while most of the nitrogen is bound in living organisms, most of the phosphorus in the aquatic ecosystem exists in inorganic combinations.

The phosphorus chemistry in water, *i.e.* the forms in which it exists and the chemical or biological conversions in the aquatic environment are not well defined. The various chemical forms of phosphorus range from inorganic phosphate ions to complex organic molecules, and exist both as dissolved and particulate phosphorus.

The major processes of the phosphorus cycle (Figure 6.2) fall into three categories: (1) addition of phosphorus in the aquatic system by input, decomposition, dissolution, etc., (2) removal of phosphorus from the system through processes such as precipitation, and (3) uptake or concentration of phosphorus by living organisms (Martin, 1970).

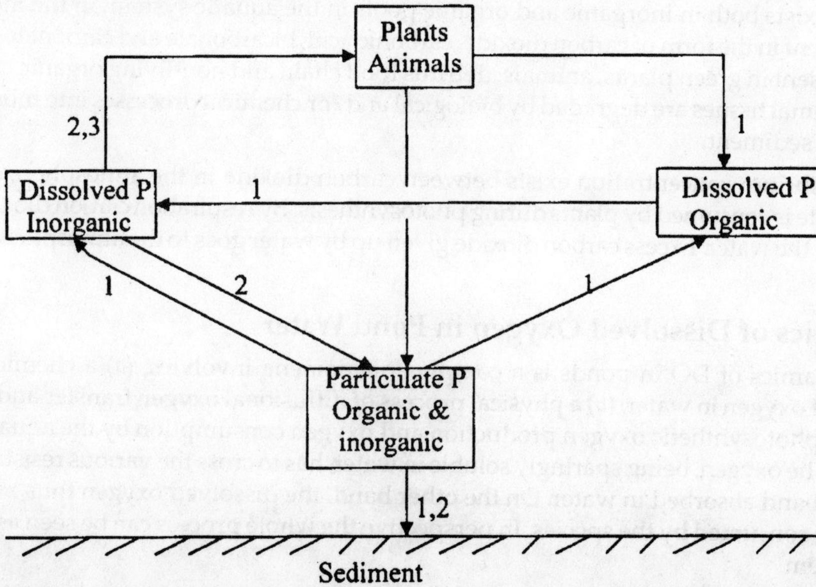

Figure 6.2: The Phosphorus Cycle (Processes: 1: Decomposition, dissolution; 2: Removal, Precipitation; and 3: Uptake by organisms)

Dissolved phosphorus exists as soluble phosphates (inorganic) and as soluble organic phosphorus compounds. The nature of the organic fraction of soluble phosphorus is not well defined, though nucleic acid materials, phospholipids and many decomposition products would be expected. The amount of dissolved phosphorus in the natural waters is maintained by dissolution of particulate inorganic phosphate or decomposition of organic phosphorus compounds and by removal through precipitation of insoluble phosphate or uptake by plants.

Particulate phosphorus exists both as inorganic and organic detritus forms. Both forms significantly affect concentration of orthophosphate by their conversion processes. Organic detritus slowly decomposes as organisms sink to the bottom loosing much phosphorus in the sediment, and inorganic phosphate goes into solution. Inorganic phosphorus compounds such as calcium phosphate, which are present in waters within high calcium concentration such as seawater, precipitate at high pH, and thereby maintain the orthophosphate concentration.

The orthophosphate present in water is utilized by bacteria and is either assimilated into their own bodies or converted into an organic form that is unavailable to green plants. Algae and rooted plants also rapidly absorb phosphorus. Studies demonstrated that phytoplankton could absorb orthophosphate very fast, for example, 41 percent of 0.3-mg/l addition of orthophosphate within 24 hours. Macrophytes, though can absorb and store large amounts of phosphorus, absorb slower than phytoplankton (Boyd, 1982).

6.1.3 Carbon Cycle

Carbon is derived primarily from the atmosphere, which contains about 0.03 percent carbon dioxide. The source of carbon is also sedimentary rocks, which are mainly carbonates, particularly calcium and magnesium carbonates.

Carbon exists both in inorganic and organic pools in the aquatic system. In the inorganic pool, carbon is present in the form of carbon dioxide, carbonic acid, bicarbonate and carbonate ions. Organic forms are present in green plants, animals, detritus food chain and non-living organic carbons. Dead plants and animal tissues are degraded by biological and/or chemical processes into inorganic carbon or into part of sediment.

An equilibrium concentration exists between carbon dioxide in the atmosphere and in water. Carbon dioxide is consumed by plants during photosynthesis. By respiration carbon dioxide generated is absorbed in the water. Excess carbon dioxide given up by water goes to the atmosphere back (Figure 6.3).

6.2 Dynamics of Dissolved Oxygen in Pond Water

The dynamics of DO in ponds is a complex phenomena involving (*a*) a chemical process of dissolution of oxygen in water, (*b*) a physical process of diffusional oxygen transfer and (*c*) biological processes of photosynthetic oxygen production and oxygen consumption by the aquatic species, or respiration. The oxygen, being sparingly soluble in water, has to cross the various resistances before it is transferred and absorbed in water. On the other hand, the dissolved oxygen thus available to the aquatic life is consumed by the species. In perspective the whole process can be seen as a supply and demand system.

In fish culture under moderate to high-density systems the oxygen demand rises due to high rate of oxygen consumption and depletion resulting from bacterial decomposition of metabolic wastes and excess feeds. The supply of oxygen by diffusion from atmospheric air and photosynthetic production cannot meet this demand, and therefore, lead to oxygen deficiencies causing stress and death of fish. Thus maintenance of DO by artificial aeration becomes a strategic factor in water quality management.

DO concentration also plays a key role in water sources for use in aquaculture purpose. Ponds, streams and coastal water where effluent and waste discharge creates water pollution face problems of low DO concentration. To that end, most regulatory programmes are largely based on understanding the DO balance in water bodies and maintaining desirable DO by artificial aeration.

6.2.1 Biological Processes

1. Photosynthetic Oxygen Production

Photosynthesis is essentially the source of all organic matter and biological available energy. Carbohydrates are synthesized in presence of solar energy and stored in green plants as chemical energy. The reaction involved in photosynthesis is:

$$6\,CO_2 + 6\,H_2O + \text{light} \rightarrow C_6H_{12}O_6 + 6O_2 \qquad \Delta F^{\circ} = 686.4 \text{ kcal} \tag{6.1}$$

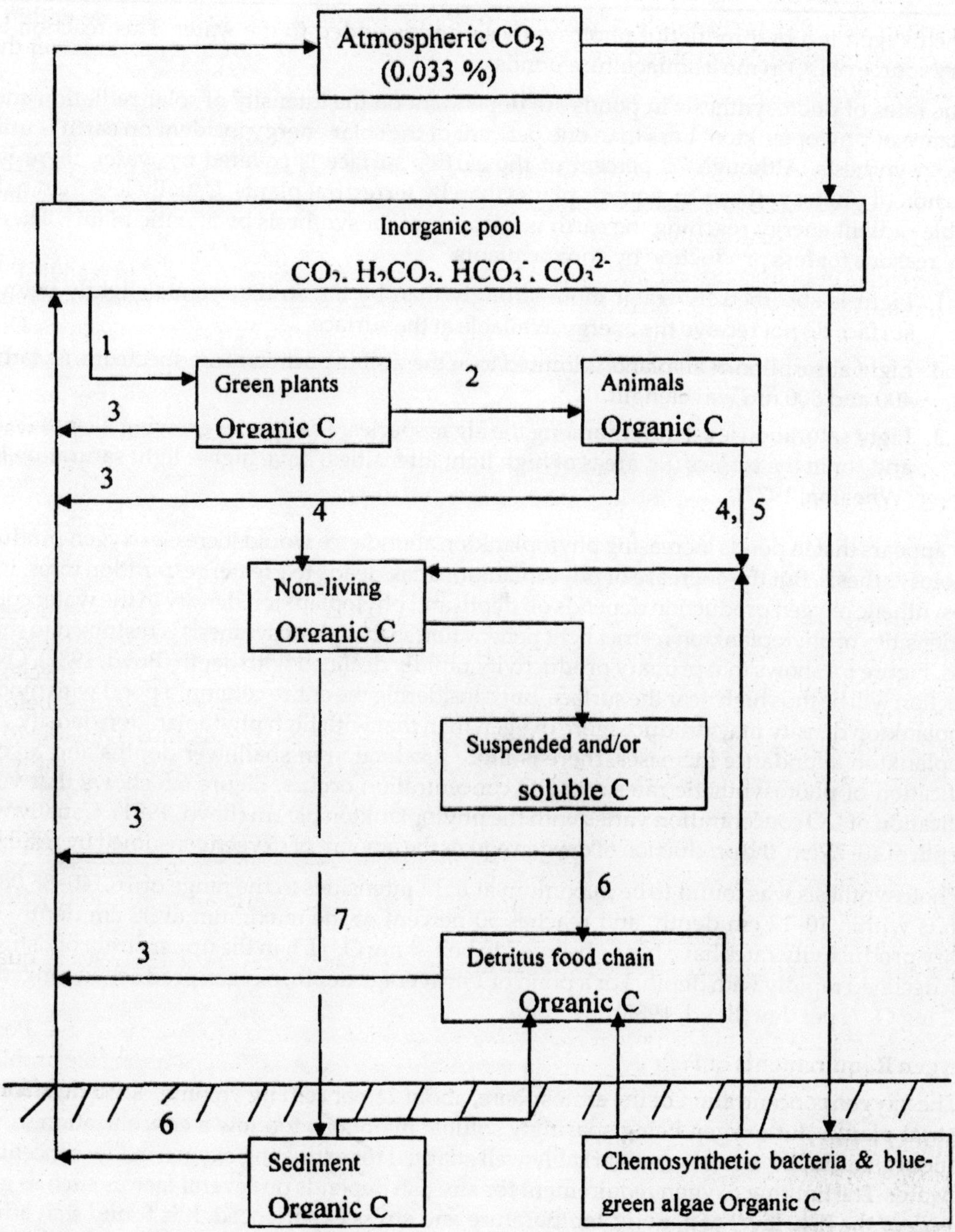

Figure 6.3: The Carbon Cycle (Processes: 1: Photosynthesis, 2: Grazing food chain, 3: CO_2 respiration, 4: Death, 5: Excretion, 6: Microbial decomposition, and 7: Sedimentation)

The oxygen is a by-product of photosynthesis and is added to the water. This reaction is the primary source of DO in most aquaculture ponds.

The rates of photosynthesis in ponds are dependent on the intensity of solar radiation and the abundance of phytoplankton. Less than one per cent of the solar energy incident on earth is utilized for photosynthesis. Although 71 percent of the earth's surface is covered by water, there is less production of photosynthesis by aquatic plants than by terrestrial plants. Usually less than half the available radiant energy reaching the earth is used for photosynthesis by aquatic plants. There are several reasons for less production by aquatic plants:

1. Light is absorbed by water more strongly than by air, so that plants slightly below the surface do not receive the energy available at the surface.
2. Light absorption of all plants is limited to in the visible portion of the spectrum, *i.e.* between 400 and 800 nm wavelength.
3. Light saturation level differs among the algae species. The species growing near the surface and/or in the tropics (*i.e.* areas of high light intensities) have higher light saturation levels (Wheaton, 1977).

It appears that in ponds increasing phytoplankton abundance would increase oxygen production by photosynthesis. But this increase of phytoplankton mass leads to greater respiration rates. In fact, photosynthetic oxygen production depends on depth and phytoplankton density in the water column. High densities of phytoplankton restrict light penetration so that photosynthesis is restricted to shallow depths. Figure 6.4 shows that primary productivity rapidly declined with depth (Boyd, 1982). Oxygen production will be thus high near the surface, but considering the entire column, a pond with moderate phytoplankton density may produce more oxygen than that with high phytoplankton density. As the phytoplankton abundance increases, there is more production in shallower depths and a vertical stratification of photosynthetic rates and DO concentration occurs. Figure 6.5 shows that vertical stratification of DO concentration varies with the phytoplankton bloom (Boyd, 1991). Usually within the depth of 40–75 cm, the production of oxygen equals the amount of oxygen consumed by respiration.

Photosynthesis was found to be maximum at light intensities in the range of 16,140–86,080 lux, which is within 10–12 cm depth, and reaches 50 percent of the maximum at 30 cm depth. Gross primary productivity rates have been observed to be 1–4 mg O_2 /l. h in the upper strata of fishponds, which declined rapidly with depth. For a pond of 1 m average depth, oxygen production ranged from 4 to 12 mg O_2 /l per day (Boyd, 1982).

2. Oxygen Requirements of Fish

The oxygen concentration of the atmosphere, about 21 per cent by volume, is too high for some terrestrial plants. But oxygen being sparingly soluble maintains too low a concentration in water. Aquatic organisms are, however, remarkably well adapted for extracting oxygen in low concentration from water. The limiting oxygen requirement for any fish depends on several factors such as genetic make-up of the fish, level of activity, temperature and stress experienced. It is found that adults of species lower in the evolutionary scale tolerate low oxygen concentrations better. Crabs and oysters, but not shrimps and prawns can tolerate low oxygen concentrations. Generally young ones of almost all species are less tolerant of low oxygen than grown-up individuals.

A hypothetical plot of oxygen concentration vs. oxygen consumption rate (Jones, 1964) in Figure 6.6 shows that at low concentration the organism cannot derive sufficient oxygen to survive. Above the basal metabolic needs the organism can survive within a range of concentrations, but will be

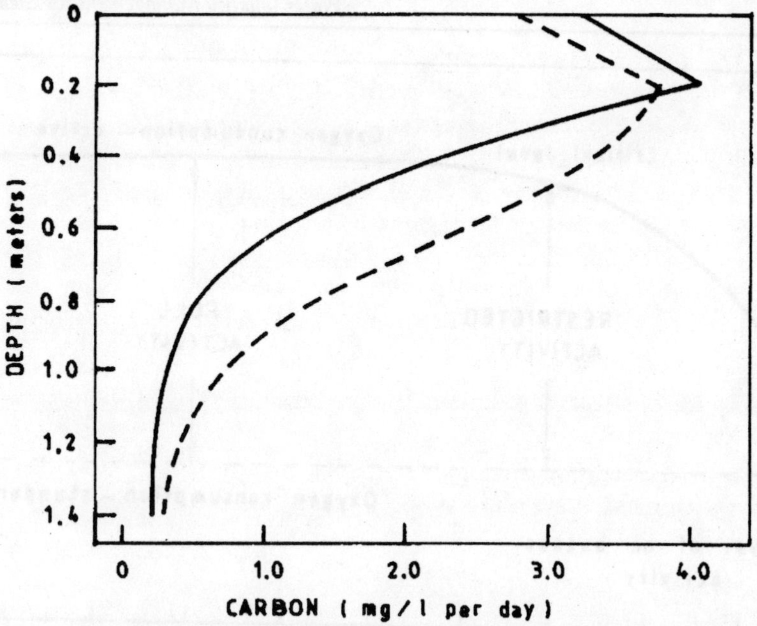

Figure 6.4: Gross Primary Productivity at Different Depths in Two Ponds Shown as Solid and Broken Lines (After Boyd, 1982)

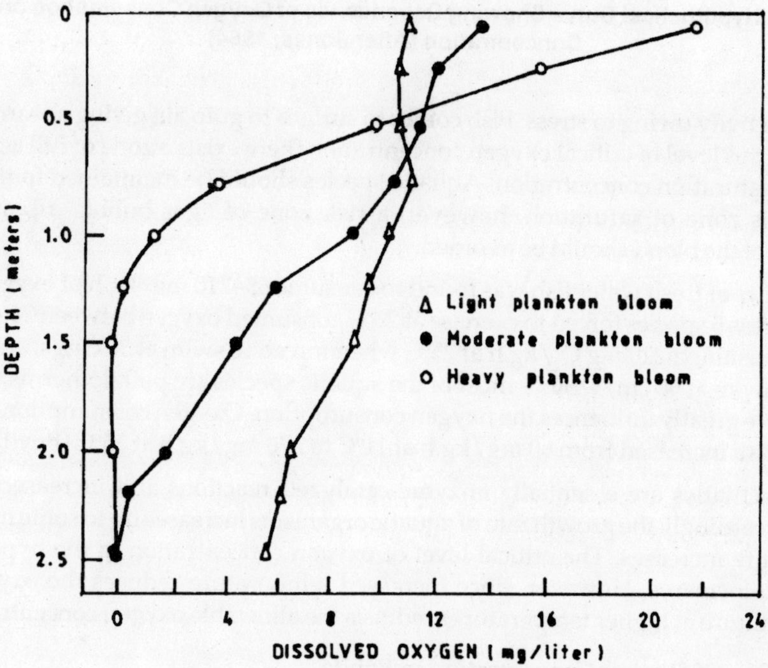

Figure 6.5: Dissolved Oxygen Concentration as Function of Depth and Plankton Density (After Boyd, 1991)

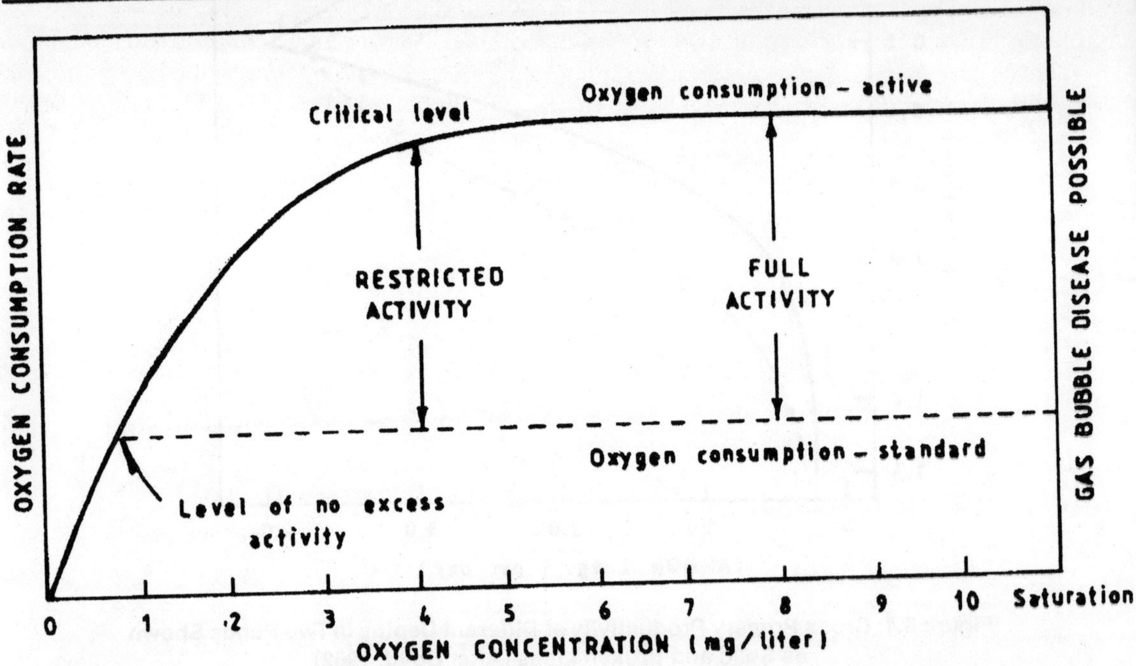

Figure 6.6: Hypothetical Curve Showing Dependence of Oxygen Consumption on Oxygen Concentration (After Jones, 1964)

under reduced activity owing to stress. Fish comes to surface to gulp air, giving a warning sign to the culturist. Above this level of critical oxygen concentration there exists a zone of full activity up to the level of oxygen saturation concentration. Aquatic species should be maintained in this full activity zone. Above this zone of saturation, however, a risk zone of "gas bubble disease" caused by supersaturating of the blood should be avoided.

Some 9 species of fresh water fish was found to consume 65–210 mg/kg.h of oxygen while at rest at 17–20°C, whereas 5 species forced to exercise at 20°C consumed oxygen between 266–888 mg/kg.h. *Tilapia nilotica* consumed 220 mg O_2/kg.h at 25°C when forced to swim at 30 cm/s, and 458 mg/kg.h when forced to swim at 60 cm/s. Since most of the aquatic species are poikilotherms, adjusting their body temperature greatly influences the oxygen consumption. Oxygen consumption by moderately active white catfish increased from 60 mg/kg.h at 11°C to 276 mg/kg.h at 25°C (Boyd, 1982).

Metabolic activities are essentially enzyme-catalyzed reactions and increased by increasing temperature. Accordingly the growth rate of aquatic organisms increases up to some maximum as the water temperature increases. The critical level of oxygen concentration in the hypothetical curve (Figure 6.6) also increases. However, since increased temperature reduces the oxygen saturation concentration, culture at higher temperatures reduces the allowable oxygen concentration range.

3. Diurnal Changes of Oxygen Concentration in Ponds

There is a daily fluctuation of DO in ponds due to biological processes and diffusional oxygen transfer. Oxygen is produced by photosynthesis in ponds in presence of light but all aerobic organisms require oxygen continuously even when it is dark. Because of these two effects DO concentration continually changes during day and night (Figure 6.7.). DO concentrations increase during the morning and usually reach a peak in the mid-afternoon. DO concentrations decline during the night, when photosynthesis stops but respiration processes continue. DO concentrations thus reach their lowest value near dawn. The daily maximum and minimum DO concentrations vary with the phytoplankton abundance in ponds. The difference between daily maximum and minimum DO concentrations in a pond tends to become greater as phytoplankton density increases (Boyd, 1991).

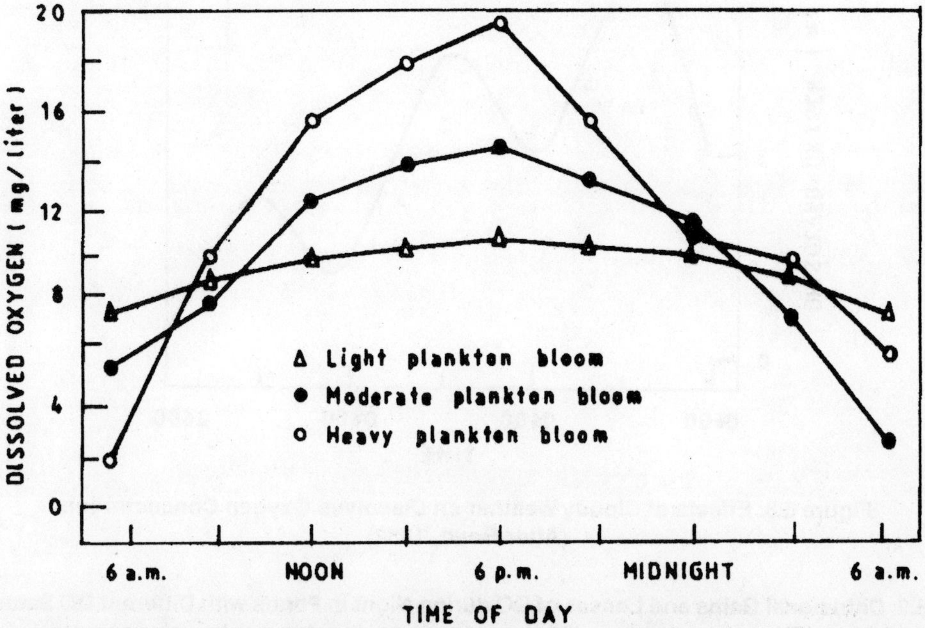

Figure 6.7: Diurnal Variation of Dissolved Oxygen Concentration in Surface Water for Different Plankton Density (After Boyd, 1991)

A typical example of DO concentration fluctuation in channel catfish pond showed that it varied between below 3 mg/l at dawn and above 15 mg/l in the afternoon (Boyd *et al.*, 1979). Cloudy weather may cause unfavorable changes in DO fluctuations. In a pond with abundant phytoplankton DO consumption rate will be more than the photosynthetic production rate during cloudy days. Few cloudy days after a clear day may cause depletion of DO below the level of optimal requirement of fish (Figure 6.8).

6.2.2 Diffusional Oxygen Transfer by Natural Aeration

Early in the morning DO concentrations may be below saturation, when oxygen diffuses into water. But later on in the daytime when super saturation condition prevails, oxygen diffuses out of water. In the early hours of night waters may be super saturated with oxygen, but later respiration of pond organisms results under saturation. Data of Schroeder (1975) on diffusional gain or loss of oxygen from pond modified by Boyd (1982) are presented in Table 6.2. It showed that if pond water were 50 percent saturated at dusk, 1.69 mg/l of oxygen would be added during night. However some quantity would be consumed during respiration. At 100 percent saturation concentration of DO at dusk only 0.44 mg/l DO would be the gain because of the deficit caused by respiration even though the water was initially saturated. The diffusion loss increases with increasing saturation.

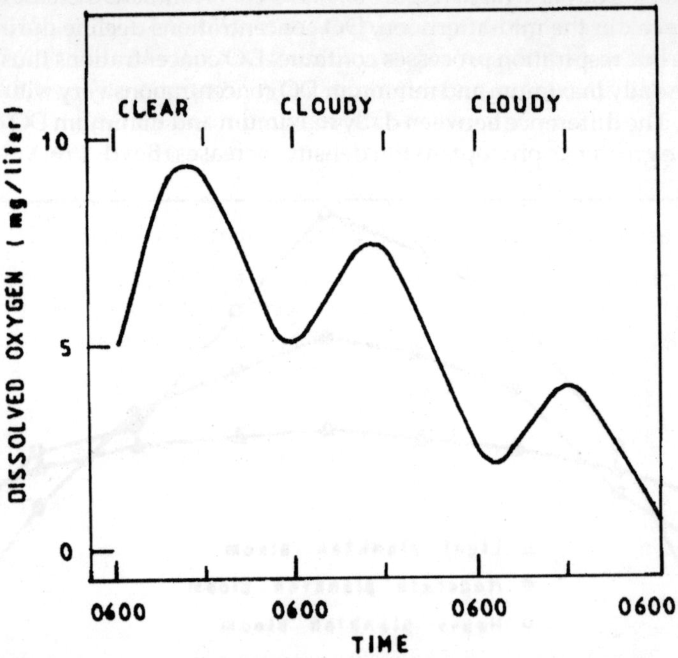

Figure 6.8: Effects of Cloudy Weather on Dissolved Oxygen Concentration (After Boyd, 1982)

Table 6.2: Diffusional Gains and Losses of DO during Night in Ponds with Different DO Saturation Values. The values represent the average change in 1-m depth water column.

DO Concentration at Dusk (% of Air Saturation)	*Gain or Loss (mg/l)*	*DO Concentration at Dusk (% of Air Saturation)*	*Gain or Loss (mg/l)*
50	+1.69	160	-1.64
60	+1.49	170	–1.82
70	+1.18	180	–1.98
80	+1.00	190	–2.11
90	+0.77	200	–2.37
100	+0.44	210	–2.42
110	+0.16	220	–2.54
120	–0.18	230	–2.67
130	–0.55	240	–2.76
140	–0.94	250	–2.91
150	–1.48		

Source: Boyd, 1982.

In natural waters accurate measurement of oxygen flux is virtually impossible, because of the continuous biological processes and complex circulation patterns in water. A value of 0.01–0.05 g O_2/

m^2.h for each 30 mm Hg or 0.2 atmosphere oxygen saturation deficit in Israel has been reported. With 99.2 mm oxygen tension of the water the deficit in saturation would be 760 x 0.21, *i.e.* 159.2–99.2 = 60 mmHg. Under this condition oxygen will diffuse into the water at 0.02–0.1 g/m^2. h or 0.02–0.1 g/m^3.h for 1 m depth of water, *i.e.* 0.02–0.1 mg/l of oxygen. Wind velocity creates turbulence in the water surface. The maximum and minimum oxygen flux reported may be due to high turbulence (strong wind) and low turbulence (calm period or little wind). For a fishpond these two values are 0.1 g/m^2.h and 0.5 g/m^2.h respectively (Boyd, 1982).

6.2.3 DO Concentration Balance in Pond Water during Culture

Several DO concentration balance equations were set up for different culture systems and species, which showed that various factors influence the DO concentration changes. The primary objective of these equations was to predict the nighttime decline in DO concentration as a water quality management strategy.

1. Channel Catfish Pond

Boyd and coworkers (1982, 1989) developed some procedures for predicting DO concentrations in channel catfish ponds. Considerable research has been conducted in channel catfish ponds, and the findings are indicative of DO variations in other fish ponds.

The oxygen balance equation showing different variables of DO concentration in catfish ponds:

$$DO_t = DO_i \pm DO_{df} - DO_{pr} - DO_{fr} - DO_{br} \qquad (6.2)$$

where, DO_t: DO concentration at time t; DO_i: initial DO concentration; DO_{df}: DO gained or lost by diffusion; DO_{pr}: DO consumed by planktonic respiration; DO_{fr}: DO consumed by fish, and DO_{br}: DO consumed in benthic respiration.

If DO_i is taken as the DO_{dusk}, which can be measured, then eq. (6.2) can be modified to eq. (6.3), so that DO depletion can be predicted any time after dusk.

$$DO_t = DO_{dusk} \pm DO_{df} - DO_{pr} - DO_{fr} - DO_{br} \qquad (6.3)$$

DO_{df} by gain or loss may be estimated from Table 6.2 of values (in percent of air saturation) against the DO_{dusk}. This equation can be solved, provided DO_{fr} and DO_{pr} can be estimated from the relations developed as given below. DO_{br} was assumed to be 61 mg/m^2.h. This is equivalent to DO reduction of 0.72 mg/l during 12 hours of darkness in a pond with an average depth of 1 m provided the entire bottom is aerobic (Boyd, 1982).

The above procedure was successfully applied for predicting nighttime depletion of DO in two catfish ponds. The percentage difference between the measured and predicted values of DO ranged from 0.24 to 11.25 percent with an average of 6.31 percent (Boyd, 1982). In most of the cases eq. (6.3) under estimated the measured DO, which, however, was more desirable than overestimation from a management standpoint.

The oxygen consumption rate of channel catfish pond can be estimated from the equation:

$$\log O_2 \text{ consumption (mg } O_2\text{/g fish.h)} = -0.999 - 9.57 \times 10^{-4} W + 6 \times 10^{-7} W^2 + 0.0327 T + 8.7 \times 10^{-6} T^2 + 3 \times 10^{-7} WT \qquad (6.4)$$

where, W = average weight of fish (g) and T = water temperature (°C). The data taken from catfish ponds for 2–1000 g fish and from 24 to 30°C fitted with correlation coefficient 0.99. The equation was developed by multiple regression analysis of data averaged on oxygen consumption by fed and fasted species (Boyd, 1982).

The DO consumed during planktonic respiration (DO_{pr}) can be predicted from the COD values of pond:

$$O_2 \text{ consumption (mg/l.h)} = -1.006 - 1.48 \times 10^{-3}\,C - 1.25 \times 10^{-5}\,C^2 + 0.0766\,T - 1.44 \times 10^{-3}\,T^2 + 2.53 \times 10^{-4}\,CT \quad (6.5)$$

where, C = chemical oxygen demand (mg/l), T = water temperature (°C). The equation is valid for the range of C = 20–160 mg/l and T = 20–32°C (Boyd *et al.*, 1978). The COD of pond water is strongly correlated with chlorophyll a concentration, indicating that phytoplankton is a major cause of COD.

Planktonic respiration was also calculated from the data of Secchi disk visibility, as it was highly correlated to COD of pond waters (Boyd *et al.*, 1978).

$$O_2 \text{ consumption (mg/l.h)} = -1.133 + 8.331 \times 10^{-3}\,S + 1.45 \times 10^{-5}\,S^2 + 8.12 \times 10^{-2}\,T - 7.49 \times 10^{-4}\,T^2 - 3.49 \times 10^{-4}\,ST \quad (6.6)$$

where, S = Secchi disk visibility (cm), and the equation is valid within S = 20–120 cm and T = 20–32°C.

The success of prediction of DO depletion depends largely on the contribution of planktonic respiration and reliability of the method of its estimation. It has been found that for a representative channel catfish pond the percentage contributions to nighttime DO loss were as follows:

DO Loss Due to	%
Plankton respiration	82
Fish respiration	9
Benthic respiration	5
Diffusion	4
Total	100

Romaire and Boyd (1979) developed one daytime equation, which can estimate the change in DO concentration during daytime (photoperiod). The following regression equation was adjusted to remove the influence of DO consumed by fish respiration:

$$\Delta DO_{day} = -3.24 + 1.24 \times 10^{-2}\,Rad - 1.2 \times 10^{-5}\,(Rad)^2 + 5.82 \times 10^{-2}\,Ch_a - 1.44 \times 10^{-4}\,(Ch_a)^2 + 8.58 \times 10^{-2} Sat + 6.51 \times 10^{-4} (Sat)^2 - 5.2 \times 10^{-5}\,(Rad)(Ch_a) - 3.55 \times 10^{-4}\,(Ch_a)(Sat) \quad (6.7)$$

where, ΔDO_{day} = adjusted day time DO change (mg/l); Rad = solar radiation (langleys/day), 60–600 langleys/day, Ch_a = chlorophyll a (µg/l), 8–280 mg/l and Sat = percentage DO saturation at dawn (10–115 per cent).

The equation is applicable within the range of values noted above. However, this equation cannot be used for prediction of DO at times other than that at dawn and dusk. DO_{dusk} can be obtained by using eqs. (6.2) and (6.7), but assuming DO_{df} and DO_{br} as constant:

$$DO_{dusk} = DO_{day} + DO_{dawn} - DO_{fr} \quad (6.8)$$

Of course, the daytime equation has little practical value, as it is dependent on the prior value of solar radiation.

Example 6.4

Calculate for a channel catfish pond of area 1 ha and water depth 1 m the quantity of oxygen consumed in mg O_2/g.h during nighttime from the following data:

Production = 1000 kg of 400 g size, DO_{dawn} = 2 mg/l, DO_{dusk} = 10 mg/l, water temperature = 30°C, COD = 75 mg/l (Data adapted from Boyd, 1982).

Solution

1. DO diffusion: At 30°C solubility of O_2 = 7.54 mg/l, and per cent saturation = 132.6, which yields DO_{df} =–0.66 mg/l, from Table 6.2.
2. Benthic respiration = 61 mg/m² O_2.h = 61 × 12 × 1 mg/m³ = 0.73 mg/l
3. Planktonic respiration = 0.384 mg/l × 12 = 4.608 mg/l, from eq. (6.5)

Therefore DO consumption from (i) to (iii) = 5.998 mg/l.

From eq. (6.3), DO_{fr} = 10–5.998–2 = 2.002 mg/l = 20020 × 10,000 × 10³/1000 × 10³ × 12 = 1.67 mg/ O_2 g fish.h

2. Trout Pond

Eq. (6.3) was also successfully applied in predicting the nighttime decline in DO. DO_{fr} was, however, obtained by an equation developed by Muller-Feuga *et al.*, 1978:

Oxygen consumption (mg O_2/ kg fish.h) = $AW^B \times 10^{CT}$ (6.9)

where, W = fish weight (g), T = water temperature(°C), and the values A, B and C for different temperature ranges are as follows:

Constant	4–10°C	12–22°C
A	75	249
B	-0.196	-0.142
C	0. 055	0.024

Planktonic respiration equation was also different:

DO_{pr} (mg/l.h) = 0.00848 + 0.00264 C, where C = COD (6.10)

Example 6.5

Oxygen budget for trout pond–Production = 3,500 kg of size 250 g. Pond area = 1 ha of water depth 1 m. Water temperature = 22°C, COD = 75 mg/l. DO_{dusk} = 12 mg/l. Calculate DO 6 h from dusk. Solubility of oxygen in water at 22° C = 8.74 mg/l.

Solution

1. DO_{df}. At 22°C DO saturation = 137 per cent, and DO loss = 0.124 mg/l.
   So DO_{df} =–0.94 =–0.47 for 6 h.

2. DO_{fr}. A = 249, B =–0.142, C = 0.024
   O_2 consumption (mg O_2/kg.h) = $249 \times 250^{-0.142} \times 10^{0.024 \times 22} = 386$
   For 3500 kg fish consumption of O_2 for 6 h = $386 \times 3500 \times 6 \times 10^{-3} \times 10^{3}$
   = 8113 g = 0.811 mg/l
3. DO_{pr}–For a trout pond:
   $DO_{pr} = 0.00848 + 0.00264\ C = 0.00848 + 0.00264 \times 75 = 0.21$
   = 1.26 mg/l
4. DO_{br}–61 mg/m².h = 366 mg/m² × 1 m = 0.366 mg/l
5. Total DO consumption = 0.47 + 0.81 + 1.26 + 0.67 = 3.68 mg/l = 3.21 mg/l
   ∴ DO after 6 hours = 6–3.21 = 2.79 mg/l

3. Warm Water Fish

Eq. (6.3) was similarly applied warm water fishes such as tilapia. Only difference was to apply a general equation for respiration of warm water fish at 20–30°C is:

$$Y = 0.001\ W^{0.82} \tag{6.11}$$

where, Y = oxygen consumption (g O_2 / h per fish) and W = fish weight (g).

The procedure outlined for nighttime depletion of DO is very complex for use in commercial aquaculture farms. A simple procedure has been developed by using the relationships between COD and Secchi disc visibility, and between average fish weight and total fish weight in a channel catfish pond. A table of values of Secchi disc visibility for predicting the DO_{dusk} corresponding to 2 mg/l of DO_{dawn} has been furnished (Table.6.3).

Table 6.3: Minimum Secchi Disk Visibility (cm) to Maintain a DO Concentration Above 2 mg/l at Dawn in a Catfish Pond of depth 1-m Containing 1000 kg Fish/ha

Temperature (°C)	*DO Concentration of Water at Dusk (mg/l)*									
	2	*3*	*4*	*5*	*6*	*7*	*8*	*9*	*10*	*11*
20	37									
21	58	26								
22	79	42	21							
23	90	58	32	16						
24	*	69	53	37	21					
25	*	79	53	37	21					
26	*	85	63	48	32	16				
27	*	90	69	53	37	26				
28	*	95	74	58	45	32	21			
29	*	95	79	63	53	40	29	18		
30	*	*	85	69	58	45	34	26	16	
31	*	*	87	74	63	50	40	32	21	
32	*	*	90	79	66	55	45	37	29	18

*: DO concentration below 2 mg/l. *Source*: Boyd, 1982.

Another simple graphical procedure, as reliable as using eq. (6.3), is based on the linear relationship between the DO concentration and time. Two measurements are made of DO concentration, one early in the night and again after few hours. A line projected through the two points and extrapolated predicts DO concentration at other times during the night or at dawn (Figure 6.9). This method often predicts DO concentration lower than the actual value, because during nighttime the rate of respiration by phytoplankton tends to decline and input of DO by diffusion from air increases. The variation between the projected and measured values fluctuated between 0.13 and 20.83 percent, the average being 10.01 percent (Boyd *et al.*, 1978; Boyd, 1982).

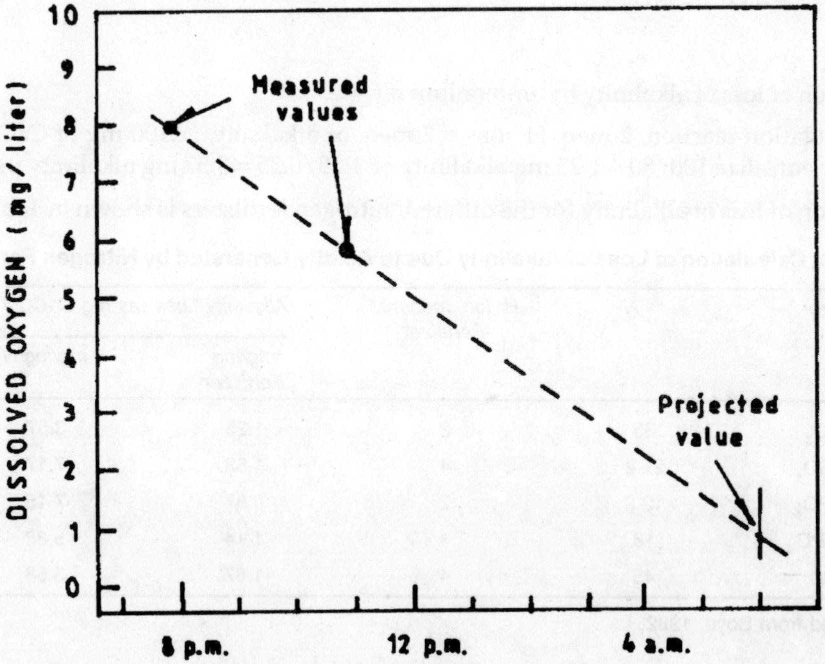

Figure 6.9: A Graphical Method of Prediction of Nighttime Decline in Dissolved Oxygen Concentration in Aquaculture Ponds (After Boyd *et al.*, 1978)

6.3 Dynamics of Fertilized Pond

Application of fertilizers is an important pond treatment practice that has been adopted from agriculture, the primary objective being the maintenance and improvement of the ponds in order to increase productivity, typically in extensive and semi-intensive farming systems. Considerable changes take place in the water chemistry and biological characteristics of the water as well as of pond bottom following application of fertilizers.

6.3.1 Effects of Fertilization on Pond Dynamics

1. Changes in Acidity due to Nitrogen Fertilizer

Acidity of the pond bottom and water results from the nitrification reactions carried by the bacteria *nitrosomonas* and *nitrobacter* present in soils and water:

$$2\,NH_4^+ + 3\,O_2 \rightarrow 2\,NO_2^- + 4\,H^+ + 2\,H_2O$$

$$2\,NO_2^- + O_2 \rightarrow 2\,NO_3^-$$

The hydrogen ions react with bicarbonate ions neutralizing the alkalinity of the pond water.

When a chemical fertilizer such as ammonium nitrate is added, it is oxidized to form nitric acid as follows:

$$NH_4NO_3 + 2\,O_2 \rightarrow 2\,HNO_3 + H_2O$$

$$2\,HNO_3 \rightarrow 2\,H^+ + 2\,NO_3^-$$

Example 6.6

Calculation of loss of alkalinity by ammonium nitrate.

In the oxidation reaction, 2 meq. H^+ ions ≡ 2 meq. of alkalinity, *i.e.*100 mg of $CaCO_3$. So 1 mg NH_4NO_3 can neutralize 100/80 = 1.25 mg alkalinity or 1.25/0.35 = 3.57 mg alkalinity per mg N.

Calculation of loss of alkalinity for the different nitrogen fertilizers is shown in Table 6.4.

Table 6.4: Calculation of Loss of Alkalinity Due to Acidity Generated by Nitrogen Fertilizers

Fertilizer	*% N*	*H^+ ion, meq/mM Fertilizer*	*Alkalinity Loss (as mg $CaCO_3$)*	
			mg/mg Fertilizer	*mg/mg N*
NH_4NO_3	35	2	1.25	3.57
$(NH_4)_2SO_4$	21.2	4	1.52	7.17
$NH_4H_2PO_4$	12.2	2	0.87	7.15
$(NH_4)_2HPO_4$	18	4	1.14	5.38
Urea	45	4	1.67	3.58

Source: Extracted from Boyd, 1982.

In pond systems the actual loss of alkalinity per unit weight of ammonium fertilizer would be less than the theoretically calculated values. Nitrogen added to ponds in the form of fertilizers is consumed in various other ways: by being absorbed by plants and in mud, lost in seepage or outflow, and lost through ammonia volatilization or denitrification (Boyd, 1982). However, losses of total alkalinity resulting from nitrification of ammonium fertilizers measured in laboratory mud-water systems showed that the result agreed well with those calculated from stoichiometry (Table 6.5). The results verified the calculation of theoretical acidity of ammonium fertilizers, which would be useful for calculation of requirements of fertilizers for neutralization of alkalinity in ponds.

2. Effects of Fertilization on Phosphorus Cycle

Increase in Orthophosphate Concentration

Mostly inorganic phosphorus is sufficiently applied in ponds to increase filtrable orthophosphate by 0.1–0.5 mg/l as P if the fertilizer is completely soluble and mixes thoroughly with water (Figure 6.10). While the ortho-phosphate concentrations approach potential values with liquid fertilizer, it will be very much low with solid fertilizer applied by broadcasting, because the solid particles quickly settle on pond bottom and get absorbed (Boyd *et al.*, 1981; Boyd, 1982).

Table 6.5: Comparison of Experimental and Calculated Loss of Alkalinity

Material	*Experimental*	*Calculated*
Ammonium nitrate	10.9	14.3
Ammonium sulphate	26.2	28.7
Monoammonium phosphate	28.6	28.5
Diammonium phosphate	16.6	21.5
Urea	12.2	14.3
Ammonium polyphosphate	23.1	

Note. Experimental values were measured 30 days after treatment of mud-water system with fertilizer.

Source: Boyd, 1982.

Absorption of Orthophosphate by Aquatic Organisms

Immediately after fertilization, different plant species compete with each other to absorb phosphorus. A rapid fall in filtrable orthophosphate concentration will occur which is enhanced by the absorption of phosphorus by mud (Figure 6.10).

Absorption of Phosphorus by Mud or Sediment

Phosphorus is partly absorbed by the plants and partly by mud. In fact the phosphorus not absorbed by plants will be absorbed by mud. Mud absorbs phosphorus rapidly. Roughly 90 percent of

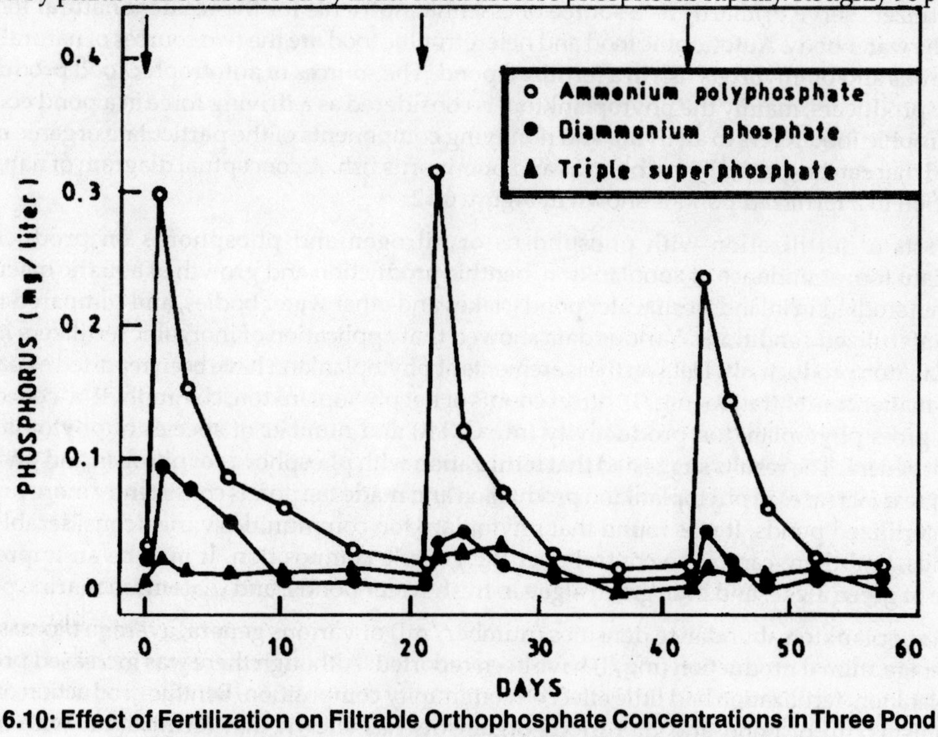

Figure 6.10: Effect of Fertilization on Filtrable Orthophosphate Concentrations in Three Pond Waters. Fertilizer application rate at 9 kg/ha of P_2O_5 (After Boyd *et al.*, 1981)

the orthophosphate added to undisturbed mud-water systems was absorbed by mud within 4 days (Boyd, 1982). In general strongly acidic or alkaline mud absorbs phosphorus more readily than slightly acidic or neutral mud. It was demonstrated that orthophosphate concentration declined logarithmically with time in a laboratory mud-water system. These results were similar to those observed in earthen pond and plastic pool (Figure 6.11, Boyd and Musig, 1981; Boyd, 1982).

Dynamic Equilibrium Between Phosphate in Mud and in Water

A dynamic equilibrium state exists between phosphate in mud and in water above it. If phytoplankton absorbs phosphate from water, more phosphate will be released from mud, because iron, aluminium and calcium phosphates in aerobic sediments are slightly soluble. This phosphate release is sufficient to maintain the phytoplankton growth. It has been demonstrated that this release of phosphate is an important source of the nutrient in unfertilized ponds. After fertilization this release was 4.5–5.5 times greater than that released from mud in unfertilized pond (Boyd, 1982). Thus the higher rate of phosphate release results in higher productivity of phytoplankton due to frequent fertilizer application during intensive fish production. The mud of fertilized pond acts as phosphorus sink, accumulates phosphorus over the years, and shifts the equilibrium in favor of water. However, phosphorus must be added even after long years of fertilization to maintain the high productivity.

Rooted macrophytes can absorb phosphorus from the sediment. Anaerobic conditions exists at a few centimeters depth of the sediment and concentration of phosphate is more in the interstitial water, which helps the roots of macrophytes to draw upon considerable quantity of phosphorus.

3. Plants and Invertebrates

Fertilizers serve primarily as a source of essential nutrients for the resident natural food chain within the water body. Autotrophic food and heterotrophic food are the two sources of natural food for herbivorous and omnivorous fish in a fertilized pond. The sources of autotrophic food production by primary producers, mainly the phytoplankton, is considered as a driving force in a pond ecosystem. Heterotrophic food refers to all living and nonliving components of the particulate organic matter in the pond that can be grazed by herbivorus and omnivorus fish. A conceptual diagram of natural food production in a fertilized pond is shown in Figure 6.12.

Effects of fertilization with phosphorus or nitrogen and phosphorus on productivity of phytoplankton, abundance of zooplankton, benthic production and growth of aquatic macrophytes have been studied in inland fresh water ponds, lakes and other water bodies, and compared with that under unfertilized conditions. Various data showed that application of inorganic fertilizers increased phytoplankton productivity. Data on measusements of phytoplankton have been reported as particulate organic matter concentration (mg/l), direct counts of net phytoplankton, chlorophyll 'a' concentration (mg/l), gross phytoplankton productivity (mg C/l.h) and number of species of phytoplankton of different genera. The results suggested that fertilization with phosphorus or nitrogen and phosphorus caused great increases in phytoplankton production and made the ponds consistently more productive than unfertilized ponds. It was found that phytoplankton communities varied considerably among the ponds, and there can be no control over the generic composition. It may be an unpredictable mixture of green algae, and blue green algae in fresh water ponds, and diatoms in marine ponds.

On zooplankton abundance densities (number/ml) of various genera, average biomass (mg/l) and average animal production (mg/l) have been reported. Although there was increased production of zooplankton, fertilization had little effect on community composition. Benthic production of number of organisms/m^2 of pond and of various organisms has shown that fertilization with inorganic

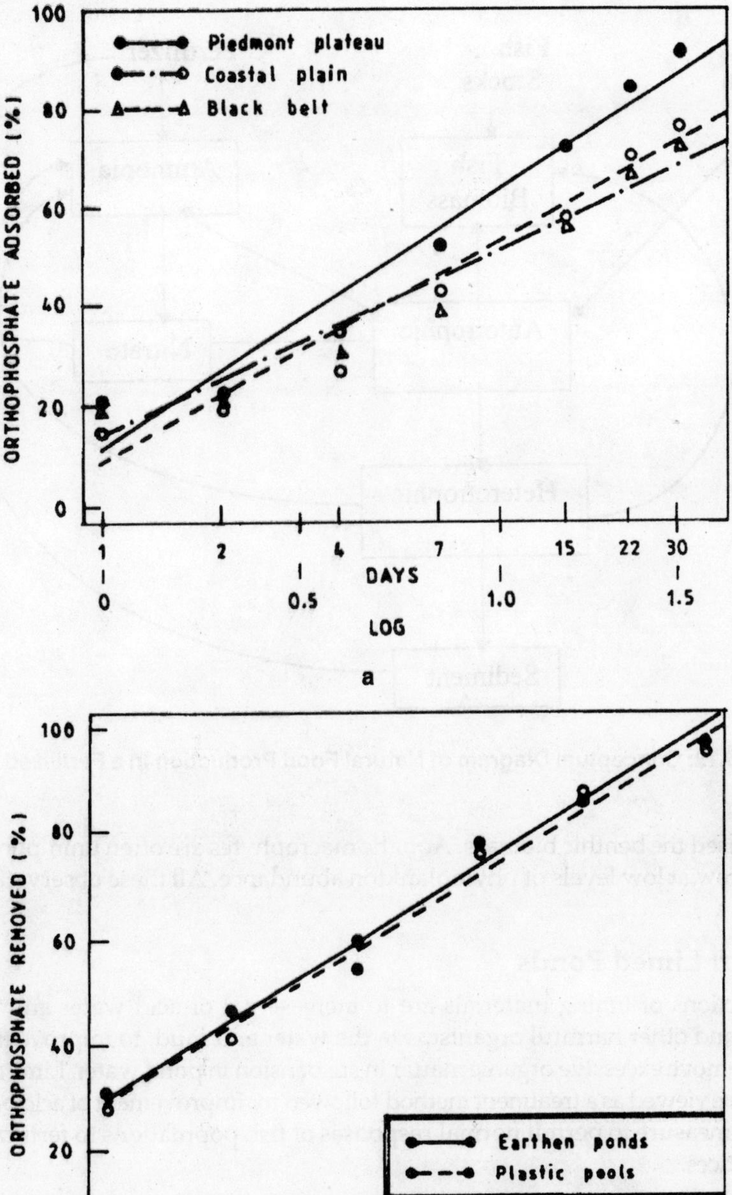

Figure 6.11: Phosphate Uptake by Mud; (a) In laboratory mud-water systems with mud taken from five different ponds and (b) In ponds and in mud water systems in outdoor, plastic pools (After Boyd and Musig, 1981)

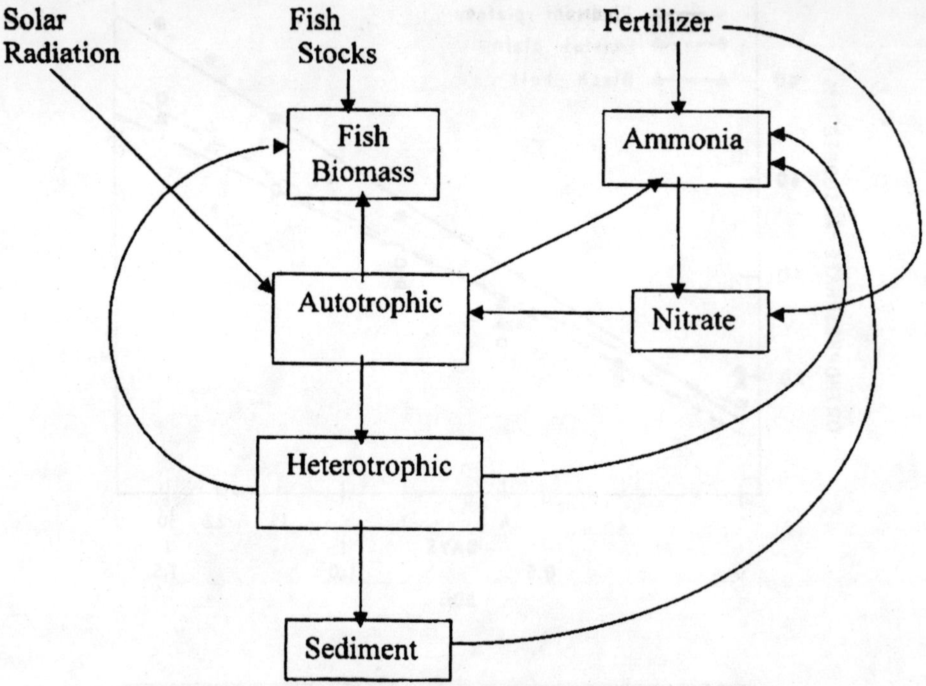

Figure 6.12: Conceptual Diagram of Natural Food Production in a Fertilized Pond

fertilization increased the benthic biomass. Aquatic macrophytes are often unimportant in fertilized ponds, but they grow at low levels of phytoplankton abundance. All these observations are given in Table 6.6.

6.4 Dynamics of Limed Ponds

The varied actions of liming materials are to increase pH of acid water and its alkalinity, to destroy parasites and other harmful organisms in the water and mud, to improve the pond bottom condition and to remove excessive organic matter in suspension in pond water. Liming is not a type of fertilization. It can be viewed as a treatment method followed for improvement of acid base relationships and as a remedial measure to permit normal responses of fish populations to fertilization and other management practices.

6.4.1 Effects of Liming on Pond Dynamics

1. Increase in Total Alkalinity

Compounds commonly used as liming materials contain calcium or calcium and magnesium carbonate, hydroxide or oxide. These compounds are contained in (a) agricultural limestone, calcite ($CaCO_3$), dolomite [$CaMg(CO_3)_2$], (b) hydrated lime or slaked lime [$Ca(OH)_2$], and (c) quick lime (CaO).

Table 6.6: Summary of the Effects of Fertilizer on Phytoplankton, Zooplankton and Benthos Production

Item	*Condition*		*Remark*
	Unfertilized	*Fertilized*	
I. *Phytoplankton:*			
a. Concentration of particulate organic matter (mg/l)	4.4, 6.0, 8.1 (av. 5.9)	8.0–31.1 (av. 22.3)	No. of unfertilized ponds = 3 No. of fertilized ponds = 11 Fertilized with various combinations of fertilizer, all received P.
b. Net phytoplankton of several genera (count/ liter)	32,700	80,400	Unfertilized, one lake. Fertilized, 3 lakes.
		>10,000,000	Fertilized with N, P.
	>1,000,000	>10,000,000	Fertilized in ponds
c. Chlorophyll 'a' (μg/l)	8.8–115.5	103.4–212.3	Fertilized with N, P (data of Israel)
	5–30	20–130	Fertilized with N, P.
	3780	11,120	Fresh water prawn ponds fertilized with distiller's dried grains with solubles
d. Gross productivityL			
i. as mg/l.h of carbon	0.18	1.76	Values measured between 10-00 and 14-00 hours.
ii. as g/m^2.day of carbon	–	1–3	
		3.3–6.4	Data of Israel
II. *Zooplankton*			
a. Density (No./liter)			
i. Rotifers	686	1826	Fertilized with P only.
	10,000	1,36,000	Fertilized with N, P.
ii. Crustacea	263	484	Fertilized with P only.
	100	1000	Fertilized with N, P.
b. Average biomass (μg/l)	150	881	Treated with high levels of N, P.
c. Average annual production (mg/l)	2767	7447	Treated with high levels of N, P.
III. *Benthos*			
Organisms/m^2	425	694	Average of two ponds, unfertilized
	371	640	and hatchery pond

Source: Extracted from Boyd, 1982.

Reactions of liming materials with acidity release calcium and magnesium ions as illustrated below:

$$CaCO_3 + 2H^+ \rightarrow Ca^{2+} + H_2O + CO_2$$

$$CaMg(CO_3)_2 + 4H^+ \rightarrow Ca^{2+} + Mg^{2+} + 2\,H_2O + 2\,CO_2$$

$$Ca(OH)_2 + 2\,H^+ \rightarrow Ca^{2+} + 2\,H_2O$$

$$CaO + 2\,H^+ \rightarrow Ca^{2+} + H_2O$$

Reactions of liming compounds such as calcium hydroxide and calcium oxide with carbon dioxide form calcium carbonate, which further reacts with carbon dioxide to form calcium and bicarbonate ions. The increase in alkalinity is reflected in the results of lime application in ponds

(Figure 6.13). The degree of acidity of water and mud and the quantity of liming materials added decide the increase in alkalinity and hardness (Arce and Boyd, 1975; Boyd, 1982).

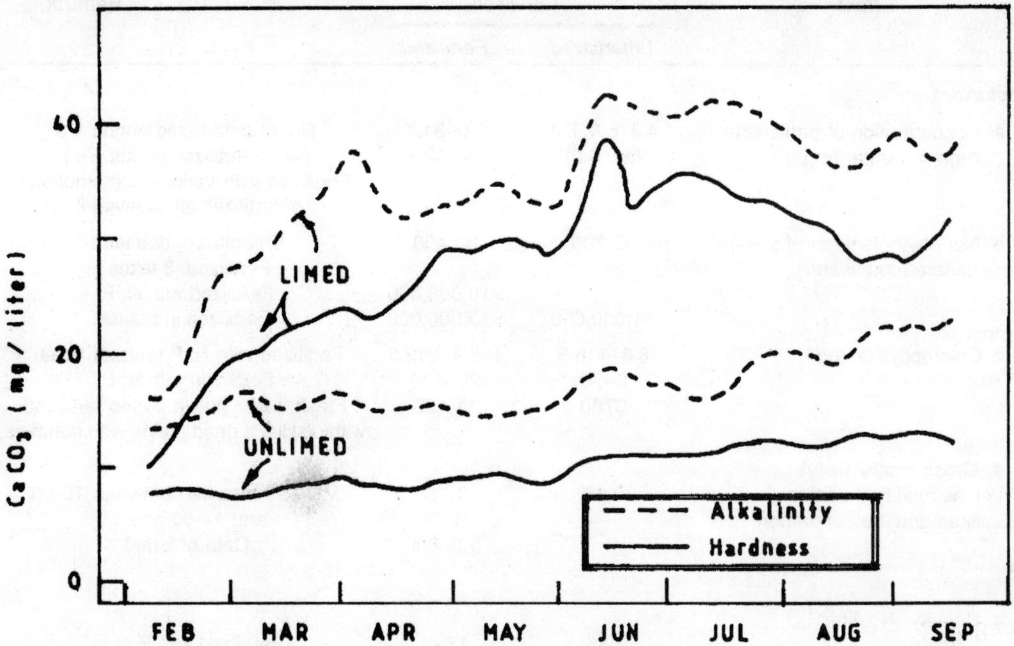

Figure 6.13: Effect of Liming with Agricultural Limestone on Total Alkalinity and Total Hardness (After Arce and Boyd, 1975)

As a result of neutralization of the acidity, the pH of water and mud will increase following lime application. However, if fertilizer is added, there will be daytime removal of carbon dioxide due to phytoplankton abundance with likely rise in pH. Usually the waters of low alkalinity are poorly buffered. When lime is added, morning pH value is increased creating an alkaline reserve and buffering against pH change. Therefore lime application may actually decrease afternoon pH in fertilized ponds.

2. Increase in Concentration of Total Available Carbon Dioxide

Due to increase in alkalinity during neutralization of acidity liming increases the availability of carbon or carbon dioxide for photosynthesis. The liming compounds react with carbon dioxide and so compete with plants for use of carbon dioxide with likely reduction of photosynthetic rates. In fact, liming materials trap carbon dioxide and prevent it to escape to the atmosphere. The carbon dioxide released in the water due to liming is removed from the water by reactions. Carbon dioxide released from the decomposition of organic matter in the mud reacts with excess lime settled in the mud. Further any carbonate in the water reacts with carbon dioxide diffused in from the air. Thus the overall effect becomes the higher concentration of total available carbon dioxide within a few weeks after liming.

3. Increase in Total Hardness

Liming increases concentration of calcium ions and also calcium and magnesium ions in water with corresponding increase in total hardness. Planktonic algae commonly found in fish ponds require 5 mg/l of Ca^{2+} and 2 mg/l of Mg^{2+} for their maximum growth. These concentrations of Ca^{2+} and Mg^{2+} correspond to total hardness of 20.7 mg/l. Liming increases total hardness to about 30 mg/l.

Calcium and magnesium ions play another important role in precipitating the colloidal particles and thereby decreasing the turbidity. Data revealed that liming reduced humic stains and lessen turbidity, which depend on the lime application rate. Because of lessening of turbidity depth of adequate light penetration for photosynthesis increases. Liming of a lake increased SD visibility from a depth of 2 m to 5 m and increased the depth at which DO was 4 mg/l from 2 m to 6 m (Boyd, 1982).

4. Effect on Activity of Microorganisms

As in agricultural soils, liming also increases the pH of water and increases the rate of microbial activity in the mud. The effect of bacterial decomposition of organic matter is indicated by carbon dioxide release. It has been found that in a laboratory mud water system at pH 4.7 and 5.4 loss of carbon dioxide was 0.65 and 0.99 percent, respectively over 38 days under unlimed condition. For limed mud this loss was 1.37 percent of its initial carbon when the pH was 6.5–7.2 (Boyd, 1982).

Parasites and other undesirable organisms can be destroyed when calcium oxide or hydroxide lime is applied in drained pond. Bacteria are also killed due to rise in pH when caustic lime is applied. As a result mud is sterilized and no decomposition takes place until pH falls.

5. Increase in the Availability of Mud Phosphate

Increasing pH of mud increases the availability of mud phosphate. Study on laboratory culture of particular algae with both limed and unlimed mud showed that algal growth was favored at high pH between 6.3 and 6.6. In unlimed mud the phosphate release was favored by mud at higher pH and higher growth of alga was observed (Table 6.7). Several workers have reported the influence of pond liming on phosphorus dynamics. Data on filtrable orthophosphste concentrations from limed and unlimed ponds have substantiated the results of laboratory studies (Boyd, 1982).

Table 6.7: Growth of Algae in Cultures where Limed and Unlimed Mud Served as the Only Sources of Phosphates

Algae (Thousands/ml) at 100 mg/kg P Addition		*P Addition (mg/kg)*	*Algae (Thousands/ml)*
Limed (pH 6.4)	*Unlimed (pH 5.5)*		
450	200	0	150
850	550	10	300
700	150	50	500
		100	700

Source: Extracted from Boyd, 1982.

6.4.2 Effect of Liming on Plankton and Invertebrates

Data on liming on the production of plankton and invertebrates are limited in comparison with those for fertilization. In ponds and lakes with water heavily stained with humic substances and mud containing decaying organic matter planktonic abundance decreases initially after liming. Probably this decrease is due to loss of available carbon dioxide and high pH. Planktonic communities reappeared after the pH declined from near 11 to about 8. Both phytoplankton and zooplankton increse several years after liming. Since these types of ponds and lakes are not fertilized, the planktonic abundance is probably due to change in the chemical modification of the environment.

In some lakes fertilization fails to produce plankton blooms due to slightly acidic (pH 5.5–6.8) water and low hardness (0–15 mg/l). Following liming pH and hardness increase and response of plankton and benthos to fertilization improves. Gross productivity of unlimed pond, 5.8 g C/m^2.day increased to 6.5 g C/m^2.day by liming the fertilized pond. However, liming had no influence in the generic composition of phytoplankton (Arce and Boyd, 1975).

6.5 Dynamics of Fed Pond

In general, artificial feed is one of the major inputs in semi-intensive and intensive fish production systems. These feed range in quality and forms from raw ingredients such as agricultural and industrial by-products supplementing natural foods to pelleted feeds supplying balanced nutrition to the animals. However, maintenance of water quality becomes aggravated in ponds where regular feeding is practiced. Uneaten feed and metabolic wastes create problems of phytoplankton abundance and oxygen demand. As feed rates increase in ponds, water quality becomes poorer and finally reduces the fish production, if scientific pond management practices are not followed.

6.5.1 Types of Feeding and Feeding Options

Feeding consists of supplying the animals being cultured with 'artificial' food. This is the food, which is not 'natural'. Natural foods, grown in the pond, are those low in the food chain or at a low trophic level, such as unicellular and filamentous algae, benthic blue green algae and diatoms, bacteria and fungi, zooplankton and insect larvae. The artificial food or feed may consist of single (individual) raw materials such as rice bran or groundnut cake, simple mixtures of ingredients, or complex formulated 'compound' feeds compacted by hand or by machine. Feeds for aquaculture may be regarded as 'supplementary' or 'complete' depending on the farming practice. Supplementary feeds supplement the natural food available in the pond. The culture animals receive the part of their nutrient requirement from natural food and part from the artificial feed. A complete feed must supply all the nutrient requirements of the animal that is cultured in intensive systems, where natural food is available not at all or in small proportion in relation to the total required.

The efficiency of a feed is usually measured by feed conversion ratio (FCR), which is the amount of feed necessary to produce a unit weight of fish. It is the unit weight of feed given, divided by the live weight (or wet weight) of animal produced.

$$FCR = \frac{weight\ of\ food\ presented}{weight\ of\ animal\ produced}$$

For example, if 1250 kg of fish were harvested from a pond to which 2500 kg of artificial feed has been presented during the culture period, the FCR would be 2500/1250 = 2.0. A feed with lower value of FCR will be more efficient. The above expression of FCR is often presented in an undefined form as

$$\text{FCR} = \frac{\text{Feed intake}}{\text{Weight gain}}, \text{ or } \text{Conversion efficiency} = \frac{\text{Weight gain}}{\text{Feed intake}}$$

The feed intake should not be taken as an expression of what is "offered" to the fish, but restricted to the "feed eaten" by the fish. Further, the weight gain may not represent a weight increase proportional to growth of all organs, but may be a deposition of fat, either connected with the viscera or in the meat (Utne, 1979).

Moisture content in the animal or in the feed is not normally taken into account when the FCR is calculated. This means that wet or moist feeds automatically have much higher FCRs, but not necessarily the less efficient than dry feeds. But comparison of feeds should be done on like basis.

Example 6.7

Two feeds, one moist with 35 percent moisture content, and the other a dry feed with 10 percent moisture content, have FCR 3.5 and 2.0, respectively. Compare the two feeds on like basis, *i.e.* as dry feed.

Solution

$$FCR = 3.5 \times (100-35)/(100-10) = 2.53$$

Therefore, the dry feed with an FCR of 2.0 is more efficient than the moist feed with FCR of 2.53. However, the strict comparison should be on dry matter basis.

The FCR calculated above is the true FCR in a farming system where the artificial feed is the only or complete feed that the animals are receiving. Even in very intensive systems of aquaculture the natural productivity of the water may have been stimulated, either by true fertilization or because of waste feed, and natural food becomes available. So the true FCR can be calculated when no feed other than the feed given is available to the animal, and the FCR calculated should be called 'apparent FCR' (AFCR). AFCR is also a way of comparing two feeds used under the similar situation.

6.5.2 Supplementary Diet Feeding

In semi-intensive or modified extensive pond production systems stocking density and standing crop is such that the natural productivity of the water alone is not adequate for growth of the species. The dietary nutrients requirements of the species are met through the production and consumption of natural live food organisms within the pond by employing fertilization and/or supplementary diet feeding. Here the supplementary feed is a direct 'supplementary' source added to supply deficiencies in dietary nutrients for the species. In fact, the role of supplementary feeds in the nutrition of cultured species is closely interlinked to the dynamics of the pond system.

All types of pond food organisms are rich source of protein, with overall composition being 50-60 percent (dry matter basis). The mean composition of protein, carbohydrate and lipid are 52.1, 27.3 and 7.7 percent, respectively with a mean calorific value of 3.9 kcal/kg (De Silva, 1995). On elemental basis, natural food organisms consist mainly of carbon, nitrogen and phosphorus, *e.g.* phytoplankton grown in a nutrient rich medium contains 45–50 percent carbon, 8–10 percent nitrogen and 1 percent phosphorus on dry basis. Therefore, the growth of the food organisms depends on the supply and cycling of carbon, nitrogen and phosphorus within the natural pond ecosystem. The natural productivity of the water body can be increased with careful management, through controlled addition of chemical inorganic fertilizers (for autotrophs), and/or organic manure (for heterotrophs). Data on proximate composition of some of the common food organisms show that the nutritional value of natural food organisms is adequate to meet the dietary requirements of most of the species cultured under semi-intensive conditions in Asia (De Silva, 1995).

The relative importance of natural food organisms and supplementary diet feeding within various culture systems is best shown in Figure 6.14 (Tacon, 1988). According to this figure, semi-intensive

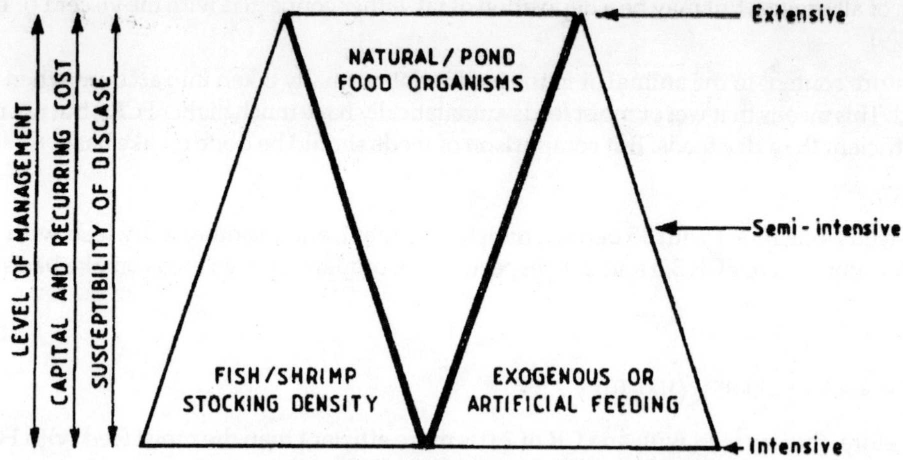

Figure 6.14: The Role of Natural Food Organisms and Supplementary Feeding within Various Culture Cystems (After Tacon, 1988)

aquaculture would constitute a significant share between extensive and intensive systems, but differs in practices in the degree of management, provision of feed, stocking density and yields.

Supplementary feeding practice leads to complex interactions between the natural food organisms and the feed, as shown in Figure 6.15. The qualitative interactions shown in Figure 6.15 indicate that the two factors: standing crop (SC) *i.e.* total biomass of the species, and availability of natural food organisms within the water body per animal or species are important. As the difference between SC of the cultured species and the critical standing crop (CSC) of natural food organisms increases, the deficit in natural protein supply also increases. Under such conditions supplementary feeding with protein rich diets becomes necessary to maintain optimum growth (De Silva, 1995). However, the importance of natural food organisms in the overall nutritional budget of pond reared fish/shrimp cannot be under stressed, especially during the early growth cycle and at low fish/shrimp stocking densities when total fish/shrimp biomass per m^2 is low and natural food availability per animal is high (Figure 6.15). The dietary energy is generally the first limiting factor at low densities (when food availability is high), whereas at high densities and standing crop dietary protein and other essential nutrients become limiting and therefore have to be supplemented. Therefore the farming practice should be to increase progressively the protein and nutrient content of the diet with increasing fish / shrimp biomass and standing crop and decreasing natural food availability. In fact, these criteria are in sharp contrast to a complete diet feeding strategy in intensive culture, where dietary nutrient levels usually decrease with increasing crop weight and age (Tacon, 1988).

Selection of feed and formulation of a supplementary diet is dependent on the culture conditions, *i.e.* the standing crop and availability of natural food organisms. Table 6.8 shows that no difference of growth was observed in several culture conditions like rations with varying protein content, without feed or fertilizer, when the standing crop and body weight remain the same.

The above observations and discussions indicate that the interactions between natural food supply and supplementation make feeding more complex than that with a complete diet. The aquaculturist has to be alert about the type and quantity of the feed to be used.

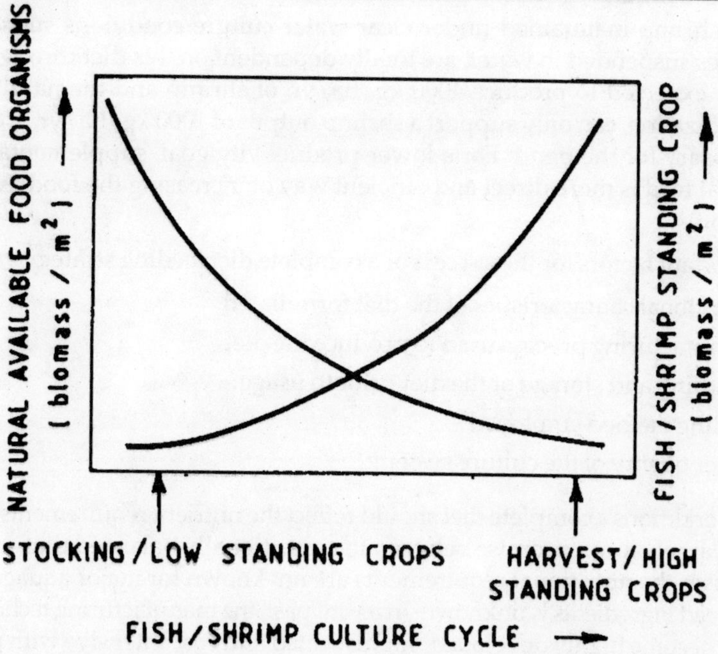

Figure 6.15: Relationship Between Fish/Shrimp Growth and the Natural Food Availability Within a Semi-intensive Culture System (After Tacon, 1988)

Table 6.8: Comparison of Culture Conditions with Respect to Growth

Species	*Feed*	*Culture Condition*	*Difference in Growth*
Common carp (*Cyprinus carpio*)	Cereal grain (sorghum) Formulated ration (22.5% protein)	Up to a standing crop 800 kg/ha Earthen ponds in Israel	No difference in growth
Common carp (*C. carpio*)	Ration (22.5% protein) Ration (27.5% protein)	Up to 1400 kg/ha Earthen pond in Israel	No difference with 2 diets
Prawn (*M. rosenbergii*)	Diet: 35 per cent protein Diet: 15 per cent protein Broiler starter feed	5 animals/m² Outdoor concrete ponds in Thailand	No difference within 3 diets
Shrimp (*P. vannamei*)	No fertilizer/feed Fertilizer only Commercial diet	7.1–9.4 animals per m² Earthen ponds in Hawaii	No difference
Tilapia (*Oreochromis nilotica*)	Diet: 9 per cent protein 19 per cent protein	Cage rearing (within pond) Stocking density 400, 600, 800 fish/ m²	No significant difference in growth rate

Source: Extracted from Tacon, 1988.

6.5.3 Complete Diet Feeding

In contrast to extensive and semi-intensive farming systems, complete diet feeding involves the external provision of a nutritionally 'complete' high quality diet containing a pre-determined nutrient profile. Fish and shrimp maintained under clear water culture conditions such as cement tanks, raceways and cages suspended in water, are totally dependent on this diet throughout their culture cycle. If a pond is expected to produce 4000 kg/ha/yr. of shrimp and the natural food available, stimulated by fertilization, can only support a shrimp output of 1000 kg/ha/yr., a complete artificial food will be necessary for the pond. For a lower productivity goal, supplementary feed would be sufficient. Artificial feed is more direct and efficient way of increasing the food available to the fish than by fertilization.

The five important factors for the success of a complete diet feeding strategy are:

1. The nutritional characteristics of the diet formulated
2. The manufacturing process used to produce the diet
3. The handling and storage of the diet prior to usage
4. The feeding method employed
5. The water quality of the culture system

Under ideal conditions a complete diet should reflect the nutrient requirements of the species. But a practical diet is largely a compromise between the nutritionally balanced diet and the economical factors. In most cases the nutritional requirements are not known for major aquaculture species and availability of the feed ingredients is unknown. In recent past, the manufacturing technology of complete pelleted feeds has become highly developed, sophisticated and cost-intensive with proprietary rights. Further, aquaculture production systems, feeding habits and feed rations of the species, and economic justification with respect to high market value of the species are some of the important factors of the complete diet formulation. Finally, water quality control and management of the culture systems become the most vital factor of complete diet feeding practice.

6.5.4 Feed Conversion, Utilization and Waste Production

The objective of aquaculture is to convert energy and nutrients in feed or fertilizer into energy and nutrients in fish, with minimum loss in the conversion. Feed conversion and utilization, and also waste production depend on the relationship between energy and nutrients (nitrogen and phosphorus) fed and those expended in fish (Bergheim and Asgard, 1996). The practice of aquaculture shows that the waste in intensive and semi-intensive culture systems as well as in hatcheries is predominantly from feed, and includes uneaten feed (*i.e.* feed waste), and undigested feed residues and excretion products (*i.e.* from eaten feed). It has been shown that energy and nutrients are lost in both types of waste. Factors such as feed quality or FCR and feeding practice greatly influence the total waste production. However, energy and nutrient retention in fish and losses in the waste are more reliable estimate of waste production, and FCR should be used judiciously for the estimate.

1. Material Balance of Feed Utilization

Utilization of feed in fish weight gain, and production of feed-derived wastes from eaten and uneaten portions can be illustrated in the form of overall material balance based on an average fish production, quantity of feed applied and FCR.

Example 6.8

Make a material balance on feed utilization from the following data:

Channel catfish pond–Net production = 5000 kg/ha, with FCR = 1.5

Feed analysis: Dry matter = 90 per cent, N = 6 per cent and P = 1.5 per cent

Fish analysis: Dry matter = 25 per cent, N = 11.0 per cent and P = 4.0 per cent (all on dry basis). Data adapted from Boyd, 1982.

Solution

Feed applied = 7500 kg. Dry matter in feed = 6750 kg, and in fish = 1250 kg

Total waste produced = 6750–1250 = 5500 kg, about 73 per cent of the total feed applied.

Nitrogen balance: N in feed = 405 kg, N utilized by fish = 137.5 kg, and N in the waste = 267.5 kg, which is 66 per cent of the total N in feed.

Phosphorus balance: P in feed =101.3 kg, P utilized by fish = 50.0 kg and P in the waste = 51.3 kg, which is 50.6 per cent of the total P in the feed.

All the values are in kg/ha.

Depicting the values in the form of a block diagram, we get the material balance shown in Figure 6.16.

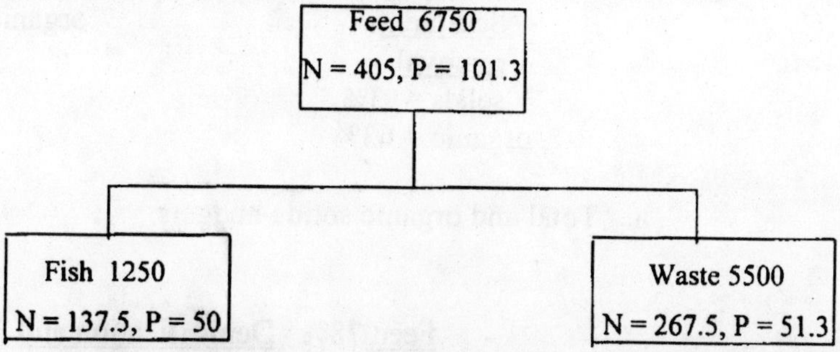

Figure 6.16: Overall Material Balance of Feed Utilization

Calculation shows that about 66 percent of the total nitrogen of feed reaches the pond as metabolic waste and uneaten feed, and the rest goes to the fish. Although much of the waste is converted to carbon dioxide during bacterial decomposition, conversion to nitrogen and phosphorus are in considerable quantities. The nitrogen added to the pond is much greater than phosphorus and is nearly twice that added in fertilizer for maximum production of the particular variety. Since the nutrients reach the pond by continuous input of feed, they are more effective in increasing plankton growth than that received by intermittent fertilizer addition.

2. Nutrients and Solids Budget

The above material balance shows an overall picture, and did not consider several factors. The example of nutrients and solids budgets for intensive shrimp ponds in Thailand was derived for solids, particulate organic matter, nitrogen, and phosphorus, and are shown in Figure 6.17 (Funge-Smith and Briggs, 1998). The types of ponds studied were one-year and two-year old ponds (stocking density 50–60 per m^2) and one-year old ponds with higher stocking density (80–100 per m^2). The budgets show that the major source of solids (88–93 percent) and organic matter (40–60 percent) was erosion of pond soil. The feed contributed little solid (4–7 percent) to the system, but was a significant source of organic matter (31–50 percent). In intensive shrimp ponds the bulk of the solids generated endogenously remain as accumulated sediment (91–94 percent), which becomes the sink of 58–70

Figure 6.17: (a) Total and Organic Solids, (b) Nitrogen, and (c) Phosphorus Budgets for Intensive Shrimp Ponds in Thailand (After Funge-Smith and Briggs, 1998)

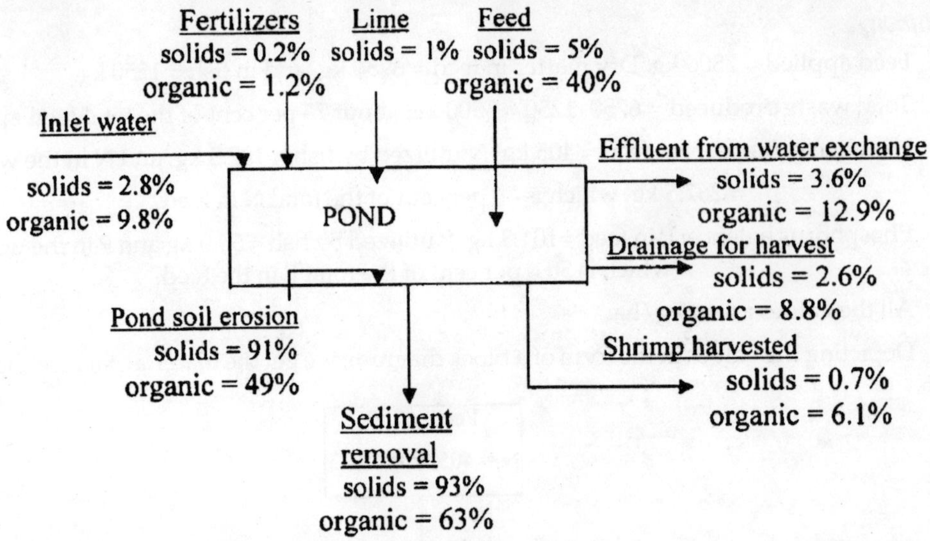

a. Total and organic solids budgets

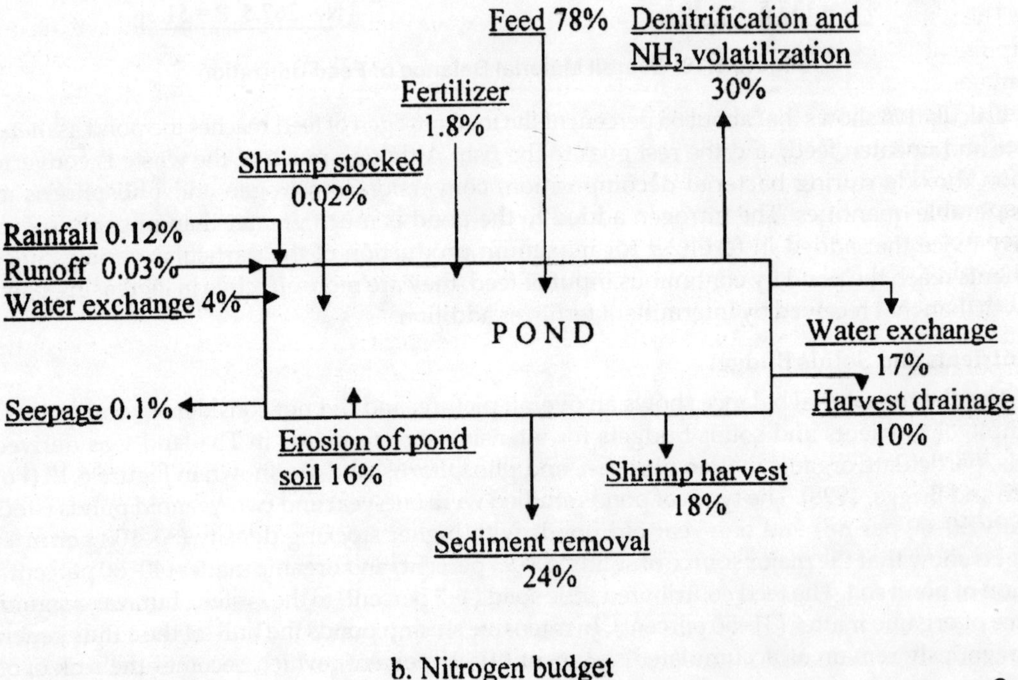

b. Nitrogen budget

Contd...

Figure 6.17–Contd...

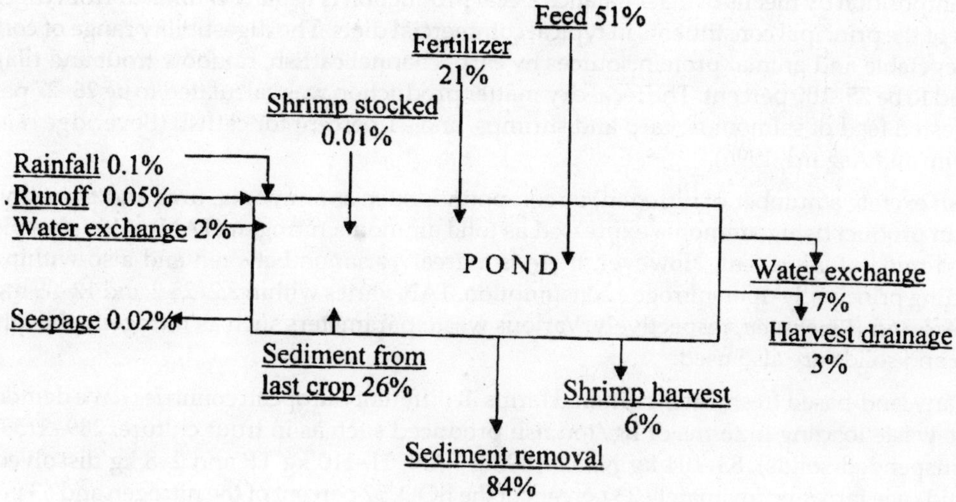

c. Phosphorus budget

percent of the organic matter in the system. This is unlike the extensive ponds, where influent water is a major source of sediments. Therefore the character of a pond soil will have a significant effect on the water quality and production of a shrimp pond.

The nitrogen and phosphorus budgets reveal in more detail the sources and sinks of the organic components. The major nitrogen input to the ponds was feed (78 percent), but the pond soil erosion, the major contributor of solids to the pond, contributed only 16 percent of nitrogen added to the system. The nitrogen sinks were the sediments (24 percent), shrimp (18 percent), and discharged water (27 percent). Rest of the nitrogen is about 30 percent, which is unaccounted and assumed to be lost to the atmosphere as nitrogen or ammonia through microbial decomposition process. For phosphorus, the principal source was the feed (51 percent), followed by pond bottom sediment (26 percent) contributed from the previous culture operations by erosion. The above material balance also shows that the actual amount of nutrients assimilated into shrimp biomass is a small fraction of the total applied as feed. Only 18–27 percent of nitrogen and 6–11 percent of phosphorus applied to the pond was assimilated on the basis of FCR 1.65–2.40, and there is a considerable wastage as nutrients, incorporated into plankton biomass, volatilized or trapped in the sediment.

3. Waste Components

If the waste components are considered separately, a different material balance on waste production can be seen. Some proportion of the feed distributed is not consumed. From the feed eaten, some proportion remains undigested, and together with mucus, sloughened intestinal cells and bacteria is excreted as feces. The digested portion is absorbed and metabolized.

Estimates of feed-derived waste production, in salmonid farming, show that the proportion of uneaten feed can vary from about 1 percent to as much as 30 percent. Data suggest that feed losses from cages are considerably larger than those from ponds (Beveridge *et al.*, 1991). The main components of waste produced from undigested feed can be estimated from feed type, feed quality, digestibility and fecal composition by means of mass balance. Fecal production is usually estimated from digestibility studies of the principal constituents in typical commercial diets. The digestibility range of commonly used vegetable and animal protein sources by carp, channel catfish, rainbow trout and tilapia was reported to be 75–100 percent. The fecal dry matter production was calculated to be 26–27 percent of the ingested feed of salmonids, carp and shrimps, and 41 percent for catfish (Beveridge *et al.*, 1991; Bergheim and Asgard, 1996).

Fish excrete a number of nitrogenous compounds such as ammonia, urea and trimethylamine, the main product being ammonia expressed as total ammonia nitrogen (TAN). Also phosphorus as excreted mineral is present. However, there is a great variation between and also within species depending principally upon nitrogen consumption. TAN varies within 2.2–24.2 and 12–39 mg N/kg. h for carp and *P. monodon*, respectively. Various waste parameters such as nitrogen, phosphorus or carbon and solids are also used.

Many land-based freshwater salmonid farms in different European countries have demonstrated specific waste loading in terms of kg/ton fish produced such as in trout culture: 289–2153 kg TSS (total suspended solids), 83–104 kg N, 37–180 kg TAN, 11–110 kg TP and 2–8 kg dissolved P. In a salmonid cage farm approximately 23 percent of the BOD, 52 percent of the nitrogen and 63 percent of phosphorus of the total supplied feed load the environment. This is equivalent to 360 kg oxygen as BOD, 32 kg nitrogen and 7 kg phosphorus per ton of fish produced. In intensive carp ponds approximately 20–30 percent of the total feed supplied nutrient are used for growth, which is 11 percent of nitrogen, 32 percent of phosphorus, and 69 percent of organic carbon. These values indicate that70–90 percent of the total nutrient input is released into water. The fraction actually loading the environment depends on the pond system. In shrimp ponds of Thailand, percentage of nitrogen and phosphorus going into the environment are 78–79 and 92–95 percent, respectively and in others, 67–78 and 76–86 percent, respectively (Bergheim and Asgard, 1996).

Figure 6.18 shows some estimates of waste production considering faecal production. A more detailed picture emerges for trout by incorporating major data and a number of assumptions. From uneaten waste and eaten waste total loss of nitrogen to the environment is approximately 98 kg, i. e 76.6 percent. Although such mass balance calculations have been arrived at for fish, none are available for shrimps and prawns (Beveridge *et al.*, 1991).

4. COD Balance

Besides nutrient (N and P) enrichment in the aquatic environment organic carbon enrichment takes place. Boyd (1982) has shown that much of the carbon fixed by photosynthesis is wasted in pond getting the regular feed supply. A COD balance in channel catfish ponds is illustrated in the following example.

Example 6.9

Four ponds receive an average of 2800 kg/ha of feed for average net production of 2100 kg/ha of fish. Considering the mean rate of carbon fixation by photosynthesis = 2.55 g of C/m^2.day, and assuming that phytoplankton generated by photosynthesis contains 48 percent carbon on dry weight basis, make a material balance on COD. Assume the COD values of phytoplankton, feed and fish as 1.29, 1.14 and 1.4 kg O_2/kg. dry weight, respectively (Data adapted from Boyd, 1982).

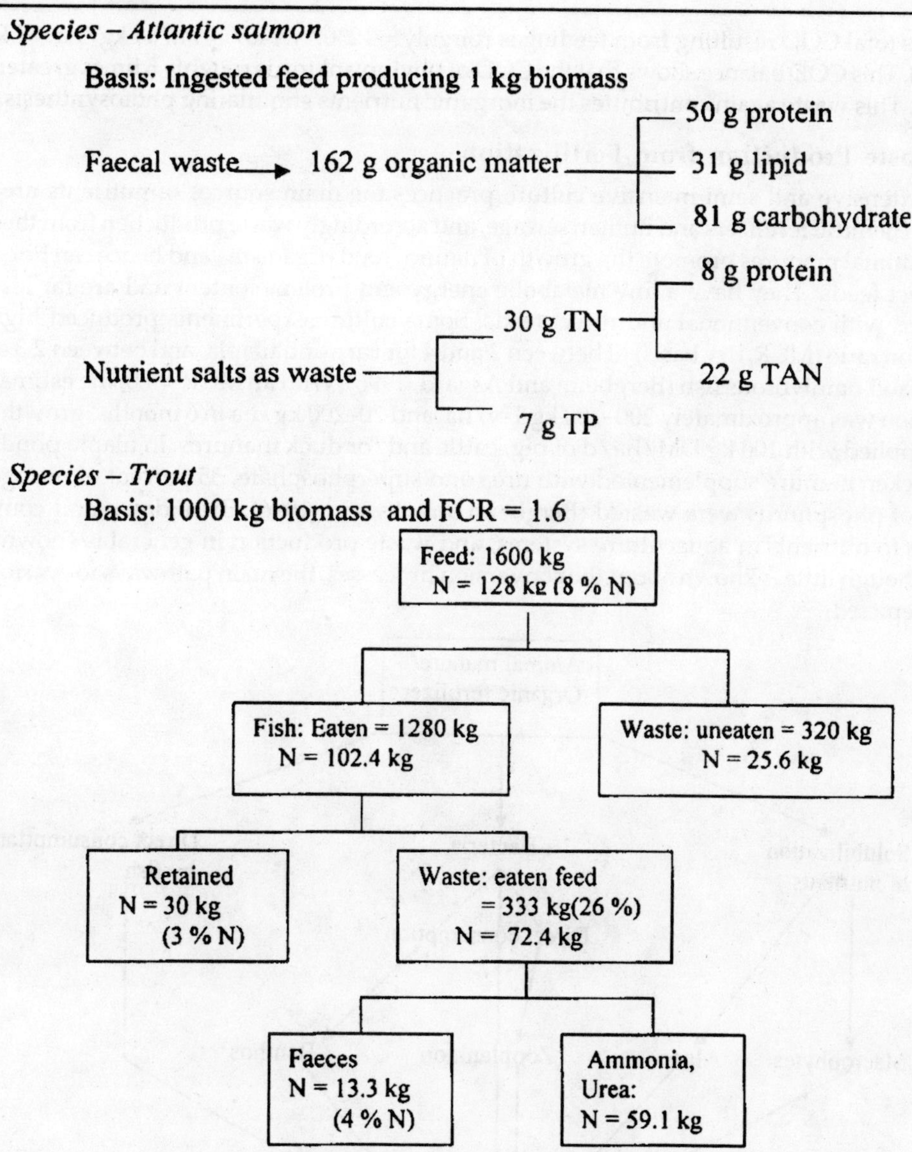

Figure 6.18: Mass Balance on Waste Production
(After Bergheim and Asgard,1996; Beveridge *et al.*, 1991)

Solution

Phytoplankton production = 2.55 / 0.48 = 5.3 g/m^2.day = 53 kg/ha. day = 9540 kg/ha. dry matter.

COD production by phytoplankton = 1.29 × 9540 = 12,307 kg/ha. = 4.4 kg/feed.

COD in feed = 1.14 × 2800 = 3192 kg/ha.

COD in fish = 1.4 × 2100 × 0.25 = 735 kg/ha (25 per cent dry matter).

COD in waste = 3192–735 = 2457 kg/ha.

Thus total COD resulting from feeding is roughly = 12307 + 2457 = 14,764 kg/ha = 7 kg/ha per kg of fish. This COD balance shows that the COD of phytoplankton is roughly 5 times greater than that of waste. This waste again contributes the inorganic nutrients stimulating photosynthesis.

6.5.5 Waste Production from Fertilization

In extensive and semi-intensive culture practices the main sources of nutrients are livestock manure, chemical fertilizers and human sewage, and accordingly waste production from these sources varies. Animal manures promote the growth of natural food organisms and hence can be considered as indirect feeds. They have a low metabolic energy and protein content and are far less efficient compared with conventional and direct feeds. Some culture experiments produced high manure conversion ratio (MCR, dry basis) of between 2 and 4 for carp and tilapia, and between 2.3 and 6.2 for filtering and omnivorous fish (Bergheim and Asgard, 1996). With an MCR = 3.0, the estimated waste production was approximately 200–700 kg TN/ha, and 70–200 kg/ha in 6 months' growth of carp in ponds applied with 100 kg DM/ha/d of pig, cattle and/or duck manures. In tilapia pond fertilized with chicken manure supplemented with urea and superphosphate, 85 percent of nitrogen and 92 percent of phosphorus were wasted (Bergheim and Asgard, 1996). Breakdown and conversion of manures to nutrients in aquaculture systems, and waste production in general is shown in Figure 6.19. Although little is known about the conversion processes, the main pathways for various species can be depicted.

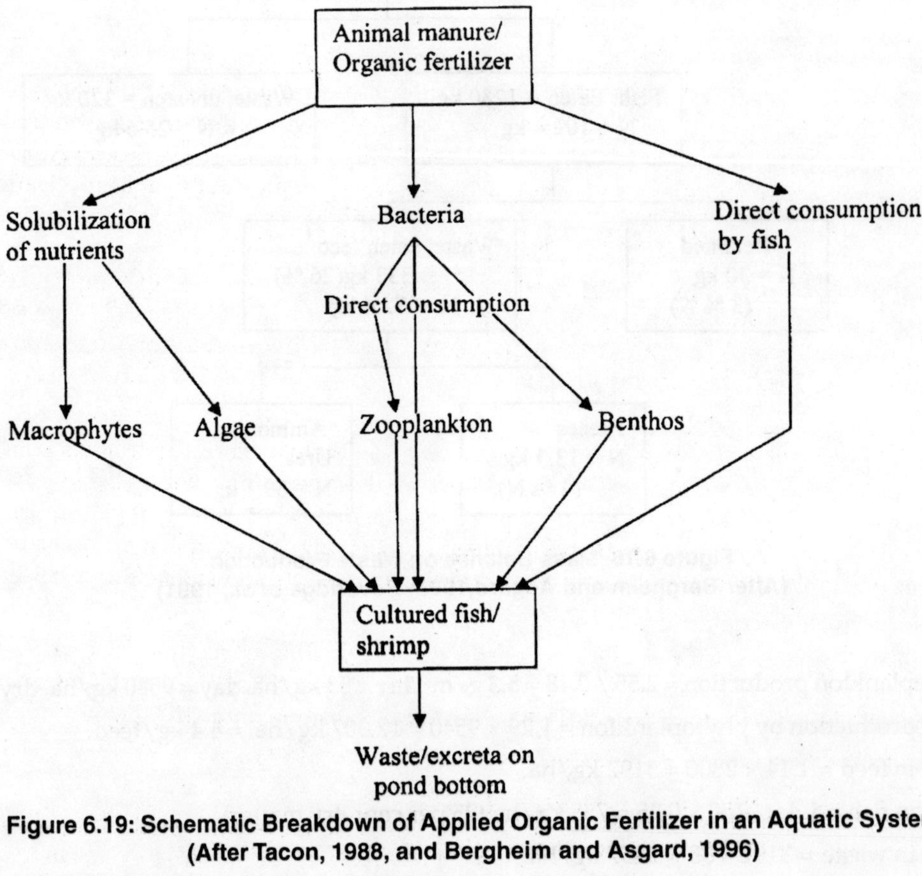

Figure 6.19: Schematic Breakdown of Applied Organic Fertilizer in an Aquatic System (After Tacon, 1988, and Bergheim and Asgard, 1996)

In ponds fertilized with inorganic fertilizers and with water exchange requirement dosage of fertilizer has to be adjusted according to the nutrient content of the water and soil, and grazing intensity. Chances of over-fertilization and normal discharge of water only at the time of harvest can result in loss of nutrient. In tidal ponds with frequent water exchange benthic algal community dislodged from the bottom can accumulate and decompose on the pond banks unless removed regularly. Static water systems with little water exchange create minimal nutrient leakage to the surrounding open waters (Pillay, 2004). Fertilization of shrimp ponds may discharge nutrients to the surrounding waters, although such effects have not been quantified (Bergheim and Asgard, 1996).

Sewage-fed aquaculture, an age-old practice in some Asian countries, uses human sewage usually after some form of pretreatment. The practice is not basically different from the application of manures and wastes. The process involves fish culture and simultaneous aerobic decomposition of wastes to remove the pathogens. The environmental problems relate to algal blooms increasing BOD in the effluents and nutrient uptake by algal blooms resulting in eutrophication of receiving waters (Pillay, 2004).

In most of the aquaculture practices involving organic fertilizer and human wastes, the organic load of the effluents discharged from the ponds is usually insignificant. Quality of effluents also depends on the water retention time and outflow of water. Effluents are discharged only at the time of harvest and retention time is usually not less than a month. Therefore adequate biological decomposition and waste utilization within this time can meet the effluent quality standards and prevent adverse environment impacts (Pillay, 2004).

6.5.6 Residues of Chemicals

A wide range of chemicals is used in aquaculture to control diseases, aquatic vegetation and other nuisance organisms; to induce spawning or promote growth; and to treat the water for water quality improvement and productivity increase (Beveridge *et al.*, 1991; Bergheim and Asgard, 1996). Table 6.9 shows some of the commonly used chemicals with examples and field of application. Some therapeutics such as antibiotics are used in feed particularly in intensive aquaculture of salmonids and shrimps. The antibiotics are commonly used in penaeid hatcheries and grow-out ponds. Many antibiotics are under regulations. Formalin and malachite green are widely used in hatcheries and farms. Although they are toxic to algae and zooplankton in low concentrations, they are potentially harmful to both pond and surrounding waters. Chlorinated hydrocarbons are also highly toxic and actual threats to the shrimp culture itself and the surrounding environment (Bergheim and Asgard, 1996).

The plant products tea seed cake and rotenone are widely used under tropical conditions at concentrations up to 20 mg/l but toxicity ceases after 8-12 days. Many selective and non-selective toxins used in pond farms for pest and predator control either decompose within hours or result in discharge of small quantities of the toxins into the receiving waters. Further, occasional waste discharges from tropical farms and long time after chemical application make the persistence of the toxicity of the chemicals unlikely (Pillay, 2004).

There are few data on the quantities of chemicals used in different countries. A survey of 148 in the U. K. shows that out of 23 chemicals the most common are malachite green and formalin. The most common antibiotic is oxytetracycline (OTC). In Norway the salmon industry is heavily dependent on chemicals, particularly the antibiotics. In tropical countries the prophylactic use of antibiotics is also widespread (Beveridge *et al.*, 1991).

Table 6.9: Chemicals Used in Aquaculture

Type	Example	Application
Therapeutics: Parasitic treatment, Fungicides	Malachite green, formalin, methyline blue, dichlorvos, potassium permanganate	Hatcheries and tropical inland, marine, and mariculture of salmonids
Algicides (antifoulants)	Copper oxide, copper sulphate	Mariculture of salmonids
Antibacterials (comprising antibiotics and antimicrobials)	5–8 types such as tetracyclines, chloramines, sulphonamides and phenicol nitrofurans Potassium permanganate	Salmonid and tropical marine shrimp culture, hatcheries, tropical inland farms
Piscicides, molluscicides	Tea seed cake, rotenone, nicotine, chlorinated hydrocarbons	Tropical marine shrimp farms
Disinfectants	Formalin, hypochlorite, chloramines, sodium hydroxide	Widely used
Water and soil treatment chemicals	Oxidizers (potassium permanganate), pH regulators (lime), zeolite	Widely used

Source: Beveridge et al., 1991; Barg, 1992; Bergheim and Asgard, 1996.

Chemical loss and release in to the environment depends on the method of treatment such as by bath, injection or incorporation in feed. Loss of therapeutics administered in feeds occurs both during leaching and feed loss. In mariculture 70 percent of the total feed-supplied antibiotics are likely to be lost to the recipient. Leaching of OTC from feed has been shown to depend on temperature, pH and pellet surface area: volume ratio, with up to 20 percent being leached in 15 min. The few data available show that the chemicals are present in effluent in low concentrations such as 0.61 and 15.2 mg/l of malachite green and formalin, respectively, in the U.K. fresh water fish farm effluent (Beveridge *et al.*, 1991).

6.5.7 Effects of Wastes on Culture Environment

As the excessive use of feed (and fertilizer also) carries the risk of increased nutrient and organic matter loading, it will alter water and sediment quality in ponds resulting in over-population in algae and build-up of anoxic sediments. Nutrient and organic wastes are generally characterized by an increase in suspended solids, BOD, COD, and content of carbon, nitrogen and phosphorus. In general effluents discharged reflect water quality conditions in the pond. Wide ranges have been observed for effluent water quality parameters during a five-month grow-out period at intensive shrimp ponds in Thailand (Table 6.10).

Table 6.10: Effluent Water Quality in Intensive Shrimp Pond in Thailand (Pond size: 0.48–0.56 ha and Depth: 1.5–1.8 m)

Water Quality Parameter	Values
Salinity, ppt	10–35
Temperature, C	22–31
pH	7.5–8.9
Total P, mg/l	0.05–0.4
Total N, mg/l	0.50–3.4
Total ammonia, mg/l	0.05–0.65
Dissolved oxygen, mg/l	4.0–7.5
Chlorophyll a, mg/l	20–250
Total suspended solids, mg/l	30–190
Water exchange, per cent per day	5–40

Source: Barg, 1997.

From a study on the effect of environmental parameters on the productivity of modified extensive and semi-intensive shrimp farms in some coastal states of India, it was found that problems due to feeding and oxygen depletion affected most of the farms (Anonymous, NEERI, 1995). The principal abiotic characteristics of the farms are shown in Table 6.11.

Table 6.11: Salient Abiotic Features of the Ponds in the Coastal Shrimp Farms in India

Farm	DO		BOD		SS		Dissolved P		NH_3-N	
	Intake	Pond	Intake	Pond	Intake	Pond	Intake	Pond	Intake	Pond
Modified Extensive										
Farm 1	5.4–6.6	6.9–7.7	4–5	4–6.2	100–246	90–149	0.15	0.1–0.2	0.72–0.74	0.7–0.9
Farm 2	4.6	4.3–4.7	5	2–6	480	470–510	0.14	0.15–0.17	0.86	0.86–1.04
Farm 3	4.1	4.0–4.3	5	7.2–12.0	153	207–374	0.05	0.01–0.02	1.5	0.5–1.1
Semi-intensive										
Farm 2	2.9	2.6–4.2	7.0	3–8	180	180–220	0.02	0.12–0.16	0.82	0.50–0.68
Farm 3	4.4–4.6	4.6–4.8	2.4–7.2	6.4–13.6	89–253	88–207	0.08–0.15	0.02–0.17	0.28–0.52	0.60–1.74

Source: Adapted from Bal *et al.*, 1998.

6.5.8 Relationship of Water Quality with Feeding Rate

Studies were conducted extensively in channel catfish ponds by Boyd and co-workers (1982) in USA on effects of feeding rate on water quality. Five number of ponds with surface area = 0.89–8.9 ha and average depth = 1.5–2 m were stocked with 7,400 fingerlings/ha. Culture season was from March to October. Feed was floating type pellet (35 percent crude protein) with feeding strategy of 3 percent body weight/day, 6 days/week. Feeding rate was reduced in September to control the ideal marketable size of fish. Their observations were as follows:

1. Temporal Changes

Water temperature gradually increased and then remained constant.

SD visibility, chlorophyll 'a' concentration and number of phytoplankton generally increased.

Total ammonia N increased in excess of 1 mg/l by late summer.

pH was usually between 8.5–9.5 (afternoon), and 6.5–7.5 (early morning).

Filterable orthphosphate was quite low, not more than 0.01 mg/l as P.

CO_2 increased with time, averaged > 4 mg/l in the early morning.

COD was > 50 mg/l.

DO showed strong vertical stratification, with depths of 0 mg/l = 1–2 M in all ponds.

2. Relationship to Feeding Rate

Ponds with lowest maximum daily feeding rate usually have the poorest water quality as shown below:

1. When the feeding rate is higher, the ponds have the supply of more nutrients as metabolic wastes, and hence the planktonic abundance and COD becomes higher.

2. Average total NH_3-N concentrations are always below 0.5 mg/l in low feeding rate. Unionized NH_3 concentration fluctuates daily.
3. DO seldom fall below 4 mg/l in low treatment, but fall below 2 mg/l in high treatment. Oxygen depletion results in 7–32 percent mortality of fish in ponds.
4. If maximum daily feeding rates do not exceed about 35–40 kg/ha, oxygen depletion and other water quality problems do not occur. At maximum daily feeding rates at about 60 kg/ha, some water quality problem may occur. At feeding rates of 80 kg/ha or above, problems with DO and unionized ammonia will be severe.
5. Fish production data (Table 6.12) show that medium treatment is clearly superior to low treatment. The average yield was greater, but cost of production was more and size of fish smaller than with the medium treatment. This may make the high feed treatment economically less attractive than medium treatment.

Table 6.12: Fish Production Data with Different Feeding Rate

	Feeding Rate (kg/ha/day)		
	Low (34)	*Medium (56)*	*High (78)*
Harvest weight (kg/ha)	2990	4100	4860
Survival (per cent)	99	93	83
Mean weight of individual fish (g)	601	443	284
FCR	1.3	1.7	2.5

Source: Tucker *et al.*, 1979.

References

Akiyama, D. 1991. Future consideration for the aquaculture feed industry. In *Proceedings Aquaculture Feed Processing and Nutrition Workshop held in Thailand and Indonesia, 19-25 September, 1991*, D. M. Akiyama and R.K. H. Tan, eds. American Soybean Association, Singapore.

Alabaster, J. S. 1974. The development of water quality criteria for marine fisheries. *Ocean Management*, 2, 101-115

Anonymous. 1995. *Environmental Management for Prawn Farming Activity*. Report for MPEDA, Kochi, prepared by NEERI, Nagpur.

Alabaster J. S. and Llyod, R. 1980. *Water Quality for Freshwater Fish*. Butterworth, London.

APHA. 1981. American Public Health Association, American Water Works Association, and Water Pollution Control Federation, *Standard Methods for the Examination of Water and Wastewater*, 16th ed. American Public Health Association, Washington D. C.

APHA. 1989. American Public Health Association, American Water Works Association, and Water Pollution Control Federation, *Standard Methods for the Examination of Water and Wastewater*, 17th ed. American Public Health Association, Washington D. C.

Arce, R. G. and Boyd, C. E. 1975. Effects of agricultural limestone on water chemistry, phytoplankton productivity, and fish production in soft water ponds. *Trans. Amer. Fish. Soc. 104*, 308-312.

Bailey, J. E. and Ollis, D. F. 1977. *Biochemical Engineering Fundamentals*. McGraw-Hill, New York.

Bal, A. S., Ghosh, T. K. and Pandey, R. A. 1998. Environmental status of shrimp farming in India, p. 14-42. In *Conservation and Management of Aquatic Resources*, A. Gautam, ed. Daya Publishing House, Delhi.

Barg, U. C. 1997. Guidelines for the promotion of environmental management of coastal aquaculture development. *FAO Fisheries Technical Paper. No. 328*. Rome, FAO. 1992. Daya Pub. House, New Delhi.

Bergheim, A. and Asgard, T. 1996. Waste production from aquaculture. Ch. 3, p. 50-80. In *Aquaculture and Water Resource Management*. B. J. Baird, M. C. M. Beveridge, L.A. Kelly, and J. F. Muir, eds. Blackwell Science, Oxford.

Beveridge, M. C. M., Phillips, M. J. and Clarke, R. M. 1991. A quantitative and qualitative assessment of wastes from aquatic animal production, p. 506-533. In *Aquaculture and Water Quality, Advances in World Aquaculture*, Vol. 3. D. E. Brune and,J. R. Tomasso, eds. World Aquaculture Society, Baton Rouge, Louisiana.

Boyd, C. E. 1982. *Water Quality Management for Pond Fish Culture*. Elsevier, Amsterdam.

Boyd, C. E. 1989. *Water Quality Management and Aeration in Shrimp Farming*. Fisheries and Allied Aquacultures Departmental Series No. 2. Alabama Agricultural Experiment Station, Auburn University, Auburn.

Boyd, C. E. 1991. Empirical modeling of phytoplankton growth and oxygen production in aquaculture ponds, p. 363-395. In *Aquaculture and Water Quality, Advances in World Aquaculture*, Vol. 3. D. E. Brune and J. R. Tomasso, eds. World Aquaculture Society, Baton Rouge, Louisiana.

Boyd, C. E., Romaire, R. P. and Johnston, E. 1979. Water quality in channel catfish production ponds. *J. Environ. Qual. 8*, 423-429.

Boyd C. E. and Musig, Y. 1981. Orthophosphate uptake by phytoplankton and sediment. *Aquaculture* 22, 165-173.

Boyd, C. E., Musig, Y. and Tucker, L. 1981. Effect of three phosphorus fertilizers on phosphorus concentrations and phytoplankton production. *Aquaculture* 22, 175-180.

Boyd, C. E., Romaire, R. P. and Johnston, E. 1978. Predicting early morning dissolved oxygen concentrations in channel catfish ponds. *Trans. Amer. Fish. Soc. 107*, 484-492.

De Silva, S. S. 1995. Supplementary feeding in semi-intensive aquaculture systems. In Farm-made Aquafeeds. M. B. New, A. G. J. Tacon and I. Csavas, eds. *FAO Fisheries Technical Paper No. 343*, Rome, FAO, 1994.

Funge Smith, S. T. and Briggs, M. R. P. 1998. Shrimp rearing: stocking density, growth, impact on sediment, waste output and their relationships studied through the N budget in rearing ponds. *Aquaculture* 164, 117-133.

Glasstone, S. 1942. *Introduction to Electrochemistry*. Van Nostrand Reinhardt, New York.

Handbook of Chemistry and Physics. 1984. R. C. Weast, ed.. CRC Press, Boca Raton.

Harvey, H. W. 1966. *The Chemistry and Fertility of Seawater*. Cambridge Press.

Jones, J. R.E. 1964. *Fish and River Pollution*. Butterworths, London, cited in Wheaton, 1977.

Levenspiel, O. 1972. *Chemical Reaction Engineering*. 2nd ed. John Wiley and Sons, New York.

Martin, D. F. 1970. *Marine Chemistry. Theory and Applications*. Vol. 2. Dekker, New York.

Muir, J. F. 1982. Recirculated water systems in aquaculture, p.357-447. In *Recent Advances in Aquaculture..* J. F. Muir and R. J. Roberts, eds. Croom Helm, London.

Muller-Feuga, A., Petit, A. and Sabbaut, J. J. 1978. The influence of temperature and wet weight on the oxygen demand of rainbow trout (*Salmo gairdneri R.*) in fresh water. *Aquaculture* 14: 355-363.

Parkins, E. J.1976. The evaluation of biological response by toxicity and water quality assessment. In *Marine Pollution*. R Johnston, ed. Academic Press, London.

Peppler, H. J., ed. 1967. *Microbial Technology*. Reinhold Pub. Corp., New York

Pillay, T. V. R. 2004. *Aquaculture and The Environment*, 2nd Ed., Blackwell Publishing, Oxford.

Poxton, M.G. and Allouse, S. B. 1982. Water quality criteria for marine fishries. *Aquacultural Eng.* 1, 153-191.

Romaire, R. P. and Boyd, C. E. 1979. Effects of solar radiation on the dynamics of dissolved oxygen in channel catfish ponds. *Trans. Amer. Fish. Soc. 108*, 473-478.

Schroeder, G. L. 1975. Nighttime material balance for oxygen in fish ponds receiving organic wastes. *Bamidgeh 27*, 65-74. Cited in Boyd, 1982.

Smith, J. M. and Van Ness, H. C. 1975. *Introduction to Chemical Engineering Thermodynamics*. 3rd ed. McGraw-Hill, New York.

Stickney, R. R. 1979. *Principles of Warmwater Aquaculture*. John Wiley, New York

Tacon, A. G. J. 1988. The Nutrition and Feeding of Farmed Fish and Shrimp–A Training Manual 3. Feeding Methods. *FAO Field Document No. 7 GCP/RLA/075/ITA*. FAO, Brazil.

Tucker, L., Boyd, C. E. and McCoy, E. W. 1979. Effects of feeding rates on water quality, production of channel catfish, and economic returns. *Trans. Amer. Fish. Soc. 108*, 389-396.

Utne, P. 1979. Standard methods and terminology in finfish nutrition, p. 438-444. In *Proc. World Symp. on Finfish Nutrition and Fishfeed Technology*. Hamburg, 20-23 June, 1978, Vol. 2, J. E. Halver and K.Tiews, eds. Berlin.

Walas, S. M. 1959. *Reaction Kinetics for Chemical Engineers*. McGraw-Hill, New York.

Weatherly, L. R. 1984. Rate models for a marine biological filter. *Aquacultural Eng. 3*, 15-29.

Wheaton, F. W. 1977. *Aquacultural Engineering*. Wiley-Interscience, New York.

Wickins, J. F. 1981. Water quality requirements for intensive aquaculture: A review. *Proc. World Symp. on Aquaculture in Heated Effluents and Recirculated Systems*. Stavanger 28-30 May, 1980, Vol. 1, Berlin.

Part III:
WATER QUALITY MANAGEMENT

7

Introduction

7.1 Culture Systems

7.1.1 Types of Culture Systems

Several approaches are taken in aquaculture practice to maintain the environmental requirement of the species. Three basic systems are: open, semi-closed, and closed water bodies (Wheaton, 1977). There are some variations or modifications to these systems, as shown in Table 7.1.

1. Open System

A natural system without any human interference or modification is an open system. Management of an open system depends on the culture practice and the desired output. In the simplest open system, if harvesting is not done too often, nature does all the management necessary. However, the open systems can be modified by adopting several techniques of management:

1. Fertilization
2. Elimination or exclusion of predators of the cultured species.
3. Culture under confinement within enclosures such as net-pen structures, or in floating systems such as raft, rack and tray culture.
4. Disease control, although difficult in open systems, can be achieved by salinity control and selective pesticides.

Open systems are used widely for aquaculture almost in all the countries, ranging from strictly natural systems to those operated by adopting management techniques.

2 Semi-closed System

In a semi-closed system, culture is carried out in static ponds with water exchange facilities, or by passing water only once through the system and returning it directly to the natural body of water. In static ponds acceptable water quality is maintained by complex processes occurring in fluctuating environment of natural water bodies. However, water quality management can be improved by controlling the parameters such as temperature, predation and diseases, and by utilizing the phytoplankton of natural water as a source of food.

Table 7.1: Culture systems

Sl.No.	Nature of System	Treatment Option	Environmental Impact Consideration
1.	Open water body – Natural system	Natural processes maintain the water quality	No impact
	– Modified natural system (in enclosures)	Natural processes maintain the water quality	Impact in culture sites and locality
2.	Semi-closed water body –Static ponds	Water quality maintained primarily by natural processes, and also by adopting techniques for improvement of the system	Large volumes of water are taken for initial filling and for water exchange. Effluent may have polluting effect on the surrounding water.
	–Flow-through or once-through system	Water quality is managed by several techniques.	Large quantities of surface water are required. Water quality is usually lower than that of the supply water.
3.	Closed water body (recirculating system)	Environmental parameters are precisely controlled.	Limited impact

Source: Adapted from Tomasso and Brune, 1991.

In general, the semi-closed systems are operated in facilities constructed for the purpose. This involves more capital cost than the open system. Skilled operators must be required because of better management practices to be followed, which increases the operating cost also. However, the control measures for water quality management are much more than pay for these increased costs by getting a good return. Semi-closed systems are widely followed for trout, catfish in the U. S. for many years, and for carp and shrimp in several countries in recent years.

3. Closed System

In a closed system, also called recirculating or water reuse system, the culture water is seldom exchanged with sources outside the system, but is circulated by pumping within the system. Environmental parameters such as water temperature, concentration of metabolites and dissolved oxygen, and several other parameters can be precisely controlled. The major advantage of this system is the maintenance of constant optimal environmental conditions for the species of culture.

Closed systems can be developed technically, but difficulty lies in making it a commercially viable system. Two primary disadvantages are: high capital cost, and high operating costs for use of good management practices. These are cost of equipment for water filtration and purification, pumping and aeration cost, and technical and managerial expenses.

7.1.2 Basic Approach of Closed System

The basic approach of closed system is that water is reused and thus conserved by recycling it through the culture unit. Water is treated in one or more units that are specifically designed for the purpose. The treatment processes are very specific, and followed usually for intensive and monoculture systems. Further, water reuse and recirculating systems are designed for high-density aquaculture, where supply of water of suitable quality and quantity is a major input and constraint.

Figure 7.1 shows the conceptual diagram of a water reuse system with the relationship between water flow (Q) and concentrations (C) of water quality parameters derived from a simple static mass balance (Muir, 1982). The water quality parameters limit the acceptability of the water, and the defined treatment efficiencies, output of waste, and flows may provide recycle rates. Alternatively, the existing recycle rate may be evaluated for the operating efficiency. The definitions used in Figure 7.1 are:

Recycle ratio (X) = Q_R/Q_T, per cent Recycle = $(Q_R/Q_T) \times 100 = (1\text{-}Q_F)/Q_T \times 100$, and per cent per day = Q_F/V, where V = system volume = $Q_F/\theta\, Q_T$, and θ = system residence time.

Recirculation efficiency = (1-EX-X)

Figure 7.2 shows water reuse advantages relative to per cent recycle, and for given per cent recycle relative to treatment efficiency (Muir, 1982).

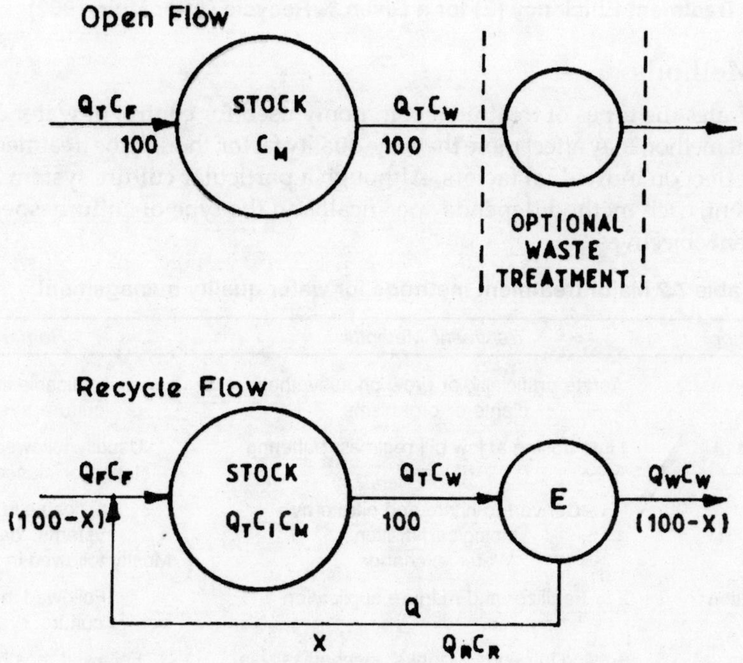

Figure 7.1: Conceptual Diagram of a Water Reuse System (After Muir, 1982)

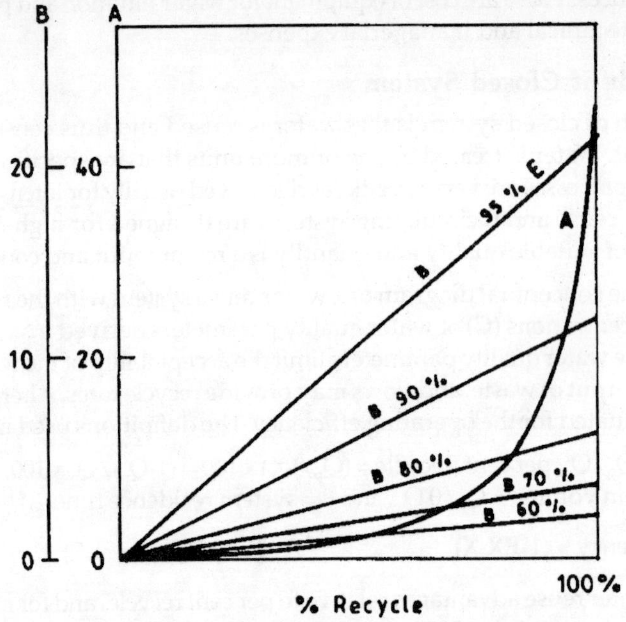

Figure 7.2: Water Reuse Advantages Relative to % Recycle (A) and to Treatment Efficiency (E) for a Given % Recycle (After Muir, 1982)

7.2 Treatment Methods

Table 7.2 illustrates the types of treatment commonly used for control of water quality factors. Individual treatment method may affect more than one quality factor, though the treatments are grouped according to their effect on individual factors. Although a particular culture system adopts several methods of treatment, each method depends specifically on the type of culture, species, the water source, and treatment objectives.

Table 7.2 Major treatment methods for water quality management

Water Quality Factor	*Treatment Methods*	*Remark*
Dissolved oxygen	Aerate artificially, or grow photosynthetic plants or organisms	Applicable in most culture systems
Carbon dioxide and pH	Lime dosing at low pH regimes, buffering of water	Usually followed in semi-closed or closed systems
Ammonia	Convert to nitrite and nitrate by biological filtration Water exchange	Applicable in closed systems, hatchery Mostly followed in pond culture
Nutrient concentration	Fertilizer and manure application	Followed in most culture systems
Suspended solids	Settling in ponds or tanks, mechanically in settling basins or by filters	Followed mostly in semi-closed or closed systems
Bacteria or other pathogens	Disinfection by chlorination, or any related methods	Applicable in closed, semi-closed systems, hatchery

7.2.1. Pond Management Methods

Continuous monitoring of water quality factors is essential in high density cultures. Key water quality factors and suggested frequency of measurement for shrimp culture are shown in Table 7.3.

Acceptable water quality can be achieved by adopting the following techniques (Clifford, 1992):

1. Controlled water exchange.
2. Maintenance of a stable, steady-state populations of algae and an autotrophic pond ecosystem by coordinating water exchange and fertilization programs.
3. Avoidance of over-feeding by proper feeding tray management.
4. Artificial aeration and water circulation.
5. Periodic or continual removal of accumulated organic material.
6. Maintenance of a high density bacterial population combined with water circulation and aeration.

Table 7.3: Key Water Quality Factors and Suggested Frequency of Measurement

Factor	*Frequency*
Water temperature	2 x daily
Dissolved oxygen	2 x daily
Turbidity (Secchi disc)	2 x daily
Salinity	1 x daily
pH	1 x daily (After noon)
Alkalinity/1	1 x daily
Total ammonia nitrogen	2 x weekly
Unionized ammonia nitrogen	2 x weekly
Nitrate nitrogen	2 x weekly
Nitrite nitrogen	2 x weekly
Total nitrogen	1 x weekly
Total phosphorus	1 x weekly
Chlorophyll-a	1 x weekly

Source: Clifford, 1992.

7.2.2. Recirculating Methods

In water reuse and recirculating methods, the water is treated usually by processes such as screening, settling and sedimentation, mechanical filtration, biological filtration, aeration and disinfection. Although treatment options are many, and have not been mentioned here, some are still in experimental stage of development and not yet proved commercially viable in aquaculture systems (Lawson, 1997). A typical recirculating unit is shown in Figure 7.3. The unit uses a 90 percent daily recycle rate with a 10 percent water exchange. Water is recycled 90 percent for reuse and replaced 10 percent with fresh water. Thus the total water volume can be replaced once in 10 days. Normally 3–5 percent water exchange per day is operated well for less waste load in culture water, but an approach to 100 percent recycles increases technical and management difficulties (Lawson, 1997).

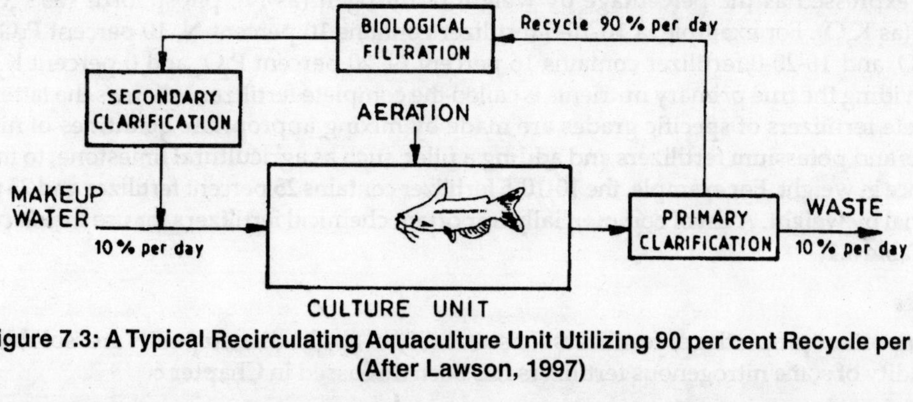

Figure 7.3: A Typical Recirculating Aquaculture Unit Utilizing 90 per cent Recycle per day (After Lawson, 1997)

8

Fertilization of Ponds

8.1 Fertilizers

8.1.1. Types, Properties and Sources of Fertilizers

1. Types and Sources

Fertilizers commonly used for aquaculture ponds are classified as chemical or inorganic fertilizers and organic fertilizers or manures. These are similar or identical to those used for agricultural crops. The chemical fertilizers are used to improve soil fertility and increase agricultural crop yields. In fish ponds fertilizers stimulate phytoplankton production, which increases fish yields.

Commercially the fertilizers are designated in terms of the primary nutrients nitrogen, phosphorus, potassium expressed as the percentage by weight of nitrogen (as N), phosphorus (as P_2O_5) and potassium (as K_2O). For example, a 10-10-5 fertilizer contains 10 percent N, 10 percent P_2O_5 and 5 percent K_2O, and 16-20-0 fertilizer contains 16 percent N, 20 percent P_2O_5 and 0 percent K_2O. The former providing the true primary nutrients is called the complete fertilizer, whereas the latter is not. The complete fertilizers of specific grades are made by mixing appropriate quantities of nitrogen, phosphorus and potassium fertilizers and adding a filler, such as agricultural limestone, to make up the difference in weight. For example, the 10-10-5 fertilizer contains 25 percent fertilizer and 75 percent filler material by weight. A list of commercially important chemical fertilizers for use in fish ponds is shown in Table 8.1.

2. Properties

Two important properties of fertilizers for suitability of application in ponds are: solubility and acidity. Acidity of some nitrogenous fertilizers has been discussed in Chapter 6.

Table 8.1: Some Common Commercial Fertilizers

Material	*% Nutrient*		*% Nutrient Solubility/1*	
	N	*P_2O_5*	*N*	*P_2O_5*
Urea	45	0	–	–
Calcium Nitrate	15	0	98.7	–
Sodium Nitrate	16	0	61.7	–
Ammonium Nitrate	33	0	98.8	–
Ammonium sulphate	21	0	85.9	–
Superphosphate	0	20	–	4.6
Triple Superphosphate	0	46	–	5.1
Monoammonium Phosphate	11	48	5.1	7.1
Diammonium phosphate	18	48	11.7	16.8

Source: Extracted from Boyd, 1982.

1. After settling through a 2 m water column at 29°C.

Fertilizers should dissolve quickly in the water before being settled at the bottom. An experimental measurement of solubility criteria in a 2-m water column at 29° C showed that the percentage dissolution of phosphorus for fertilizers varied between 4.6 and 16.8 percent, whereas that of nitrogenous fertilizers ranged between11.7 and 98.8 percent. Thus fertilizers containing phosphorus do not go into solution appreciably during settling (Table 8.1).

8.1.2 Requirement of Fertilizers

1. Principle

For use of fertilizers in fish ponds for increase of productivity the concept of diminishing return with increasing fertilizer rates applicable for agricultural crops has been followed. The basic principle regulating the benefits of fertilizers was developed by Mitscherlich and Spillman in the early 1900's for the agricultural crops. The principle states that when a single growth factor is limiting the growth of a crop, the increase in growth with each equal successive addition of growth factor is progressively smaller (Tisdale and Nelson, 1956). The relationship between the yield (Y) produced by a given quantity of growth factor (X) and the maximum yield possible (A) is:

$$\log (A-Y) = \log A - 0.30\,X \tag{8.1}$$

Starting with no growth and no yield, X = 0 and Y = 0. With one unit of growth (X = 1), and A = 100, so log (100–Y) = log 100–0.30 (1), or Y = 50.

This means that 50 percent of the maximum yield is obtained by addition of one unit of growth factor. It can be shown by calculation that there will be little increase on yield by addition of growth factors beyond 3 or 4 units (Table 8.2).

The economic implication of the above relationship is that at a point the cost of an additional unit of fertilizer is greater than the value of the increment of the crop yield. The concept was demonstrated in fish production to show that fish production did not increase in direct proportion to fertilizer application, as in agricultural production, and did not increase above a certain level with further increase in fertilizer (Table 8.3, Hickling, 1962).

Table 8.2 Results of Calculation by Equation (8.1)

Units of Growth Factor (X)	*Yield (Y)*	*Increase in Yield (%)*
0	0	0
1	50	50
2	75	25
3	87.5	12.5
6	98.44	1.56
10	99.90	0.10

Source: Extracted from Tisdale and Nelson, 1956.

Table 8.3: Application of Spillman's Concept in Fish Production

Pond	*Yield, kg/ha*	*Increase*
Unfertilized	97	–
Fertilized with 22.4 kg/ha P_2O_5	315	219
Fertilized with 44.8 kg/ha P_2O_5	418	102

Source: Hickling, 1962.

There are several factors influencing the action of chemical fertilizers (Tacon, 1988):

Sunlight

A pond will respond to fertilization by turning green, mostly within 24 hours. Primary productivity reaches a maximum value due to the presence of adequate inorganic nutrients. The gross primary productivity in tropical waters = 17.8 g of carbon fixed per m^2 per day (upper limit) or 47 g of oxygen, *i.e.* 2.6 g of oxygen produced per g of carbon fixed. The amount of solar energy penetrating the pond water dictates the upper limit of autotrophic production. Continued application beyond a certain level will not result in increased primary productivity.

Water Exchange

Flushing out of the pond can also flush fertilizers and nutrients out of the pond. Water retention time in the pond should be at least 3-4 weeks, equivalent to 5 percent per day of water exchange.

Water Chemistry

High calcium concentration (hard water) and elevated pH should be avoided in order to prevent loss of phosphate. Phosphate application should never be made at the same time or within one week of liming.

Natural Soil Fertility

Ponds on fertile pasture soils and with rich alluvial soils with a high organic matter require lower fertilizer application rates.

Previous Pond Management

Newly constructed ponds generally require higher initial fertilizer application rates than ponds with a history of fertilization and accumulated bottom sediments.

Aquatic Weed Infestation

Weeds compete with plankton for available nutrients and sunlight. Weeds also shade the water surface and prevent sunlight penetration, which is essential for phytoplankton growth.

Algal Taxonomic Composition

Nutrient concentrations favoring predominance and growth of specific algal groups differ. For example, diatoms require 20-30:1, and 10-20:1 as N: P, whereas phytoplankton, in general, requires 4:4:1, and 4:1:0 as N:P:K.

Fertilizer Solubility

This is not generally a problem for nitrogen fertilizer, the majority being very soluble. Phosphate fertilizers vary in solubility depending on their particle size and chemical composition.

Fertilizer Application Method and Frequency of Application

This factor will be discussed in the following pages.

2. General Guidelines for Fertilizer Requirement

Fish and shrimp production can be greatly increased by application of fertilizers. In fact, fertilization generally cause large increases in production compared to unfertilized ponds. There has been considerable research on fertilization of fish ponds (Boyd, 1982, 1989), but not much studies on shrimp ponds Based on the practical experience, some general guidelines have been suggested for fertilizer requirement.

1. Waters with high concentrations of nitrogen and phosphorus, *e.g.* polluted water or sewage fed ponds, do not require as much fertilizer as waters with low concentrations of these nutrients.
2. Nitrogen fertilization does not appreciably increase the potential for fish production in fresh water ponds. A small amount of nitrogen may benefit in increasing fish yield in new ponds not previously fertilized.
3. Phosphorus is the most effective nutrient in pond fertilization and increases considerably the fish yield.
4. Nitrogen is more important as a limiting nutrient in brackish water ponds than in fresh water ponds. Nitrogen fertilizers are more suitable in supplying nutrients to diatoms than phosphate fertilizers. Since diatoms are good food organisms for shrimp, a fertilization programme for shrimp ponds should employ more nitrogen and less phosphorus than recommended for fresh water ponds.
5. Potassium is non-essential in pond fertilization.
6. Weekly fertilization is considered to be enough for increase in fish yield. Increase of nitrogen and phosphorus concentrations beyond the maximum by addition of more fertilizers would lead to wastage.
7. Frequency of fertilization to be followed for shrimp ponds is not clearly known. However, fertilization two or more times a week is necessary, if water exchange rates of 5–10 percent

per day are used. If feed is not applied, heavy fertilization must be followed in shrimp ponds, which are infertile.

8. Fertilization becomes less important as feeding rates increase.

8.2 Application of Fertilizers

8.2.1. Types of Fertilizer

Most pond fertilization research has been conducted in freshwater. Either phosphorus-only fertilizers (0-1-0) or fertilizers with a N: P ratio of about 1-3-0 have been found to be best in freshwater ponds (Boyd, 1982). Triple superphosphate or superphosphate may be used for phosphorus-only fertilizers. Diammonium phosphate (18-48-0) has a nutrient ratio of roughly 1-3-0 and is a good source of nitrogen and phosphorus in a solid fertilizer. Alternatively, a combination of nitrogen-only fertilizer, *e.g.* ammonium nitrate, sodium nitrate or urea, and a phosphorus-only fertilizer, *e.g.* triple superphosphate may be used for nitrogen and phosphorus fertilization.

8.2.2. Application Rate

Application rates usually are 2 to 9 kg/ha of P_2O_5 alone, or of both N and P_2O_5. One of the best fertilization programs for freshwater ponds consists of periodic applications of 3.9 to 7.8 kg/ha of P_2O_5 and 1.1 to 2.2 kg/ha of N. Frequency of applications normally varies between 3 and 4-week intervals.

A fertilization program for brackish water ponds should employ more nitrogen and less phosphorus than recommended for freshwater ponds. Fertilization with 30 kg N/ha and 1 kg P/ha or 15 kg N/ha and 1 kg P/ha (2.29 kg P_2O_5/ha), results in good phytoplankton blooms consisting of 50–90 percent diatoms. The ideal frequency of fertilization for shrimp ponds is not known. However, 2 or more times per week of fertilization is needed, if water exchange rates of 5–10 percent per day are used. The fertilization rate of once or twice weekly and limited water exchange is suggested when the ponds are first stocked. If feed is applied, fertilization application may be reduced and possibly discontinued as feeding rates increase. But heavy fertilization must be applied to ponds, if feed is not applied and waters are infertile (Boyd, 1989).

A large number of combinations of nitrogen and phosphorus fertilizers has been cited for shrimp ponds in different countries (Table 8.4).

Calculation of Quantity of Fertilizers for Application

The quantity of fertilizers needed can be calculated using the information from Table 8.1.

Example 8.1

A 1.0 ha pond is to be fertilized with 10-24-0 type fertilizer at a rate of 50 kg/ha using ammonium sulphate and diammonium phosphate. Calculate the quantities of these fertilizers.

Solution

Quantity of N = $50 \times 0.10 \times 1.0 = 5$ kg, and quantity of $P_2O_5 = 50 \times 0.24 \times 1.0 = 12$ kg.

So diammonium phosphate [$(NH_4)_2HPO_4$] needed to supply 12 kg $P_2O_5 = 12/0.48 = 25$ kg.

25 kg $(NH_4)_2HPO_4$ supplies $25 \times 0.18 = 4.5$ kg N.

For the remaining (5–4.5) = 0.5 kg N, ammonium phosphate [$(NH_4)_2SO_4$] to be required = $0.5/0.20 = 2.5$ kg.

Therefore, 2.1 kg of $(NH_4)_2SO_4$ and 25 kg of $(NH_4)_2HPO_4$ can be applied to pond which would be equivalent to 50 kg of 10-24-0 fertilizer.

Example 8.2

The suggested amounts of nitrogen and phosphorus for fertilization are 0.95 and 0.11 ppm respectively. Calculate the quantity of ammonium sulphate and triple superphosphate for a 1-ha pond with 1-m depth of water.

Solution

Volume of water = 10,000 m³. 1 ppm = 1 mg/l or 1 g/m³.

Amount of N to be added = 10,000 × 0.95 g = 9.5 kg

Amount of P to be added = 10,000 × 0.11 g = 1.1 kg.

Fertilizers required = 9.5 kg N/ 0.21 = kg of ammonium sulphate, and

= 1.1 kg P/ 0.46 = kg of triple superphosphate.

Table 8.4: Combinations of Nitrogen and Phosphorus Fertilizers in Use in Shrimp Ponds

Fertilizers (kg/ha)	*N:P*	*Remark*
U (15) + DAP (4)	9:1	In Columbia
U (16) + TSP (8)	4:1	In Ecuador for pond preparation
U (20) + TSP (7)	6:1	In Ecuador in extensive pond
U (5.6) + TSP (2.2)	6:1	In Ecuador
U (100) + TSP (50-100)	2-4:1	In Indonesia for pond preparation
U (25) + TSP (10)	6:1	In Indonesia
U (50) + AP (50)	7:1	In Philippines for extensive pond preparation
U (7) + DAP (65)	1:1	In Philippines for extensive pond preparation
–	20-30:1	Universal, for pond preparation
U (7.9) + TSP (2.6)	7:1	Universal, for pond preparation

U: Urea, DAP: Diammonium phosphate, TSP: Triple superphosphate, AP: Ammonium phosphate

Source: Clifford, 1992.

8.2.3 Method of Fertilizer Application

Usually solid granular fertilizers are broadcast over the pond surfaces in shallow depths of water. If the fertilizers are broadcast over the deeper waters, they will sink to the hypolimnion, and nutrients will not be available to phytoplankton. Phosphates are strongly absorbed by mud and much of it never reaches the water. Fertilizers can be applied in several ways to keep them out of bottom mud.

1. Platform Method

A platform or table may be built of wood, bamboo or galvanized iron sheets, which can rest 30–40 cm underwater (Figure 8.1 a). One platform of about 4 m² area is adequate for 2–4 ha pond area (Boyd, 1982). ASEAN (1978) suggests that the base of the platform should be 15–20 cm below the water surface. It should be located near the pond water inlet or at the end of the pond from which the prevailing wind comes. A single platform is sufficient for ponds up to 7 ha, when plankton is grown.

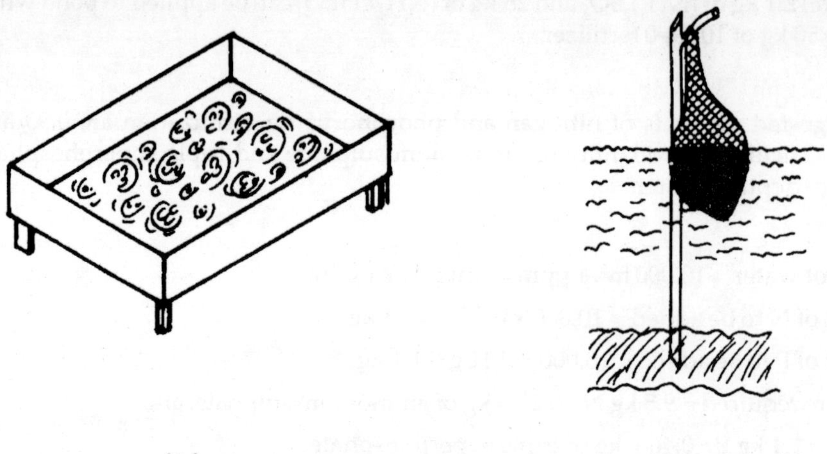

(a) (b)

Figure 8.1: Fertilizer Application Methods: (a) Underwater platform (b) Suspended Nylon/Cloth Bag (After ASEAN, 1978)

According to pond area, the platform top dimensions vary, *e.g.* 0.85 m x 0.85 m for 1-ha pond area to 2.25 m x 2.25 m for 7 ha pond area.

A two-week dose of fertilizer can be placed on top of the platform. Water currents distribute nutrients as they dissolve. Fertilizer can be added as required to maintain the desired phytoplankton abundance. This practice can be followed when the platform is seen through clear water.

Platform method is effective in preventing wastage of solid fertilizers in the bottom mud. The method often reduces fertilizer requirements by 20–40 percent (Boyd, 1982).

2. Nylon or Cloth Bag Method

Nylon or cloth bags, used to transport fertilizers or food materials, may be filled with the required quantity of fertilizer and tied to a post below the water surface (Figure 8.1 b). Dissolved nutrients diffuse through the bag into the water. Large ponds may require more than one bag.

3. Application of Liquid Fertilizers

There are several methods of liquid fertilizer application. The required quantity of fertilizer may be dissolved in buckets and splashed over the entire pond surface. Nutrients get dispersed into the water column faster than platform and bag methods. By this means daily application of fertilizers can give best results. The methods are similar to organic manure application.

There are mechanical means of liquid fertilizer application by using irrigation pump, hand sprayer, power sprayer, and outboard motor. Boyd (1982) evaluated these devices such as a centrifugal pump with 400 l/min capacity, modified to discharge liquid fertilizer; a compression-type hand sprayer, 10-l capacity; a power sprayer operated by a 3.75 kW gasoline engine; and an electric trolling motor, 5 kg thrust, on a boat. All the four methods were found to be effective in appreciably increasing average filterable orthophosphate concentrations in 24-h after application in ponds varying from 0.38 to 5.71 ha and from 1 to 2 m in average depth. The method of application, however, depends on the equipment available.

8.3 Organic Manures

8.3.1. Types of Manures

Organic manures are all plant and animal materials, in decomposed, dried, or even as fresh, which can be used as fertilizers to increase the production of natural live food organisms in fish and shrimp ponds.

The types and sources of organic manures are:

1. Livestock manures, fresh or dried, *e.g.* farm animal dung, with or without urine.
2. Farmyard manures, *e.g.* mixture of animal dung and urine with crop residues, usually straw or sawdust, and occasionally soil.
3. Plant residues, fresh or dried, *e.g.* straw, husks, leaves, vegetable wastes, grass cuttings, tree by-products, seaweed.
4. Compost of partially decomposed mixture of animal and/or vegetable materials.
5. Human waste, *e.g.* night soil, sewage and sewage effluents

Table 8.5 gives the average nutrient composition of animal and plant manures commonly used as organic fertilizers in aquaculture (Tacon, 1988). The nutrient composition of organic manures is highly variable, and each manure source must be considered as being unique, and accordingly analyzed.

8.3.2. Method of Straight Manual Application of Manures

1. Frequency of Application

The effect of animal manure on pond productivity will depend largely on the quantity and frequency of application. There is a maximum amount of manure that a pond can aerobically digest/unit area / unit time. It was found that when soft fresh animal manure such as cow dung was mixed with water and repeatedly spread over the entire pond area, sufficient amounts of carbon compounds were liberated to achieve a high primary productivity. Liquid cow manure contains approximately 30 percent of the total dry matter in colloidal state, which acts as an ideal substrate for bacterial and protozoan growth on the pond bottom and within the water column, and gets rapidly decomposed by bacteria. Further, approximately 90 percent of the coarse organic matter of fresh dung settles to the pond bottom after one to two hours, and the sediment deposit of more than a few mm results in the development of anaerobic conditions (Moav et al., 1977; Schroeder, 1980; Edwards, 1982). Therefore, addition of manure over the maximum level is to be avoided. For Israeli pond conditions, a pond can digest without any adverse effects the maximum amount of 100–200 kg manure (dry weight)/ha/day or 70–140 kg organic matter/ha/day. These amount of manure could be produced from 100–200 pigs weighing 100 kg each/ha/day, 15–30 cows weighing 500 kg each/ha/day, or 2000–4000 poultry each weighing 2 kg/ha/day (Schroeder, 1980; Edwards, 1982).

Based on the above field data, the following guidelines have been suggested:

1. Manures should be applied as frequently as possible, at least daily, on a little and often basis.
2. Manures should be applied during mid-morning when oxygen levels are increasing rapidly through photosynthesis. This would counteract the oxygen demand of the bacteria during decomposition of manure in the pre-dawn hours.

Table. 8.5: Average Elemental Composition of Organic Manures

Manure	*C:N Ratio*	*% by Weight on Moisture-free Basis*		
		N	*P*	*K*
Animal				
Cattle dung	19	1.91	0.56	1.40
Pig dung	13	2.80	1.36	1.18
Poultry manure	9	3.77	1.89	1.76
Duck manure	10	2.15	1.13	1.15
Cattle urine	–	9.74	0.05	7.78
Bone meal	8	3.36	10.81	–
Fish manure	4.5	7.5	2.82	0.8
Plant				
Rice straw	105	0.58	0.1	1.38
Groundnut straw	19	0.59	–	–
Grass	20	0.41	0.03	0.26
Tree leaves (general)	60	1.0	0.3	0.57
Water hyacinth	18	2.04	0.37	3.4
Azolla spp.	–	3.68	0.2	0.15
Marine seaweeds, dry	–	0.66	0.32	1.2
Neem seed cake	4.5	5.21	0.46	1.19
Mustard oil cake	–	5.93	0.53	0.65
Soybean oil cake	–	6.95	2.88	1.02
Mixed				
Farmyard manure	–	0.8	0.21	0.68
Rural compost	–	1.1	0.29	1.37
Sewage sludge	9	4.0	1.4	0.3

Source: Adapted from Tacon (1988).

3. Since the manure requirement of a pond depends upon the live food availability for the fish/shrimp biomass, the manuring rate should be increased with increasing biomass of the standing crop.
4. The manuring rates are always pond and culture specific, and so the pond fertilization program of a particular system should be developed. For example, use of manures in intensive shrimp ponds is not necessary and so not recommended because of excessive organic loading and increased BOD/COD.

2. Methods

Three basic methods, usually employed for the distribution of animal manure to ponds, are (Tacon, 1988):

1. Distribution of the diluted manure by hand is carried out either from the shore or from a small boat. This is usually followed for small ponds.
2. Distribution from a boat is carried out with the help of a basket, made of parallel iron rods, approx. 2–2.5 cm apart. The basket attached to the side of the boat is suspended 10–20 cm below the water line. The manure gets dispersed as the boat moves and forces water into the basket.
3. Distribution from a boat is carried out with the help of a pump. The manure is shoveled into a hopper, diluted with the pumped water and sprayed out into the pond through a flexible hose.

3. Manure Application through Integrated Farming of Livestock

Integrated livestock and aquaculture production system necessitates that the livestock living area be located adjacent or above the fish pond. The wastes resulting from livestock feeding, *i.e.* uneaten feed residues can be directly consumed by the cultured stock. Several animals such as pig, duck, chicken and goose, have been successfully employed in these integrated systems. Tacon (1988) has cited many references on the integrated livestock-aquaculture farming with these animals. Figure 8.2 shows a diagrammatic representation of a typical integrated farming.

Table 8.6 shows the amount of manure normally produced by the different livestock animals per year and the stocking density of animals usually kept per ha of pond water. It must be admitted that the composition of the animal manure depends on total live weight and age of the animal; feed types; climate; and management, and that stocking density of fish depends on the farming practice, culture species, and the water quality tolerance of the species.

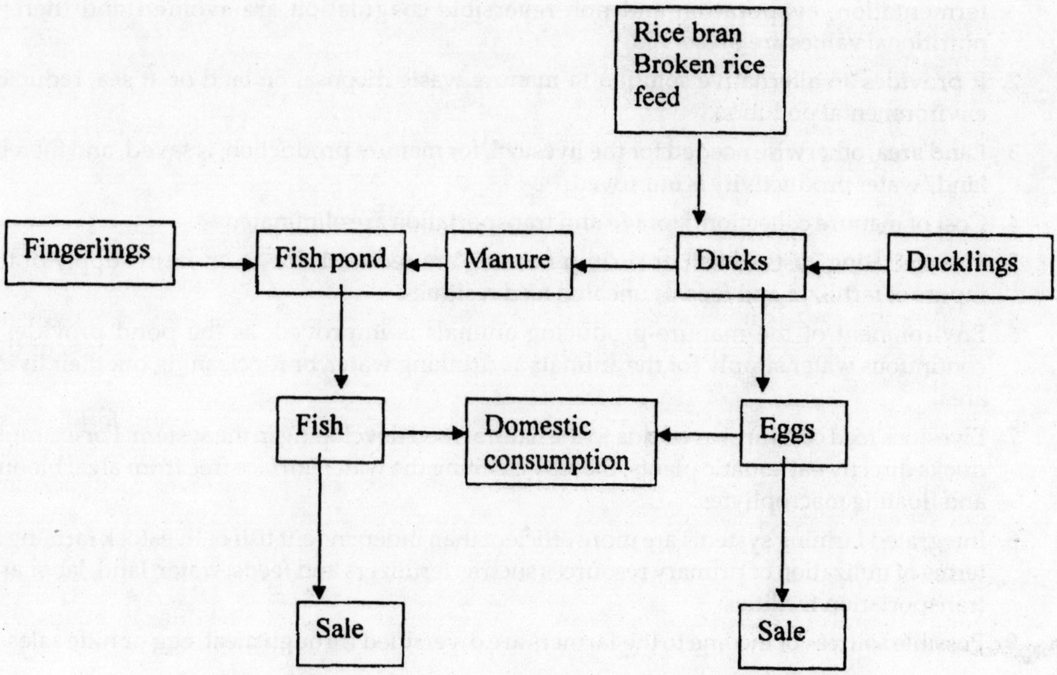

Figure 8.2: Diagrammatic Representation of a Duck-fish Integrated Farming System in Thailand (After Tacon, 1988)

Table 8.6: Livestock Manure Production Rate and Recommended Fish Stocking Density in Integrated Aquaculture

Animal	*Manure Production (kg dry matter/ animal/yr)*	*Stocking Density (No./ha of pond water surface)*	*Production of Fish (kg/ha)*
Pig	178	15–300	–
		40–60	1950 over 90 days of polyculture with mainly carp and tilapia
		103	3549 over 180 days of polyculture
Ducks	20.6, 24–32	150–13,125	–
		750–1250	1690 over 90 days of polyculture with mainly tilapia and carp
Chicken	6–11, or 25 (as fresh)	1000–10,000	–
			1758 over 90 days of polyculture with mainly carp and tilapia
Cattle	784 (for cow)	25–50	
	1260		
Sheep/goat	290	200–400	
Goose			

Source: Adapted from Tacon, 1988.

There are several advantages of integrated production of livestock and fish as cited by Tacon (1988):

1. Losses of nitrogen and energy of the manure and feed remnants due to natural wastage, fermentation, evaporation and non-reversible coagulation are avoided and thereby nutritional values are preserved.
2. It provides an alternative solution to manure waste disposal on land or at sea, reducing environmental pollution.
3. Land area, otherwise needed for the livestock for manure production, is saved, and thereby land/water productivity is improved.
4. Cost of manure collection, storage and transportation are eliminated.
5. The operating costs of fish or shrimp farming are reduced due to on-farm supply of the inputs of fertilizer and feed as uneaten feed residues.
6. Environment of the manure-producing animals is improved, as the pond provides a continuous water supply for the animals as drinking water, or for cleaning out their living area.
7. Livestock feed costs are saved due to the natural food developing in the system. For example, ducks directly eat aquatic plants, thereby cleaning the water surface free from algal blooms and floating macrophytes.
8. Integrated farming systems are more efficient than independent fish or livestock farming in terms of utilization of primary resources such as fertilizers and feeds, water, land, labor and transportation facilities.
9. Possible sources of income to the farmers are diversified through meat, egg or hide sales.

9

Liming of Ponds

Liming is one of the major treatments of pond for its maintenance. Liming has been adopted from agricultural practices like fertilization. However, liming is not a type of fertilization. It can be viewed as a treatment method followed for improvement of acid-base relationship and as a remedial measure in some ponds to permit normal response of fish populations to fertilization and other management methods. In freshwater aquaculture, where alkalinity and hardness is too low, and bottom mud is acidic, lime is added to increase their concentrations, and so the mud pH. In brackish water, alkalinity and hardness are usually high. So control of these parameters is seldom followed in management strategy. However, ponds with acidic soils require lime treatment.

According to Boyd (1982), the basic types of ponds that respond favorably to liming are: (a) dystrophic ponds, where water is highly stained with humic substances and mud rich in slowly decaying organic matter. Such waters have pH 5–6, alkalinity 1–5 mg/l, and acidity zero; (b) ponds with moderately acidic soils on watershed, where pH is 5.5–7, alkalinity 3–15 mg/l, and acidity zero; and (c) dystrophic ponds with acid sulphate soils on watersheds, with pH 2–4.5, alkalinity zero and acidity 10–250 mg/l. Low alkalinity ponds are more common than zero alkalinity ponds with acidity.

9.1 Lime Requirement and Liming Rate

Lime requirement is decided on the total hardness and alkalinity of the pond water. There is generally a positive response to lime application by ponds with waters in the hardness range of 10-20 mg/l. When the hardness goes below 10 mg/l, the response is more striking. It has been observed that ponds with hardness less than 20 mg/l should be limed for maximum response to fertilization. If the hardness is nearly 20 mg/l, lime application may not be necessary. However, total alkalinity is considered a much better indicator of lime requirement than hardness. There are waters where hardness may be quite high and alkalinity quite low. So the requirement of lime on the basis of hardness may be misleading. In fact, analysis of both alkalinity and hardness should be encouraged.

It appears that the amount of liming materials could be calculated as the amount of calcium carbonate necessary to raise the total hardness and total alkalinity of a pond to a specified concentration. For example, alkalinity of a 1-ha pond of water depth of 1-m can be increased from 10 mg/l to 20 mg/l by adding the total quantity of 100 kg calcium carbonate. In fact, this amount of calcium carbonate would react with the acidic mud and be lost from the water. A much larger amount of liming material must be required. It was found that most application rates for liming of fish ponds were 1000 kg or more per ha (Boyd, 1982).

9.1.1 Calculation of Liming Rate

1. Technique Employed on Agricultural Crop

It has been recognized that the lime requirements of ponds differ in about the same manner as agricultural soils. On the basis of soil pH and soil texture a table of values for estimation of liming rates for ponds can be used for heavy loams or clays, sandy loam and sand (Boyd, 1982).

2. Technique Based on Exchange Acidity of Soil

The procedure used by soil testing laboratories was modified by Boyd (1974, 1982) for fish ponds. This is a direct calculation procedure based on cation exchange capacity and exchange acidity of soils.

Cation Exchange Capacity (CEC)

Mud and soils contain colloidal particles of organic matters and clay minerals. These particles are negatively charged and attract cations present in the interstitial water. The cations on mud colloids and in interstitial water are called exchangeable cations, and the sites of adsorption of cations on colloids are referred to as exchange sites.

The quantity of cations that is adsorbed on exchange sites of soils or mud is defined as the cation exchange capacity (CEC), expressed as milliequivalent of cations per 100 g (meq/100 g) of dry soil or mud. The CEC values of agricultural soils vary with the type of clay minerals, clay content, and organic matter content, and range between 1 and 100 meq/100 g. Usually mud with high proportion of sand has low CEC, and mud with high proportion of organic matter, clay or both has higher values of CEC. Pond mud in Alabama, USA, has values of 0.5–26 meq/100 g (Boyd, 1982).

Exchange Acidity

Exchange acidity (or potential acidity) is defined as capacity to yield hydrogen ions. Acidity of a mud sample as measured by a pH meter indicates the hydrogen ion activity of the water of mud solution, but not the capacity of the sample to yield hydrogen ions. A mud has exchange acidity resulting from acidic cations on exchange sites. Aluminium, ferric and hydrogen ions are acidic cations, and calcium, magnesium, potassium, sodium and ammonium are basic cations. The fraction of the total CEC filled by acidic cations is termed 'base unsaturation". The pH of the water surrounding mud particles, *i.e.* mud pH, is closely related to the degree of base unsaturation of the mud particles. This relationship is shown for mud in ponds of Alabama, USA (Figure 9.1). The magnitude of exchange acidity of a mud sample depends more on the value of CEC than on the degree of unsaturation. This is because of the base unsaturation which is contributed by the acidic cations only, not by the total cations on the exchange sites.

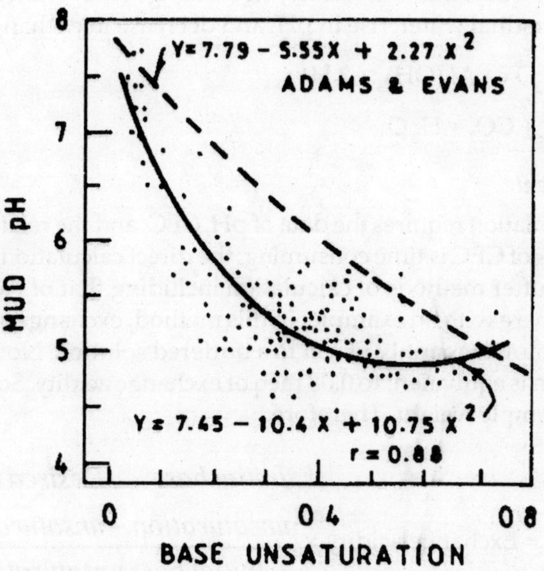

Figure 9.1: Relationship Between Base Unsaturation and pH of Mud in Alabama Ponds Broken lines for agricultural soils used by Adams and Evans (From Boyd,1982)

Example 9.1

Analysis of a mud sample shows the following ions. Calculate the CEC and base unsaturation.

Solution.

Ions	*mg/100 g*	*mg/meq*	*meq/100 g*
Ca^{2+}	80	20	4
Mg^{2+}	36.6	12.2	2
K^{+}	39.1	39.1	1
Na^{+}	11.5	23	0.5
NH_4^{+}	1.8	18	0.1
Al^{3+}	72	9	8

CEC = 16.6 meq/100 g, and base unsaturation = 8/16.6 = 0.48

Neutralization of Exchange Acidity by Lime

If a mixture of 1 part water and 1 part acidic mud is neutralized with enough bases, the acid pH will rise to 7, but it will soon become acidic again. So further addition of base is necessary to make the exchange acidity permanently neutral. For example, when calcium carbonate ($CaCO_3$) is added to aluminium mud, hydrogen ions released from the acidic portion of the mud are neutralized. As a result more aluminium ions are released in the mud and aluminium hydroxide ($Al(OH)_3$) is precipitated and calcium ions (basic) replace aluminium (acidic) on exchange sites. The net effect is neutralization of hydrogen ions in the interstitial water, rise in pH, and decrease in exchange acidity.

$$\text{Al mud} \leftrightarrow Al^{3+} + 3\,H_2O \leftrightarrow Al\,(OH)_3 + 3\,H^+ \tag{9.1}$$

$$CaCO_3 + 2\,H^+ \rightarrow Ca^{2+} + CO_2 + H_2O \tag{9.2}$$

Calculation of Lime Requirement

A direct method of calculation requires the data of pH, CEC, and the relationship shown in Figure 9.1. However, as the analysis of CEC is time consuming; the direct calculation method is seldom used. Boyd (1982) cited several buffer methods of calculation including that of Adams and Evans (1962), which has been elaborated here with an example. In this method, exchange acidity is estimated from change of pH by adding a known quantity of soil to a buffered solution. Now each pH change of 0.1 units in 20 ml buffer solution is equivalent to 0.08 meq of exchange acidity. So exchange acidity = 0.08 x units of pH change/soil sample weight. Therefore,

$$\text{Acidity to neutralize} = \text{Exchange acidity} \times \frac{\begin{matrix}\textit{Initial base} & \textit{Desired base}\\ \textit{unsaturation} - & \textit{unsaturation}\end{matrix}}{\textit{Initial base unsaturation}} \tag{9.3}$$

Example 9.2

A soil sample is to be limed from an initial pH of 5.0 to pH of 6.0. 10 g sample of the pond soil changes pH of 20 ml buffer by 0.5 pH units. Calculate the quantity of lime required for the pond soil, using the plot for the Alabama pond soil (Data adapted from Boyd, 1982).

Solution

Acidity to neutralize = 0.08 x 5 x (0.45–0.10)/0.45 = 0.311 meq/10 g = 1555 mg/kg $CaCO_3$

Assuming that lime reacts in the upper 15 cm layer of soils, which is usually the case with agricultural land, and that 1 ha of 15 cm deep soil weighs 2.2 x 10^6 kg

Lime requirement would be = 1555 mg/kg $CaCO_3 \times 2.2 \times 10^6$ = 3421 kg/ha.

Further, assuming a liming factor of 1.5 for slow and incomplete reaction of liming materials in soils, the adjusted liming rate = 3412 × 1.5 = 5132 kg /ha

Boyd (1982) modified the above method of calculation for application in ponds for lime requirement. The modification was based on the fact that lime treatment of ponds is to raise the total hardness and total alkalinity of pond water above 20 mg/l, and not to raise the mud pH to a desired level. The following suggestions were made by Boyd (1982) for estimation of buffer lime requirement:

1. Extensive studies of Alabama pond mud showed that total hardness was well correlated with mud pH and base unsaturation. At a mud sample pH above 5.9, and base unsaturation at 2.0 or less, the hardness of pond water exceeds 20 mg/l, indicating that the pond needs no lime treatment. This hardness (and alkalinity) value is considered suitable for good fish production.
2. The relationship between mud pH and base unsaturation was quite different from the one used by Adams and Evans for agricultural soils, as shown in Figure 9.1. This is because

pond mud contains more organic matter and has higher CEC values than agricultural soils used by Adams and Evans.

3. The average weight of 1.5 cm layer of mud from several ponds has been found to be 1.5×10^6 kg/ha, which gives the lime requirement about a third less than those used for agricultural soils.

   On the basis of the above modification a table for lime requirement values (Table 9.1) has been prepared by Boyd (1974).

4. The lime requirement procedure outlined above was developed for use in pond in the Southeastern U. S. Lime requirement in these ponds with initial alkalinities between 5 and 12 mg/l ranged from 1000 to 5000 kg/ha. Liming rates were in practice 2000–6000 kg/ha with pH of limed ponds ranging between 6 and 7. It was suggested that the procedure should be applicable to ponds in most warm climates. However, the relationship between base unsaturation and mud pH of a number of representative ponds in the region should be established. Based on this and the values of base unsaturation and mud pH corresponding to a total hardness of 20 mg/l, a table similar to Table 9.1 should be prepared for lime requirement estimation (Boyd, 1982).

Table 9.1: Lime Requirement of Ponds to Increase Total Hardness and Total Alkalinity of Pond Water to 20 mg/l or more (neutralizing value of $CaCO_3$ = 100 per cent)

pH of Pond Mud	*Lime Requirement as $CaCO_3$ (kg/ha) on the Basis of Mud pH in Buffered Solution*									
	7.9	*7.8*	*7.7*	*7.6*	*7.5*	*7.4*	*7.3*	*7.2*	*7.1*	*7.0*
5.7	91	182	272	363	454	544	635	726	817	908
5.6	126	252	378	504	630	756	882	1008	1134	1260
5.5	202	404	604	806	1008	1210	1411	1612	1814	2016
5.4	290	580	869	1160	1449	1738	2029	2318	2608	2898
5.3	340	680	1021	1360	1701	2042	2381	2722	3062	3402
5.2	391	782	1172	1562	1948	2344	2734	3124	3515	3906
5.1	441	882	1323	1765	2205	2646	3087	3528	3969	4410
5.0	504	1008	1512	2016	2520	3024	3528	4032	4536	5040
4.9	656	1310	1966	2620	3276	3932	4586	5242	5980	6552
4.8	672	1344	2016	2688	3360	4032	4704	5390	6048	6720
4.7	706	1412	2116	2822	3528	4234	4940	5644	6350	7056

Source: Adapted from Boyd, 1974.

9.2 Liming Methods

9.2.1 Liming of ponds

1. Liming Materials

The liming materials used in ponds are the same as applied commonly in agricultural soils. These are: (*i*) Agricultural limestone, prepared by crushing either calcite ($CaCO_3$), or dolomite ($CaMg(CO_3)_2$), (*ii*) Calcium hydroxide, $Ca(OH)_2$, known as hydrated lime, slaked lime, and (*iii*) Calcium oxide, CaO, known as quick lime, or burnt lime. All these liming materials produce similar effects by increasing the total alkalinity, when applied in equivalent amounts (Table 9.2).

However, their selection depends upon the relative advantages and disadvantages:

1. Calcium hydroxide and calcium oxide react faster than limestone, but are unpleasant to handle due to their highly caustic nature. They may raise pH to toxic levels for fish.
2. Limestone is usually the safest, most effective and cheapest material for liming.
3. Calcium hydroxide and calcium oxide are available in fine powders, and the limestone is to be ground to fine powders. Particles less than 0.25 mm react fastest, and are rated 100 percent efficient.

Table 9.2: Effects of Equivalent Application Rates of Liming Materials

Liming Materials	*Total Alkalinity (mg/l)*	
	Before Liming	*After Liming*
None	11	13
Hydrated lime	9	55
Dolomite limestone	11	42
Calcite limestone	12	35

Source: Adapted from Boyd, 1974.

2. Methods of Application

The methods vary with the pond, new and old. In the new ponds, liming should be followed before filling with water. The calculated amount of liming materials is spread, and incorporated into the pond bottom by discing. The older ponds, after drain harvest or renovation, are dried and limed in the same manner as new ponds. In the tropics and sub-tropics, time of application should be a few weeks before fertilization, and in the temperate regions, after fertilization.

In full ponds liming materials are spread from a boat with the help of a shovel, or from a raft and pump system by mixing the lime with water and discharging into the pond water. In full ponds liming becomes more difficult to spread over the entire surface. In small ponds, the small amount of liming material can be easily spread from a boat.

9.2.2 Liming of Acid-Sulphate Soils

In areas such as saline and brackish water tidal swamps and marshes, acid-sulphate soils occur. Initially heavy load of sediment emptied by rivers and runoff is deposited near shores, becomes covered with vegetation, and is gradually developed into a swamp forest. In course of time tree roots trap organic and inorganic debris, which decompose in anaerobic conditions by sulphur reducing bacteria. Sulphides accumulate in pores as hydrogen sulphide, or combine with iron forming iron sulphide precipitates. Iron sulphide undergoes further reaction to form iron pyrite (FeS_2). If ponds exist in such areas and the pyrite sediment is exposed to air and water during draining of ponds, oxidation occurs and acid (H_2SO_4) is produced. The following reactions may occur:

$$FeS_2 + H_2O + 3\tfrac{1}{2}O_2 \rightarrow FeSO_4 + H_2SO_4 \quad (9.4)$$

$$FeSO_4 + \tfrac{1}{2}O_2 + H_2SO_4 \rightarrow Fe_2(SO_4)_3 + H_2O \quad (9.5)$$

$$Fe_2(SO_4)_3 + 6\,H_2O \rightarrow 2\,Fe(OH)_3 + 3\,H_2SO_4 \tag{9.6}$$

$$FeS_2 + 7\,Fe_2(SO_4)_3 + 8\,H_2O \rightarrow 15\,FeSO_4 + 8\,H_2O \tag{9.7}$$

The production of ferric sulphate from ferrous sulphate is highly accelerated by the bacteria *Thiobacillus*, and oxidation of pyrite under acidic condition is very rapid (Sorensen *et al.*, 1980).

Ferric hydroxide crystallizes as a reddish brown material in the sediment. The sediment containing pyrite is called an acid sulphate soil or "cat's clay". As long as the sediments of pyrite are submerged and remain anaerobic, they are not oxidized and change little. Once oxidation occurs, pH becomes acidic and fall below 4.0. The pH values of dry acid-sulphate soils may decrease in the range of 2 to 3. As exchange acidity will not account for pH values less than 4–4.5, the procedure of determining lime requirement of acid-sulphate soils involves determining the buffer lime requirement, and a technique for estimating the amount of calcium carbonate that would neutralize all of the acid produced from the oxidation of pyrite. A suitable procedure for the potential acidity has been described by Boyd (1982).

The sulphide is oxidized to sulphuric acid by hydrogen peroxide treatment, and the amount of the acid is measured by titration with a standard base. The sample of soil should be passed through 20 mesh sieve, which is called the reactive fraction.

$$\text{Potential acidity (meq/100 g)} = \text{(ml of NaOH) N } \}\ 100\ \text{g/weight of sample in g}$$
$$= 500\ \text{mg/kg of } CaCO_3$$

$$\text{Liming rate} = \text{Potential acidity (meq/100 g)} \times 500 \times 1.5\ \text{(liming factor)} \times \text{reactive fraction}$$

The liming rate = buffer lime requirement + lime requirement due to sulphate. However, this procedure is not meaningful, as the pyrite in soil is not destroyed by lime, but the acid only is neutralized. The residual lime continues to react with the acid produced by subsequent oxidation of pyrite. Therefore the buffer lime requirement may be used to estimate the liming rates, and the potential acidity should not be included to avoid overliming.

10

Aeration

Maintenance of dissolved oxygen (DO) concentration in pond water is probably the most important management practice for regulating production of fish, shrimp and other species in semi-intensive and intensive aquaculture operations. Artificial aeration is currently the most effective mechanical means of increasing the DO concentration in water. Aerators are used to increase the rate at which oxygen enters water. There are several designs of mechanical aerators available, and various aeration systems are followed in aquaculture practice.

Over the past years, interest in aeration practice has increased tremendously, and many designs of aeration equipment are manufactured and used in ponds. However, farmers, manufacturers and farming enterprises have poor understanding of aeration fundamentals. Therefore this important tool of water quality management is not in proper use and gives poor performance.

10.1 Aeration Fundamentals

10.1.1 Theory of Oxygen Transfer

The oxygen is transferred from air, which is in contact with water. The air contains about 21 percent oxygen by volume, which is much larger than that dissolved in water in saturated conditions. Water at 0°C and with zero salinity will contain oxygen about 15 mg/l, *i.e.* 0.0015 percent by weight. This concentration gradient of oxygen between the air and the water is the driving force for oxygen transfer by diffusion. The driving force can also be expressed in terms of oxygen partial pressure or oxygen tension, as has been explained in the Chapter 5. The oxygen will enter the water until the oxygen tension in the water becomes equal to the partial pressure of oxygen in the air. The concentration of DO at equilibrium condition is the saturation concentration. Oxygen under supersaturated condition in water is lost to the atmosphere by diffusion.

The oxygen transfer takes place at the air-water interface through three physical steps or resistance paths (Figure 10.1), which can be depicted as transfer from a gas bubble into the liquid or through layers of gas and liquid:

1. Transfer of oxygen from the bulk gas to the gas-liquid interface through the gas film,
2. Movement through the gas-liquid interface, and
3. Diffusion through the liquid film into the bulk liquid.

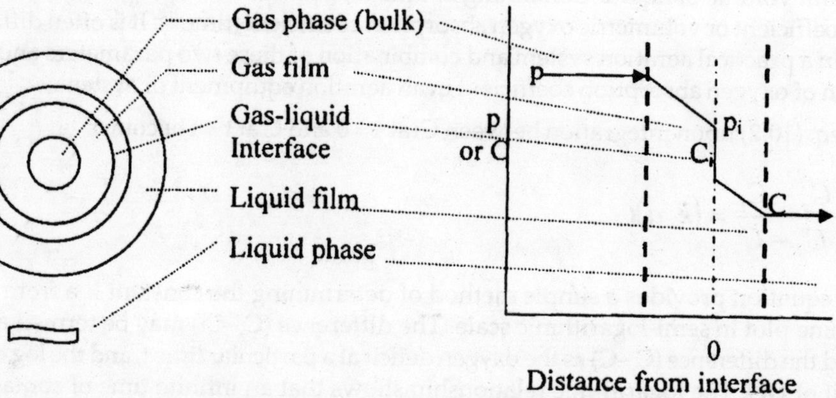

Figure 10.1: Physical Steps and Resistance Paths in Oxygen Transfer Process

In the first step gas molecules move under turbulent flow condition in bulk phase and under laminar flow condition in the gas film. It takes place by a combination of molecular diffusion and convective movement. The process is relatively rapid. Movement of oxygen through the gas-liquid interface, which is a very thin film, takes place strictly by molecular diffusion, and so it is a slow process. The third step occurs largely by convection in the bulk phase, and is also a slow process, because oxygen is sparingly soluble in water.

Under normal conditions, the predominant resistance lies in one of the two films–gas and liquid. According to "two film" or "two resistance" theory, the predominant resistance lies within the liquid film in case of sparingly soluble gases such as oxygen, nitrogen and other gases (Treybal, 1981). The third step is, therefore, the rate limiting.

According to mass transfer theory, the rate of oxygen transfer is a function of oxygen transfer coefficient and the DO concentration gradient between the equilibrium or saturation concentration (C_s) and concentration in the bulk liquid at a particular time (C). In differential form

$$\frac{dC}{dt} = k_L A (C_s - C) \tag{10.1}$$

where all C's are concentration terms, dC/dt = time rate of change of concentration, A = area of gas-liquid interface, t = time and k_L = liquid phase transfer coefficient, which is the overall transfer coefficient in case of liquid film resistance controlling process. A positive concentration gradient of the gas transfers gas into the liquid phase, and conversely, a negative gradient transfers gas into the gas phase. The transfer can be increased by increasing the value of k_L, A, or C_s.

Converting the transfer coefficient to volumetric coefficient, eq. (10.1) becomes

$$\frac{dC}{dt} = k_L\left(\frac{A}{V}\right)(C_s - C) = k_L a(C_s - C) \tag{10.2}$$

where A/V = a = ratio of the surface area to volume of the liquid being aerated and is the interfacial area per unit volume of liquid. Combining k_L and 'a', a parameter 'k_La' is defined as overall oxygen transfer coefficient or volumetric oxygen absorption coefficient (time^{-1}). It is often difficult to measure k_L and A in a practical aeration system and combination of these two parameters and volume allows evaluation of oxygen absorption coefficient in an aeration equipment or system.

The eq. (10.2) upon integration between Ci at t = 0 and C at t = t becomes

$$\ln\frac{C_s - C_i}{C_s - C} = (k_L a)t \tag{10.3}$$

This equation provides a simple method of determining the constant k_La from the slope of the straight–line plot in semi-logarithmic scale. The difference (C_s–Ci) may be termed as initial oxygen deficit and the difference (C_s–C) as the oxygen deficit at a particular time t, and the log of the ratio as the log deficit of DO. The logarithmic relationship shows that an infinite time of contact is required to reach saturation concentration, and therefore points towards a practical and economical condition of not increasing the saturation beyond 95 per cent by aeration.

1. Factors Affecting Volumetric Oxygen Transfer Coefficient (k_La)

The overall oxygen transfer coefficient depends on conditions prevailing in a specific gas-liquid contact system. The important factors are temperature, characteristics of the liquid, and turbulence.

Temperature affects viscosity, which, in turn, influences gas diffusivity, thickness of the liquid film and the area of gas transfer. k_La has been found to increase by 2.4 percent over each °C rise in temperature in clean water according to the following relation:

$$(k_L a)_T = (k_L a)_{20} 1.024^{T-20} \tag{10.4}$$

where, $(k_La)_T$ = overall gas transfer coefficient at liquid temperature T, $(k_La)_{20}$ = overall gas transfer coefficient at 20°C. In case of municipal sewage treatment the value of 1.028 rather than 1.024 was used. The values in seawater appear to be larger than 1.024, especially for temperatures greater than 20° C (APHA, 1980).

Natural waters contain substances that affect the surface tension and concentration of oxygen at saturation. Most wastewater lowers the saturation concentrations slightly. The factor α is used to represent the ratio of k_La in pond water to k_La in pure water under a common set of conditions:

$$\alpha = \frac{(k_L a)_{pondwater}}{(k_L a)_{pure\ water}} \tag{10.5}$$

Similarly, a β factor is defined as:

$$\beta = \frac{(C_s)_{pond\ water}}{(C_s)_{pure\ water}} \tag{10.6}$$

which is used to correct k_La values for aeration tests in tap water to pond water conditions.

The value of α is influenced by presence of surface-active agents such as detergents found in domestic sewage, and natural surfactants such as proteins present in pond water. These materials form a coating at the interface and reduce the oxygen transfer under less turbulent conditions. The depression of α appears to depend on the leaching of soluble compounds of feed or from compounds produced by algae. Salinity has a major effect on α. The transfer of both oxygen and air increases in seawater, because of a decrease in bubble size. The magnitude of this effect depends on the particular type of aerator used. The β factor is influenced primarily by dissolved solids and to some extent by dissolved organics and suspended solids. For pond water $\alpha = 0.8$ and $\beta = 0.98$. The values vary among ponds, such as between 0.66 and 1.07 for α in catfish ponds with average as 0.94 (Shelton and Boyd, 1983). β Values of wastewater typically range from 0.95 to 1.00 (Metcalf and Eddy, 1979). β values for aquaculture conditions are not available but it is recommended to use $\beta = 0.95$ for ponds and recycle systems and 1.00 for flow-through systems, when it is not known experimentally (Colt and Orwicz, 1991).

Turbulence increases the oxygen transfer rate by minimizing the surface film thickness, which maximizes k_L, and by renewing the fluid particles exposed to the atmosphere. The latter action reduces the transport of oxygen by diffusion (a slow process), maximizes convection, and increases the area of transfer. By breaking up the water into small particles, the effective surface area is increased. Mixing of water by bringing from lower layers to the surface for direct aeration also increases the surface area.

2. Evaluation of k_La by Aeration Experiment

The experiments were originally developed to determine oxygen transfer and water mixing rates used for wastewater treatment (APHA, 1980; Boyd and Watten, 1989). These tests have been adapted for evaluation of aeration equipment or aerators in aquaculture. Two types of testing procedures can be used: unsteady-state and steady-state procedures. In the unsteady state test, the concentration of oxygen in water changes with time during the test. In comparison, in steady state testing both the input and output oxygen concentrations are measured (Colt and Orwicz, 1991).

Unsteady-state Testing

A convenient basin or tank, rectangular in shape and 1 to 1.5-m in depth, is used for the test under standard conditions. In a typical test, water is first deoxygenated by applying a solution of sodium sulphite (Na_2SO_3) and cobalt chloride ($CoCl_3$). Theoretically, 7.9 mg/l of sodium sulphite is required to remove one mg/l of DO. In practice, 1.5–2 times the theoretical quantity is added in order to compensate some oxidation of sodium sulphite. The requirement of cobalt chloride, which catalyzes the reaction of chemical oxidation, is 0.05–0.1 mg/l (as Co).

$$Na_2SO_3 + 1/2\ O_2 \rightarrow Na_2SO_4 \tag{10.7}$$

After mixing the solution thoroughly with water in the tank by running the aerator for at least 30 minutes, the aerator is run and DO concentrations of the aerated water are measured at timed intervals while DO increases from 0 percent saturation to at about 80 percent saturation.

If the tank is larger than 50 m^3 or if the aeration does not mix the water well, the DO measurements should be made at two or more locations in the tank and the result averaged. A minimum of four sampling locations is recommended for basins containing more than 200 m^3 of water (APHA, 1980). Some test basins are as large as 3000 to 6000 m^3 and can be used to test aerators as large as 50 to 100 kW.

Example 10.1

The following data were obtained for DO concentration of water after aeration for different times. Calculate the k_La value of the aerator used. Ci = 0.1 mg/l and C_s = 8.2 mg/l.

Time, min	1	2	3	4	5	6	7	8	9	10
C, mg/l	5.4	5.4	5.6	5.6	6.0	6.1	6.2	6.4	6.6	6.7

Solution

Following eq. (10.3) the values of ln $[(C_s-C_i)/(C_s-C)]$ have been calculated and tabulated in below. $C_s-C_i = 8.1$

Time (min)	*C (mg/l)*	*C_s–C (mg/l)*	*ln Deficit*
1	5.4	2.8	1.0622
2	5.4	2.8	1.0622
3	5.6	2.6	1.1364
4	5.6	2.6	1.1364
5	6.0	2.2	1.3034
6	6.1	2.1	1.3499
7	6.2	2.0	1.3987
8	6.4	1.8	1.5041
9	6.6	1.6	1.6219
10	6.7	1.5	1.6864

The plot generated by computer graphics is shown below in Figure 10.1.1 and from the linear regression equation, $k_La = 4.4\ h^{-1}$.

3. Measurement of DO

Traditional Method

Winkler procedure involves treatment of a water sample with manganese sulphate ($MnSO_4$), potassium iodide (KI) and sodium hydroxide (NaOH), and finally with sulphuric acid (H_2SO_4) to produce iodine (I_2). The reactions are:

$$Mn^{2+} + 2\,OH^- + 1/2\,O_2 \rightarrow MnO_2 + H_2O \qquad (10.8)$$

Manganese dioxide (MnO_2) is a brown precipitate, which is dissolved in sulphuric acid and under acidic condition potassium iodide is oxidized to iodine:

$$MnO_2 + 2\,I^- + 4H^+ \rightarrow Mn^{2+} + I_2 + 2\,H_2O \qquad (10.9)$$

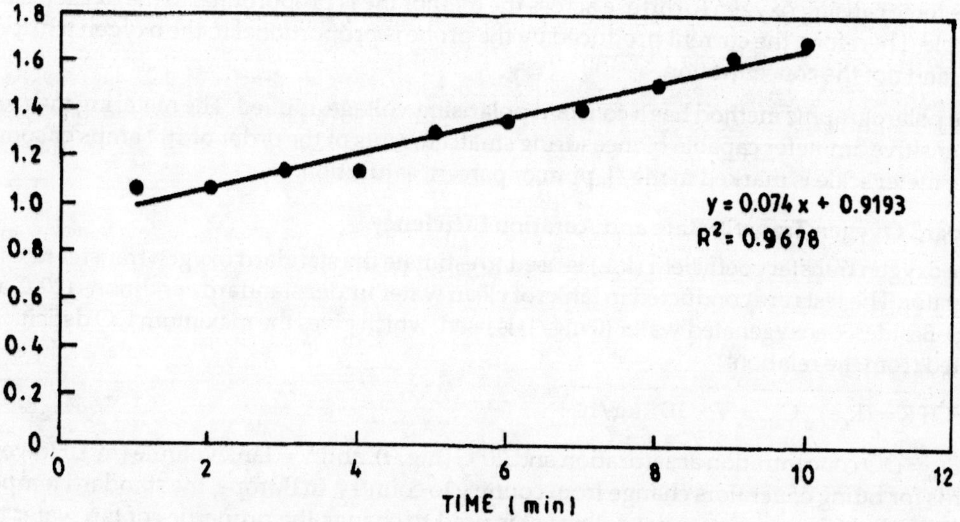

Figure 10.1.1: Linear Regression Plot of Example 10.1

The quantity of iodine released is proportional to the amount of molecular oxygen (DO) originally present. The amount of iodine is titrated with standard sodium thiosulphate solution using starch as the indicator. The blue color of the starch solution turns colorless at the end point.

$$\text{DO (mg/l)} = \text{Volume of } Na_2S_2O_3 \text{ (ml)} \times \text{Normality of } Na_2S_2O_3 \text{ (meq/ml)} \times 8\text{x } 1000/\text{sample volume (ml)}.$$

The standard Winkler procedure requires a laboratory for preparing reagents and various items of glassware for conducting the analysis. Further, the method is tedious and time consuming if several analyses are to be made. However, the method affords high precision and accuracy, and enables samples to be stored.

Polarographic Measurement

The oxygen meter consists of an electrode that produces a current proportional to the oxygen tension in the water, and instrumentation to convert the flow of current to a meter reading that indicates the DO concentration in mg/l. A typical DO electrode, often called a probe, consists of a gold cathode and a silver-silver oxide reference electrode. The electrodes are bathed in a 4-M potassium chloride (KCl) and separated by a membrane usually made of Teflon. The membrane is permeable to oxygen and the rate at which oxygen crosses the membrane is related to the oxygen tension in the sample.

When an electrical voltage (usually around 0.7 volts) is applied to the probe, oxygen diffusing across the membrane reacts with the gold cathode and is reduced to form hydroxide:

$$4\,e^- + O_2 + 2\,H_2O \Leftrightarrow 4\,OH^- \qquad (10.10)$$

Oxygen will take electrons if the potential difference between the surface and the liquid is more than 0.3 volts. A current then flows to the silver anode and the hydroxide reacts with silver to form silver oxide (Ag_2O):

$$2\,Ag^{\circ} + 2\,OH^- \Leftrightarrow Ag_2O + H_2O + e^- \qquad (10.11)$$

The force causing oxygen to diffuse across the membrane is proportional to the oxygen tension in the sample. Therefore, the current produced by the probe is proportional to the oxygen tension in the sample, and not the concentration.

The polarographic method has a constant polarising voltage applied. The metering instrument is a very sensitive ammeter capable of measuring small currents of the order of 10^{-9} amps or sometimes less. The meter scale is marked in mg/l, ppm or percent saturation.

4. Standard Oxygen Transfer Rate and Aeration Efficiency

The oxygen transfer coefficient (k_La) is used to estimate the standard oxygen transfer rate (SOTR) of an aerator. The tests are conducted in tanks of clean water under standard conditions (20°C and 760 mm Hg). Besides, deoxygenated water (0 mg/l) is used, which gives the maximum DO deficit. SOTR is calculated from the relation:

$$SOTR = (k_La)_{20}\, C_{s20} \times V \times 10^{-3} \text{ kg/h} \tag{10.12}$$

where, C_{s20} = DO concentration at saturation and 20°C (mg/l), and V = tank volume (m³). However, the standards for rating of aerators change from country to country. In Europe, the standard temperature used is 10°C. In U. K., a commercial detergent is used to change the properties of tap water to more closely approach the properties of wastewater, which may not be desirable for aquaculture purpose.

Although SOTR is an important parameter used to compare aerators, it describes the amount of oxygen transferred by a given aeration unit. Standard aerator or aeration efficiency (SAE) gives a better parameter for comparison. SAE is defined as the amount of oxygen transferred per unit of power input under standard conditions (kg/kWh), and calculated from:

$$SAE = SOTR/P, \tag{10.13}$$

where, P = power input (kW). The power input can be the measured wire power, that is, actual power used by the (a) motor, drive and blower, or (b) motor, coupling and gearbox. It has been customary to report power in terms of power applied to the aerator shaft (break power). There are several losses such as loss in the motor, in the drive, and frictional losses, and all power applied to the aerator shaft is not transferred to the water to facilitate aeration. A more realistic approach is, therefore, to express SAE in terms of power used (break power), according to Boyd and Watten (1989). However, there is a controversy in using the power input for calculation of SAE. As majority of the units in aquaculture applications are supplied with a motor, SAE values based on only the brake power would report the better efficiency than that based on wire power (Colt and Orwicz, 1991).

10.1.2 Rating of Aeration Systems under Field Conditions

The standard values of SOTR and SAE are determined at 20°C, 760 mm Hg, tap water, and zero initial DO concentration, and therefore should be modified to use for rating of actual aeration systems under field conditions. The rating will depend primarily on the required DO concentration and to a lesser extent on the temperature, pressure, and water characteristics. The rating of aerators under field conditions requires computation of values designated as field oxygen transfer rate (OTR_f), and field aeration efficiency (AE_f) (Colt and Orwicz, 1991), and also as actual oxygen transfer rate (AOTR), and actual aeration efficiency (AAE) (Boyd and Watten, 1989).

Using the eqs. from (10.4) to (10.6) and (10.12), the following relation can be derived:

$$OTR_f = SOTR\left[\frac{\beta C_{s,f} - C_w}{9.08}\left(1.024^{T-20}\right)\alpha\right] \tag{10.14}$$

where, 9.08 is the value of C_s at 20°C, $C_{s,f}$ = saturation DO concentration at local atmospheric pressure and temperature, and C_w = DO concentration in pond water during aeration. For surface aerators, $C_{s,f}$ is assumed to be the atmospheric saturation concentration, which is equal to 9.08 mg/l. The values of α and β have a significant effect on the oxygen transfer rate under field conditions. Therefore, computation of OTR_f requires α and β for the particular water condition, and local temperature. When α = β = one, temperature = 20°C, $C_{s,f}$ = zero, and C_s = 9.08 mg/l, eq. (10.14) reduces to SOTR, eq. (10.12).

Field aeration efficiency (AE_f) can be computed from the relation:

$$AE_f = OTR_f/P \tag{10.15}$$

A table of factors (Table 10.1) has been prepared for rapid solution of eq. (10.14) in channel catfish ponds (Shelton and Boyd, 1983). For this purpose, eq. (10.15) was modified to

$$OTR_f = f \times SOTR \tag{10.16}$$

where, f is the factor given in Table 10.1 and SOTR is the standard oxygen transfer rate given by the manufacturer. The method may be applicable to ponds for species other than channel catfish. For use of Table 10.1, multiply the SOTR (manufacturer's OT rating) by the appropriate factor to get the OTR for pond condition. For example, for pond DO = 4 mg/l and temperature = 25°C, f = 0.36. If the SOTR = 19 kg O_2/h, OTR_f = 0.36 x 19 = 6.84 kg/h. If P = 8.60 kW, AE_f = 6.84/8.6 = 0.8 kg O_2/kWh.

The aerator performance in brackish water must be adjusted for salinity by correcting the values of $C_{s,f}$, and also C_s at 20°C and 760 mm Hg. For example, for an aerator to be used in water of 15 ppt salinity, the value of 9.08 is to be replaced by 8.31.

Table 10.1: Factors for Rapid Solution of the Aerator Rating Equation

DO Concentration (mg/l)	*Water Temperature, °C*					
	10	*15*	*20*	*25*	*30*	*35*
0	0.71	0.71	0.72	0.74	0.76	0.79
1	0.64	0.64	0.64	0.65	0.66	0.67
2	0.57	0.56	0.55	0.55	0.55	0.55
3	0.51	0.49	0.47	0.45	0.44	0.43
4	0.44	0.41	0.38	0.36	0.33	0.30
5	0.37	0.34	0.30	0.26	0.22	0.18
6	0.31	0.26	0.21	0.17	0.12	0.06
7	0.24	0.19	0.13	0.07	0.01	0.0
8	0.17	0.11	0.04	0.0		
9	0.11	0.04	0.0			
10	0.04	0.0				
11	0.0					

Source: Shelton and Boyd, 1983.

Example 10.2

Calculate the field oxygen transfer rate of an aerator with 2 kW rating to be used for pond aeration under the following condition:

Test data on the aerator (C_s = 7.81 mg/l, water temperature = 28°C, and water volume = 80 m^3):

Time (min)	2	4	6	8	10	12	14	18	20
C (mg/l)	5.1	5.25	5.3	5.5	5.7	5.9	6.2	6.4	6.6

Assume C_i = 1.5 mg/l, tank volume and average DO concentration in the night = 4 mg/l. α = 0.90 and β = 0.98

Solution

First k_La value has been calculated by tabulating and plotting as in Example10.1. $C_s - C_i$ = 6.31

From the linear regression equation, $k_La = 2.7\ h^{-1}$, and putting the values in eq. (10.4), $(k_La)_{20}$ = $2.7/(1.024)^{28-20} = 2.23\ h^{-1}$, and from eq. (10.12)

SOTR = 2.23 x 7.81x 80 x 10^{-3} = 1.39 kg/h

$$OTR_f = 1.39\left[\frac{0.98x7.81-4}{9.08}(1.024)^8 0.9\right] = 0.61 \text{ kg } O_2/\text{h}.$$

10.2 Aeration Systems

10.2.1 Types of Aerators

Aerators used in aquaculture practices are similar to those involved in wastewater treatment. All basic types of these aquaculture aerators are, therefore, modifications of the wastewater aerators. The basic functions of aerators are that they utilize an energy input to (a) increase the liquid surface area available for oxygen transfer, and/or to (b) mix the liquid to ensure that water of low DO concentration is brought into contact with air or oxygen. Mixing increases the surface area and the effective concentration gradient, thereby causing oxygen transfer.

1. Classification

The two basic aeration methods are: (a) air contact system which increases the air water interfacial area and degree of turbulence in water, and (b) pure oxygen absorption system which brings pure oxygen and water into contact and employ high pressure, turbulence or high surface area. Air contact systems can saturate the water, and the efficiency of transfer declines as the DO concentration in water increases. If the water is super-saturated, oxygen will escape from water to air, and the device then acts as a degasser.

According to Wheaton (1977), there are four basic types of air contact aerators: gravity aerator, surface aerator, diffuser aerator, and turbine aerator. There are several aerators that combine two or more of these types. According to Colt and Orwicz (1991), aeration devices can be classified as surface, subsurface and gravity. Surface aerators break up, or agitate the surface of water, and then spray, or splash water to bring it into greater contact of air. Subsurface aerators inject air or oxygen into water in the form of bubbles, and transfer of oxygen takes place from the gas bubbles to the water by diffusion. Diffuser aerator falls under this category. Turbine aerators employ a propeller to circulate water and effect aeration at the surface. Gravity aerators are a special type of surface aerator that utilises energy released when water loses altitude to increase the air water surface area. Aerators, except gravity aerators, are powered by electric motors. Where electricity is not available, aerators may be run by the power-take-off (PTO) of farm tractors or by diesel engines.

Surface aerators create water drops with high transfer coefficients and large surface to volume ratio. They also create turbulence in water and in thin layers for better oxygen transfer. But the principal theoretical disadvantage is that water gets relatively short contact time with air. For example, a jet rising 3-m and then falling have only 2-s as contact time with air. With diffuser aerator, a bubble with little acceleration gets 10-s contact time during rise of 3-m. However, there are many practical disadvantages to diffuser aerators: cost of installation at the pond bottom, obstruction to fish harvest by the diffuser, and clogging of holes in diffuser when not in operation (Boyd, 1982). For these reasons, paddle wheel aerators are more widely used in aquaculture practices. Gravity aerators are inexpensive to operate when sufficient head is available.

2. Surface Aerators

There are several designs available, amongst which the major ones are:

1. Simple surface aerator, using a pump, which sprays water into the air as a fountain, also called pump sprayer.
2. Paddlewheel aerator, which uses rotating paddles partially submerged in water, and splashes water into the air.
3. Spray-type aerator, which draws water into a vertical tube by a propeller and pumps as a spray over the water surface, also called vertical pump.
4. Nozzle aerator, which pumps water through a nozzle at a high velocity striking the water surface and inducing strong current that favors oxygen absorption.

Figure 10.2 shows the simplified sketches of paddle wheel and vertical pump aerators.

Pump Sprayer

The pump sprayer aerator consists of a high-pressure pump that discharges water at high velocity through outlet slits attached to the pump discharge. Impeller suction pump, impeller lift pump, or turbine pump can be used. Discharge can be made directly through a manifold, which has discharge slits on top and outlets at each end. Aerator sizes range from 2 to 15 kW, and the impeller speeds are from 500 to 1000 rpm. Most are powered by PTO or electricity (Boyd, 1998).

Paddle Wheels

Paddle wheels are multi-bladed, with variations in the type of blades such as rectangular or triangular, flat or bent, plain with holes or serrated. Figure 10.2a shows a simplified sketch of multi-bladed paddle wheel aerator. The paddles are fitted in 2 or more number of rows or in a spiral arrangement in a rotating hub of varying lengths (Ahmad and Boyd, 1988; Boyd and Watten, 1989). Variations in design and operational parameters are mainly paddle wheel diameter, paddle shape, paddle wheel rotational speed, paddle tip submergence (paddle depth), and power rating. In general, paddle wheel diameters range from 50 to 90 cm, speeds range from 75 to 135 rev/min and paddle depth from 8 to 25 cm. Depending on the speed and depth, the power rating ranges from 2 to 10 kW (Boyd and Watten, 1989).

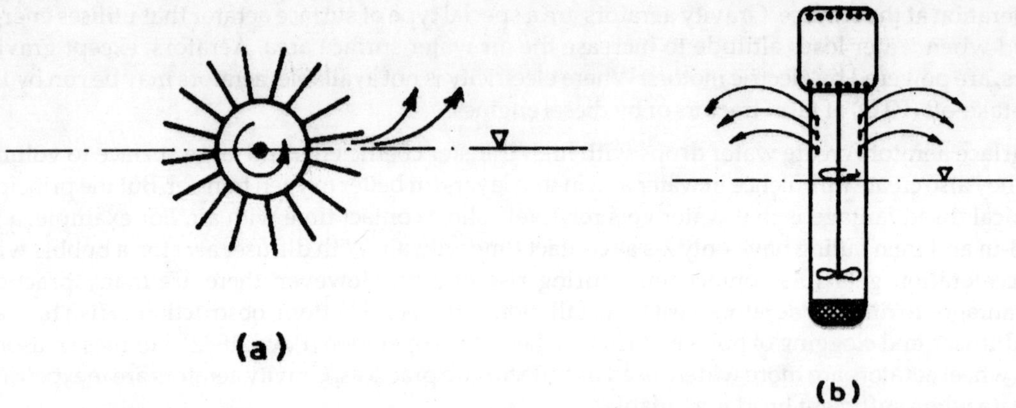

Figure 10.2: Simplified Sketches of (a) Paddle wheel aerator, and (b) Vertical pump aerator

An electric paddle wheel aerator usually consists of a motor, a speed reduction mechanism, and paddle wheel mounted on a float. Either a gear motor, or a gearbox connected to the output shaft of the motor is used. Paddle wheel shafts are fitted with bearings and mounted on a moveable metal frame, which is attached with the floating frame. Some means of raising and lowering both ends of the paddle wheel are provided for minor adjustments in paddle depth after it is installed in a pond. Installation in a pond is done by anchoring either into the pond bottom, or into the pond bank.

Paddle wheel aerators may be constructed of plastic. The popular "Taiwanese paddle wheel" is widely used in aquaculture, and a Japanese model is similar to Taiwanese aerator. An U.S. prototype aerator was constructed of polyurethane plastic. The polyurethane paddles were attached to threaded polyurethane rods and screwed into the polyurethane hub. Hubs were inserted into bearings made of solid polyurethane. The frame was made of polyurethane bars, and the flotation system consisted of plastic barrels. PVC pipes were also used for construction of aerator parts (Boyd and Watten, 1989).

Vertical Pump Aerator

The vertical pump aerator (Figure 10.2b) consists of a submersible motor, which rotates an impeller to pump surface water into the air as a spray. The motor is suspended by floats. Units ranging from less than 1 to more than 50 kW are manufactured. Units for aquaculture are seldom larger than 2 kW, and have high speed impeller, which rotate at 1730 or 3450 rpm (Boyd, 1998).

3. Diffused Air Systems

Diffused air system or subsurface aeration is a two-phase process in which oxygen transfer takes place through diffusion of oxygen from air bubbles released continuously below the water surface. Gas transfer occurs across the gas liquid interface. Diffused air system aerators use a low pressure, high volume air blower to release air through diffusers positioned at the pond bottom or suspended in the water. The minimum permissible pressure increases with increasing depth of water above diffusers, because enough pressure must be required to diffuse against total pressure at the discharge point. Water depth is normally between 1 and 2 m. For 1 m depth approximately 0.1 kg/cm^2 hydrostatic pressure is required. Therefore, air pumps for aquaculture need not develop high pressure.

A diffuser aerator consists of an air pump or air blower, air distributing piping, and diffusers. Various types of blowers are used in aquaculture such as rotary vane blower (up to 3.5 kg/cm^2

pressure), rotary lobe blower (at 0.56 to 1.06 kg/cm^2 pressure), turbine blower (0.7 kg/cm^2 pressure, 28–142 m^3/min air flow). Centrifugal blowers provide a high volume of air at very low pressure, whereas high pressure piston type compressors provide relatively low volumes of air at pressures above those usually required in aquaculture (Boyd and Watten, 1989).

Various types of diffusers have been used including porous ceramic tubing, perforated plastic pipe, ceramic dome diffusers, packed columns, carborundum airstones, U-tube and nozzle system. Commercially available diffuser tubing may be constructed by drilling small holes in PVC pipes. Large air stones attached to the pieces of tubing of manifold can be suspended in the water rather than resting on the bottom. The diffusers that release fine air bubbles usually are more efficient than those that discharge coarse bubbles (Boyd and Watten, 1989).

Diffuser with aspiration system consists of a suitably designed venturi, orifice, nozzle or any other type of head loss meter. Air is sucked into water through an opening located at the low pressure side, and bubbles thus transferred aerate the water. The system requires a liquid pump and an aspirator arrangement but not a pressurized air supply.

Figure 10.3 shows four types of diffused air systems.

4. Gravity Aerators

Gravity aerators can be categorised into three types: Natural aerators, man-made aerators, not designed for the purpose, and aerators designed for flowing water aquaculture. Natural streams, where turbulent motion continually reaerates the surface water, and waterfalls, where water breaks up into fine droplets and gets aerated, are natural gravity aerators. Dams and fish ladder are artificial aerators, usually not designed for aeration. In flowing water systems such as raceways, water is aerated by allowing it to (*a*) fall over weirs with splash board, paddle wheel, or rotating brush; (*b*) flow down various types of inclined planes (cascades); (*c*) flow down lattice (stairs step surface with holes); (d) flow down perforated trays; and (*d*) flow up a riser and fall through perforated aprons (Wheaton, 1977). Figure 10.4 shows four types of gravity aerators.

Perforated tray gravity aerators are more efficient than simple weirs, but will require cleaning. A series of screens, perforated trays, or expanded metal sheets are stacked on above the other in a box. Four to 10 aprons, with 10 to 25 cm spacing, are commonly used. A height of 1 m can provide at least 70 to 80 percent saturation of water with oxygen. With a 0.6-m head loss, 52 percent saturation with DO has been achieved (Boyd and Watten, 1989).

10.2.2. Types of Aeration

There are various types of aeration followed in practice: (*a*) emergency aeration, (*b*) supplemental or continuous aeration, (*c*) aeration to prevent thermal and oxygen stratification, and (*d*) aeration of source water. Although the most complete information on theses types of aeration is available for catfish culture practice, it may be applicable to other types of culture (Boyd, 1982; Boyd and Watten, 1989).

1. Emergency Aeration

The practice of aerating ponds when DO concentrations reach low level is commonly known as emergency aeration. It is experienced by aquaculturists and farms managers that when DO concentrations fall below 2 to 4 mg/l during the night, emergency aeration is necessary. When feeding rates exceed 50 kg/ha/day, emergency aeration may be needed frequently. This practice is initiated between midnight and dawn and continued during daytime until phytoplankton photosynthesis

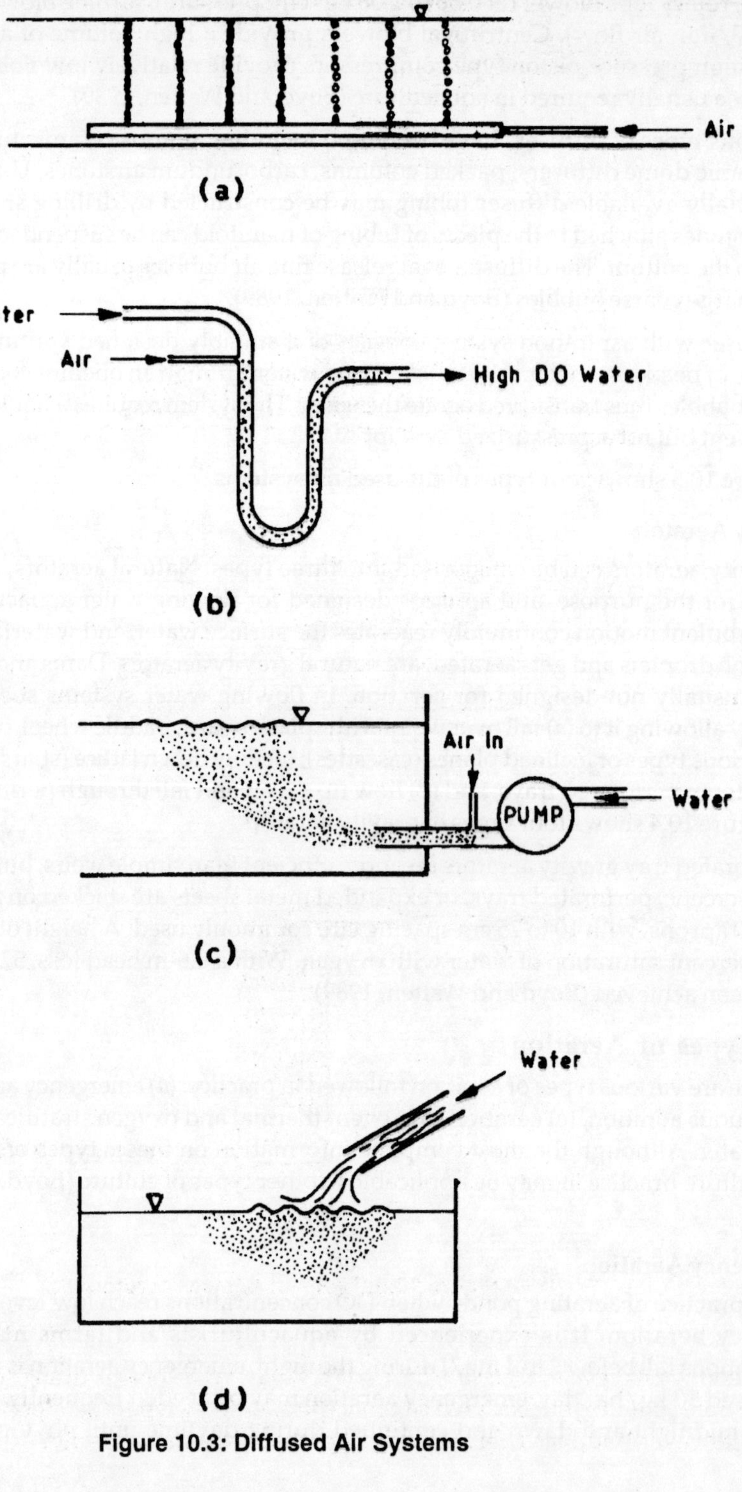

Figure 10.3: Diffused Air Systems

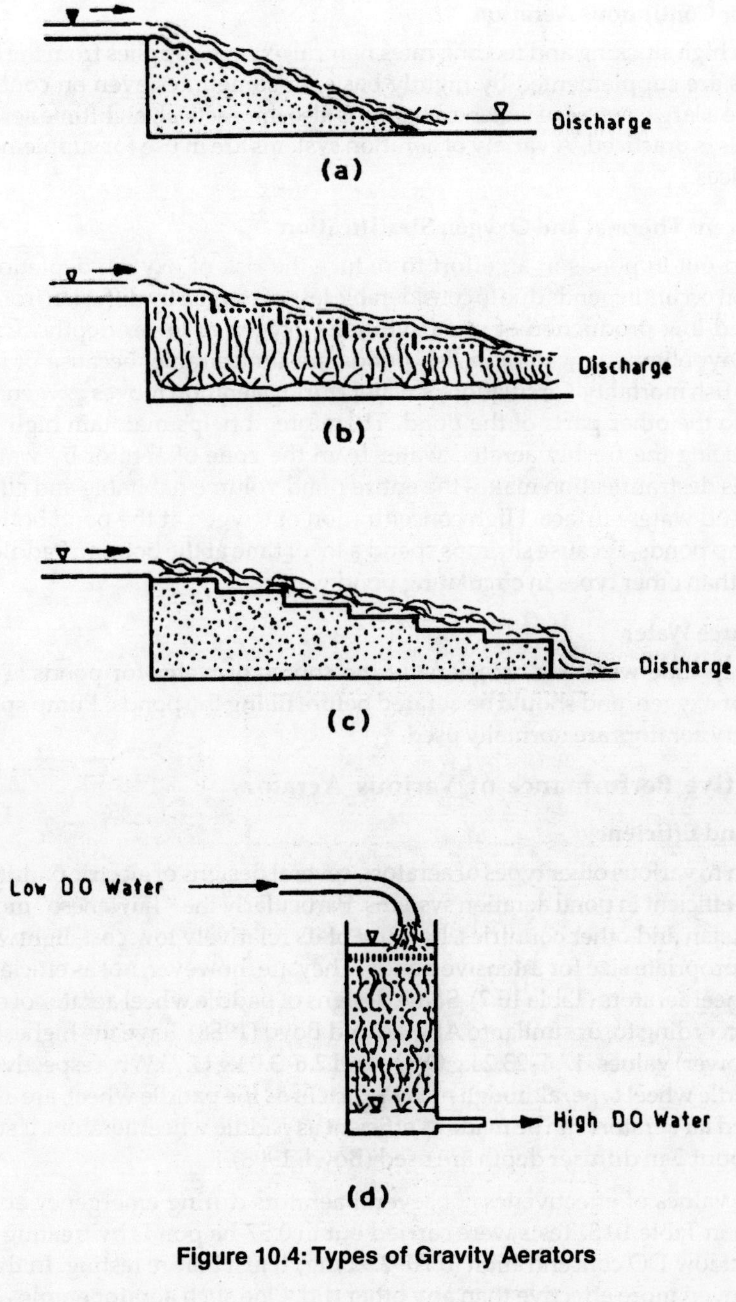

Figure 10.4: Types of Gravity Aerators

produces enough oxygen to reach the desired DO level. Some powerful aerators suitably designed for the purpose are started to quickly raise DO concentration. Any type of aerator can be used for emergency aeration. However, paddle wheel aerators are probably the most effective.

2. Supplemental or Continuous Aeration

In ponds with high stocking and feeding rates natural oxygen supplies from the atmosphere and by photosynthesis are supplemented by nightly basis, or sometimes even on continuous basis. In some ponds, where a large degree of water exchange is also impractical, nighttime aeration sometimes on continuous basis is practiced. A variety of aeration systems are in use for supplemental aeration in aquaculture practices.

3. Aeration to Prevent Thermal and Oxygen Stratification

This is carried out in ponds in an effort to reduce the risk of oxygen depletion. Thermal and oxygen stratification occurs in ponds due to considerable loss of oxygen by diffusion from supersaturated surface waters, and low production of photosynthetic oxygen at lower depths. Large volumes of oxygen-depleted hypolimnetic waters in fish ponds are undesirable, because of sudden low DO concentration and fish mortality. Circulation of water during aeration moves oxygenated water away from the aerator to the other parts of the pond. Therefore, it helps maintain high oxygen transfer efficiency by replacing the freshly aerated water from the zone of aerator by water of lower DO concentration. This destratification makes the entire pond volume habitable and eliminates oxygen depletion at the mud water surface. High concentration of oxygen at the pond bottom is especially important in shrimp ponds, because shrimps spend a lot of time at the bottom. Paddle wheel aerators are more efficient than other types in circulating pond water.

4. Aeration of Source Water

Wells and deep tube wells are occasionally the source of water for ponds. These waters are normally devoid of oxygen, and should be aerated before filling the ponds. Pump sprayers or nozzle aerators and gravity aerators are normally used.

10.2.3 Comparative Performance of Various Aerators

1. Aeration Rate and Efficiency

In comparison to various other types of aerators, several designs of electric paddle wheel aerators have become very efficient in pond aeration systems. Particularly the "Taiwanese" model, has become very popular, in Asian and other countries, because of its relatively low cost, lightweight, corrosion resistance, and appropriate size for intensive ponds. They are, however, not as efficient as some other types of paddle wheel aerators (Table 10.2). Some designs of paddle wheel aerator of capacity 2.25–7.5 kW, constructed according to, or similar to Ahmad and Boyd (1988), have the highest SOTR and SAE (based on brake power) values, 17.4–23.2 kg O_2/h, and 2.6–3.0 kg O_2/kWh, respectively. The aerators other than the paddle wheel type, although not as efficient as the paddle wheel, are also used in small ponds. The diffused air aerators can be made as efficient as paddle wheel aerators, if sufficient number of diffusers and about 1-m diffuser depth are used (Boyd, 1998).

Comparative values of effectiveness of several aerators during emergency aeration in catfish culture are shown in Table 10.3. Tests were carried out in 0.57-ha ponds by treating each pond with algaecides to effect low DO concentration (0.70–2.92 mg/l) just before testing. In these tests paddle wheel aerators proved more effective than any other did. One such aerator employed for the above tests was 12-bladed, each blade 35 cm long by 15 cm wide, immersed 25–30 cm below the water surface and rotated at about 120 rpm. It was however powered by PTO. Large paddle wheels, usually rotated by 3.7–7.5 kW electric motors, are more energy efficient than the PTO–powered aerators.

Table 10.2: Comparative Values of SOTR and SAE for Some Electric Aerators

Types of Aerator	*Number of Aerators*	*Range of SOTR ($kg\ O_2/h$)*	*SAE ($kg\ O_2/kWh$)*	
			Average	*Range*
Paddle wheel:	24	2.5–23.2	2.2	1.1–3.0
Several designs–				
Auburn University Design			2.25	
"Taiwanese" model			1.17	
"Japanese" model			1.03	
Propeller-aspirator-Pump	11	0.1–24.4	1.6	1.3–1.8
Vertical pump	15	0.3–10.9	1.4	0.7–1.8
Pump sprayer	3	11.9–14.5	1.3	0.9–1.9
Diffused air	5	0.6–3.9	0.9	0.7–1.2

Source: Boyd and Watten, 1989 and Boyd, 1998.

Note. SAE values are based on brake power, and the wide variation of SOTR is due to various sizes of the aerator.

Table 10.3: Comparative Performance of Aerators in Emergency Aeration

Type of Aerator	*Capacity*	*Oxygen Added (kg)*	*Effects of Aeration*	*Relative Efficiency (%)*
Paddle wheel	3.7–7.5 kW, tractor-powered	31.9	Raised DO by 4 mg/l in 4 h, 900 m² of O_2-enriched zone with 6.2 mg/l DO	100
Spray-type	(i) 3.75 kW	9.7	Each failed to appreciably increase DO during 4 h period	31
	(ii) 2.25 kW	7.2		23
	(iii) 0.25 kW	2.5		8
Pump-sprayer	(i) 30.5-cm, 18.9 m³/min	7.5	Raised DO by 1.46 mg/l during 4 h, 1000 m² O_2-enriched zone	24
	(ii) 15.2-cm, 3.8 m³/min	6.8	Raised DO by 0.46 mg/l during 4 h, 300 m² O_2-enriched zone	22
Pump as circulator discharging water from adjacent pond	(i) 30.5-cm, 18.9 m³/min	12.1	Discharged 7.8 mg/l DO, increasing DO in pond water by 2 mg/l in 4 h,1600 m² O_2 enriched zone	39
	(ii) 5.2-cm, 3.8 m³/min	3.8	Discharged 8.5 mg/l DO, increasing by less than 1 mg/l in 4 h,1000 m² O_2-enriched zone	12
Pump with Sprayer	30.5-cm pipe, with end made into a sprayer	19.9	Discharged 5.02 mg/l DO, increasing by about 3 mg/l in 4 h, 500 m² O_2 enriched zone	64

Source: Extracted from Boyd, 1982. The values have been corrected for increase or decrease of DO resulting from photosynthesis and respiration.

The spray-type surface aerators are most suitable for small ponds and not used for emergency aeration in large ponds. Pumps may be used by placing intakes 25–50 cm below the surface and releasing the water onto the pond surface. Effectiveness of this technique depends on pump capacity. A 10–30 cm diameter pump with discharge capacities of 4–20 m^3/min can be used. A 20 m^3/min pump would discharge the entire volume of a 1-ha pond of 1-m depth in 8.3 h. Spray type surface aerators, rotated at 0.75–3.75 kW, are used in small ponds.

Pumps working as circulator created oxygen-enriched zones with DO concentrations appreciably higher than those in other areas of the ponds. The same two pumps discharging water from adjacent ponds resulted in higher DO concentration of the discharge water and create larger zones of oxygen-enriched water. However, the pump fitted with a sprayer induced strong currents that favored absorption of oxygen, and increased DO by about 3 mg/l (Boyd, 1982).

2. Oxygen Saturation and Oxygen Transfer

For evaluation of aeration devices in supplemental aeration, the effectiveness of aeration should be studied with respect to the stocking and feeding rates, fish production, and economic returns. The evaluation parameters studied by Boyd and coworkers (Boyd, 1982) were average percentage of oxygen saturation just before sunrise, and oxygen transfer (kg/h and kg/kWh) (Table 10.4 and 10.5).

Table 10.4: Comparative Performance of Aerators in Supplemental Aeration: Effect on Oxygen Saturation

Aeration Device	*Capacity*	*% O_2 Saturation Just Before Sunrise (Average)*
Electric paddle wheel (Japanese water wheel)	(i) 0.38 kW (ii) 0.19 kW	97, 82, 99 84, 68, 60
Spray-type surface aerator	0.25 kW	67, 63
Diffuser and plastic pipe	5.8 kW, 0.08-0.25 m^3/min air flow at 460-600 mm Hg	50,34,27

Source: Adapted from Boyd, 1982.

Table 10.5: Comparative Performance of Aerators in Supplemental Aeration: Effect on Oxygen Transfer

Aeration Device	*Capacity*	*Oxygen Transfer*	
		kg/h	*kg/kWh*
Spray-type aerator	i. 0.75 kW	1.15	0.48
	ii. 0.25 kW	0.50	0.50
Air blower	12 lengths of porous tubing fixed about 10 cm from pond bottom, each tubing receiving 1.7 m^3/h air	0.29	0.96
Diffuser aerator (jet eductors)	(i) Two 3.8 cm units/pond	0.67	0.32
	(ii) One 7.6 cm unit/pond	0.79	0.35
	(iii) Two 3.8 cm units/pond One 7.6 cm unit/pond	1.48	0.34

Source: Adapted from Boyd, 1982.

Table 10.4 shows that the best results were obtained with paddle wheels and spray type aerators in seasons of 2–3 crops. Tests were carried out in ponds of 200-m^3 volume with exchange of water every 10 days. 20 percent oxygen saturation was taken as the threshold value, below which fish death might occur within 2–3 h. However, all methods of aeration gave much higher levels of fish production than could be achieved without aeration.

Performance characteristics of aeration devices with respect to oxygen transfer in a 0.06-ha pond (Table 10.5) showed that the spray type aerator was effective and economic aeration device, for small ponds, provided there was no disturbance in the bottom soil causing turbidity. The aerator is easily movable in ponds and do not need supports. With jet eductors oxygen transfer was considerable but with high energy consumption.

In this comparison, paddle wheel aerators were not included. Small paddle wheel aerators are more expensive than other types of small aerators, because of high costs of small gear motors. For aerators of 1-kW and larger, paddle wheel aerators are equal or less in cost than other types of aerators. For these reasons, paddle wheel aerators are found more efficient than other types of aerators in large ponds.

10.3 Fish Production

It is well known that aeration will increase fish and shrimp production. There is much data on effect of aeration in channel catfish ponds, but few in other fish and shrimp ponds. Table 10.6 shows the data on aeration of catfish ponds (Loyacano, 1974; Hollerman and Boyd, 1980). Effects of aeration treatment, in terms of air volume injection per min, with maximum daily ration (145 kg/ha) in unaerated ponds and 196 kg/ha in aerated ponds showed that there was no appreciable difference in pH, carbon dioxide or temperature in the ponds, but DO and turbidity increased with the degree of aeration. Average net fish production increased with rate of aeration. All unaerated ponds had fish mortality due to low DO and no oxygen-related fish kill occurred in aerated ponds. In one economic analysis unaerated treatment was found to be more profitable than all aerated treatments except that with the 40,000 fish/ha stocking rate.

According to Boyd (1989), relationships among feeding, dissolved oxygen, and shrimp production being almost identical in general features to these relationships in catfish production, information on catfish pond aeration should be applicable to shrimp ponds. Some data on shrimp ponds are shown in Table 10.7.

10.4 Aeration Process and Aerator Design

10.4.1 Computation of Oxygen Demand and Supplemental Aeration Requirement

Sizing of aeration devices requires an estimate of the total oxygen demand by the fish and other aquatic organisms, the available oxygen supplied by water flow and the supplemental oxygen requirement. The total oxygen demand of the fish depends on the number and size of the animals, and temperature of water. The total oxygen demand can be estimated from fish respiration, planktonic and benthic respiration (whole pond respiration) and oxygen balance in a pond system. The procedure has been discussed under the topic of pond oxygen dynamics in Section 6.2, Chapter 6.

A method to estimate total oxygen demand and supplemental oxygen requirement for sizing pure oxygen system of aeration was described by Colt and Watten (1988). The method defines the average and maximum daily oxygen requirements over the whole production cycle on a weekly and monthly basis.

Table 10.6: Fish Production with Aeration in Catfish Ponds (0.02–0.04 ha sizes)

Item	*Culture Conditions*	*No Aeration*	*With Aeration*
Net fish production (kg)	Aerator–Blower, 2.54 cm air line, 15 cm from pond bottom Stocking–20,000 per ha Aeration–143 days Feeding–6 days weekly, at max. daily 145 kg/ha (unaerated pond) and 196 kg/ha (aerated pond). With air volume, m^3/min per ha	2740	
	–6.9		4560
	–10.4		5510
Harvest weight (kg/ha)	Aerator–Air-lift pump moving 0.11 m^3/min of water with 0.06 m^3 per min of air, supplemented, if		
Stocking–5,000		3214	
(No./ha)–5,000	required, with spray type aerator		4,278
–10,000	(0.25 kW)		2,863
–20,000	Feeding–daily at 3 per cent for 204		10,416
–40,000	days		15,837
Harvest weight (kg/ha)	Emergency aeration with pump sprayer (5.5 kW/ha)		
Stocking–5,000	Feeding–daily, max. 28 kg/ha	2,518	
(No./ha)–18,500	(unaerated ponds), and 206 kg/ha		8,554
–25,000	(aerated ponds)		11,108
–31,000			12,846
Harvest weight (kg/ha)	Aerator–Spray type (0.25 kW) Stocking–20,000/ha Feeding–6 days /week at 3 per cent, max. 71 kg/ha (unaerated ponds), and 90 kg/ha (aerated ponds)	1,400	5,307

Source: Loyacano, 1974; Hollerman and Boyd, 1980 and Boyd, 1982.

Table 10.7: Shrimp Production with Aeration

Species	*Culture Conditions*	*Production (kg/ha)*
Penaeus penicillatus	Stocking–286 PL/m^2. Aeration with paddle wheel at 11.7 kW/ha. Water exchange uo to 30 per cent per week. 141 days culture. FCR = 1.7	12,000
P. vannamei	Stocking–45 PL/m^2. Aeration with paddle wheel at 7.5 kW/ha. Average water exchange of 17 per cent per day. FCR = 2.5	7500

Source: Boyd, 1998.

1. Average Daily Oxygen Demand

Fish consumes a certain amount of oxygen in processing and digesting feed. The oxygen requirement is based on size of the animal, ration size, feed composition and feeding rate, and digestibility of the feed components. Assuming that all the feed is converted to energy, *i.e.* no growth, the oxygen requirement is described by the oxygen-feed ratio (OFR) as:

$$OFR = M_e/Q_{ox}, \qquad (10.17)$$

where, OFR = oxygen-feed ratio, oxygen required per unit mass of dry feed ingested, kg O_2/kg dry feed; M_e = metabolizable energy, the energy available from food less exogenous fecal and non-fecal excretion, kcal/g feed; and Q_{ox} = oxycaloric equivalent, the number of calories released by a substrate per mass of oxygen consumed, kcal/g O_2.

Example 10.3

Assuming a value of 3.24 kcal/g O_2 for the Q_{ox} (Brett and Groves, 1979), calculate the OFR of feeds (17.8 per cent moisture content) containing 3.67 kcal/g of M_e.

Solution

OFR = 3.67/3.24 = 1.13 kg O_2 /kg dry feed, or 1.13/1.21 = 0.93 kg O_2/kg wet feed.

If the M_e value of the feed is lower, the OFR will be also less. With higher moisture content of the feed, OFR will also reduce. For example, with M_e = 2.6 kcal/g feed, OFR of a feed with 10 percent moisture would be 0.80 kg O_2/kg dry feed, or 0.72 kg O_2/kg wet feed, but would be reduced to 0.48 kg O_2/kg wet feed with 40 percent moisture.

Once the OFR is known, the average daily oxygen demand can be calculated from equation as:

$$\text{Average daily oxygen demand (kg/d)} = \text{OFR.R} \tag{10.18}$$

where, R = total ration, kg/d. In most production systems, feeding levels would be higher than maintenance and probably closer to maximum feeding levels. In salmon and trout culture under production system, the OFR is about 0.20 kg O_2/kg wet feed, whereas the value of 1.00 kg O_2/kg wet feed is commonly used in commercial high-density warm water fish culture. The higher values of OFR for warm water fish may be due to higher values of M_e in the feed, lower moisture levels in the feed, reduced reaeration across the water surface, higher bacterial oxygen demand in the pond, differences in activity and feeding behavior (Colt and Watten, 1988).

2. Maximum Daily Oxygen Demand

On a daily basis, the maximum oxygen consumption occurs 4-6 h after feeding (Figure 10.5a). A peaking factor of 1.44 was, therefore, suggested to compensate for the additional oxygen consumption:

$$\text{Maximum daily oxygen demand (kg/d)} = 1.44\ (\text{OFR.R}) \tag{10.19}$$

3. Oxygen Supplied by Water Flow

The amount of oxygen supplied by the flow is:

$$\text{Oxygen supplied by flow (kg/d)} = 86.4\ Q_w\ (DO_{in} - DO_{out}) \tag{10.20}$$

where, Q_w = flow in m^3/s. Detailed DO data for the system should be available for correct estimation of the oxygen supplied by the flow. For undeveloped streams and rivers, and highly oxygen-rich systems, where reliable data is lacking, DO_{in} may be estimated as 0.90 C_s. For springs and well waters, which are more variable in DO concentrations, the oxygen supplied by the flow must be assumed zero (Colt and Watten, 1988).

4. Supplemental Oxygen Demand

The amount of supplemental oxygen demand is equal to the difference between eqs. (10.19) and (10.20). The supplemental oxygen requirement increases throughout the culture cycle, as the fish biomass and feeding rates increase (Figure 10.5b). Depending on the harvest schedule and temperature,

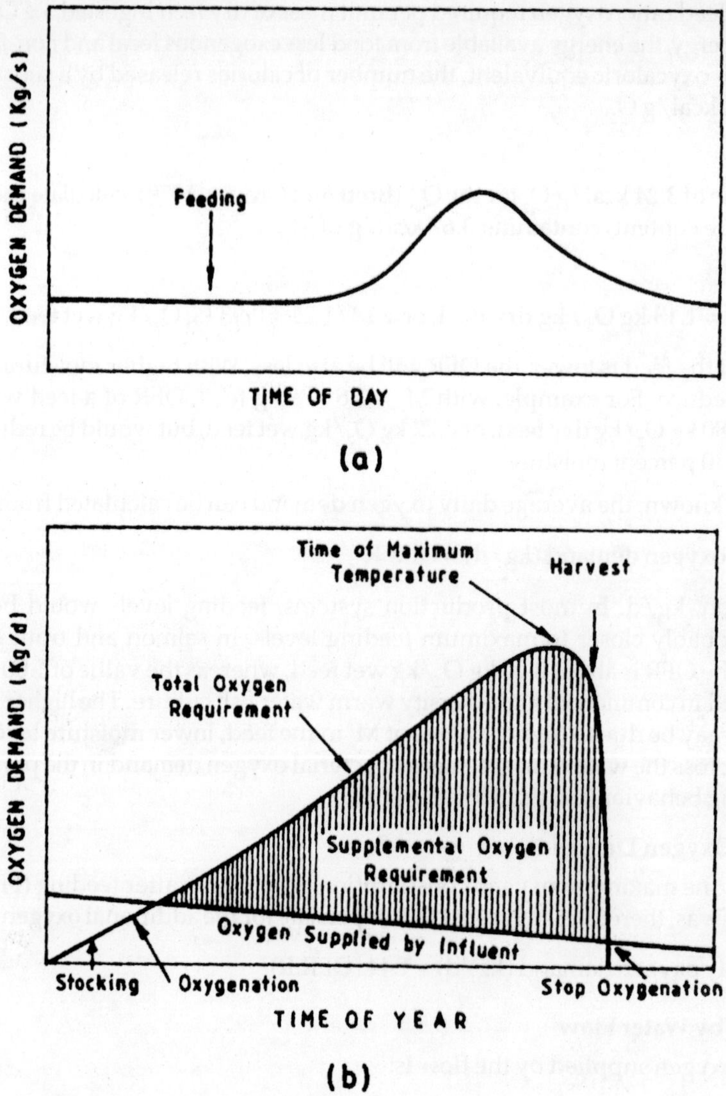

Figure 10.5: Oxygen Demand in Relation to (a) Time of feeding and (b) Culture cycle (After Colt and Watten, 1988)

maximum oxygen requirement may occur near the end of the culture cycle. Therefore, the supplemental oxygen requirement for aerator design purpose should be estimated on a weekly or monthly basis.

Example 10.4

Calculate the supplemental oxygen requirement in a fish rearing unit containing 12,000 fish having a mean weight of 0.3 kg. The fish are fed 2.5 percent of their BW daily. The water flow rate through the unit is 3 m^3/min with DO concentration of 12 mg/l. The effluent DO concentration is maintained at 6 mg/l. Use the OFR value from Example 10.3.

Solution

Total ration = 12,000 × 0.3 × 0.025 = 90 kg/d.

Average daily DO demand = 0.93 × 90 = 83.7 kg.

Maximum daily DO demand = 1.44 × 83.7 = 120.5 kg

DO supplied by flow = 86.4 × 3 × (12–6) / 60 = 25.9 kg/d

Therefore, supplemental DO demand = 120.5–25.9 = 94.6 kg/d.

The procedure for computation of OFR and supplemental oxygen requirement under pond conditions may not be particularly important, as the contribution of fish respiration on the whole pond respiration is minor.

10.4.2 Surface Aerator Design

1. Practical Approach

Surface aerators are available from several manufacturers in package units. Purchase of this equipment is often the most economical choice. However, in some cases it is desirable to design an aerator for a specific application. Most aerator manufacturers publish OTRs for their equipment. These values are usually taken under standard conditions, or are corrected to the standard conditions. These parameters are SOTR, SAE, and certain others applied for field rating of aeration efficiency, as outlined before.

The practical approach of the process design is, therefore, to determine the number and size of surface aerators required by the culture units. According to Wheaton (1977), the following steps can be outlined:

1. Determine the oxygen demand in the system.
2. Determine the volumetric oxygen transfer coefficient, k_La.
3. Evaluate a by determining k_La for pure water and for the waste under similar conditions.
4. Determine what the oxygen concentration gradient will be during actual operating condition.
5. Compute oxygen transfer per hour (OTR).
6. Divide oxygen demand by the oxygen transfer per kWh for the aerator selected to determine the total power requirement to supply oxygen.
7. Divide total power requirement by power rating of individual aerator to determine the number of aerators needed.
8. Locate aerators in the culture tank such that each aerator has an equal tank surface area in which to operate.

Example 10.5

Calculate the number of 2 kW paddle wheel aerators required for aeration of a shrimp pond with 1-ha area and 1-m depth of water under the following conditions:

Initial stocking density = 16/m², survival rate = 80 percent, maximum average weight of shrimp = 40 g, minimum DO level to be maintained = 4.5 mg/l, DO saturation level at 30°C water temperature = 7.5 mg/l, oxygen transfer rate of aerator under standard conditions = 2 kg O_2/kWh, $\alpha = 0.90$, $\beta = 0.94$

Assume for single shrimp respiration, mg O_2/h = 0.487 $W^{0.88}$, where W = shrimp weight (g), and 20 percent of the oxygen transferred by aerator is used for shrimp respiration.

Solution

Total number of shrimps = 16 × 10,000 × 0.80 = 1,28,000

Oxygen respiration by a single shrimp = 0.487 $(40)^{0.88}$ = 12.5 mg/h, and total oxygen demand by all shrimps = 12.5 × 1,28,000 = 16 × 10^5 mg/h.

Therefore, total oxygen to be transferred by an aerator = 16 × 10^5 / 0.2 = 8 kg/h

$$\text{Now, } OTR_f = 2\left[\frac{0.94x7.5-4.5}{9.08}(1.024)^{10}\ x0.9\right] = 0.64\ \text{kg/kWh}$$

Total power requirement for the total oxygen to be transferred = 8/0.64 = 12.5 kW, and the number of 2 kW aerators required for aeration = 12.5/2 = 6.25, *i.e.* 6–7 aerators.

The above calculation shows that 5,120 kg shrimp in the pond will require about 6 aerators of 2 kW each, *i.e.* each 2 kW aerator will support 853 kg shrimp, or about 426.5 kg shrimp production will be supported by each 1-kW aerator. This calculation tallies with the rough estimate of 500 kg shrimp production by each kW of aerator worked out by Boyd (1998).

2. Simulation Approach

In the field of wastewater treatment, many investigators used the theory of dimensional analysis and obtained optimal geometric similarity of horizontal rotor aerators under different conditions (Schmidtke and Horvath, 1976; Zlokarnik, 1979; Horvath, 1984; Ognean, 1993). On the contrary, aquaculture aeration studies dealt with design and manufacturing aspects of paddle wheel aerator, and correlations for prediction of aeration performance are not available. Moulick *et al.*, (2002, 2005) studied the design characteristics of a single hub paddle wheel aerator and developed generalized equations based on dimensional analysis for prediction of aeration efficiency.

The oxygen transfer rate depends on the number of air bubbles per unit volume and their size and distribution while aerating the water. The bubble parameters also depend on the physical properties of air and water, the flow parameters, and the geometric parameters of the aeration system. The operating variables of surface aerators, on which the oxygen transfer rate depend, are depth of submergence, rotational speed of the motor, rotor diameter, or diameter of the aerator, power input per unit area or unit volume of liquid being aerated, characteristics of waste being aerated, size and shape of liquid tank, oxygen concentration gradient, and aerator design. (Wheaton, 1977)

The main parameter of the oxygen transfer or absorption process, is the absorption rate coefficient, expressed as k_La_{20} x V = SOTR/ΔC, where ΔC = C_s–C_o. The functional relationship between SOTR/ΔC and the variables of a bent type paddle wheel aerator studied by Moulick *et al.*, (2002,2005) has been expressed as follows:

$$SOTR/\Delta C = f_1(h, b, \alpha, l', n, V, l, d, N, g, \rho_a, \rho_w, \sigma, \nu, D_L) \quad (10.21)$$

where, h = submergence depth of paddle blade, b = breadth of paddle blade, α = bent angle of paddle blade, l′ = horizontal projection of bent length of paddle blade, n = number of paddle blades in a hub, l =length of paddle blade, d = paddle wheel diameter, N = paddle wheel speed, g = acceleration due to gravity, ρ_a = mass density of air, ρ_w = mass density of water, ν = kinematic viscosity of water, and D_L = molecular diffusion coefficient for oxygen. By dimensional analysis Moulick et al (2002) derived the absorption correlation and power correlation, which were simplified to eq. (10.22) and eq. (10.23), respectively, as

$$Y = SOTR/\Delta C = f_2 (h/d, b/d, \alpha, l'/d, n, V/d^3, F, R, W) \quad (10.22)$$

$$\text{and } N_e = P/(\rho N^3 d^5) = f_3 (h/d, b/d, \alpha, l'/d, n, F, R, W) \quad (10.23)$$

where, $Y = SOTR/\Delta C$, Absorption number; $N_e = P/(rN^3d^5)$, Power number; $F = N^2d/g$, Froude number; $R = Nd^2/n$, Reynolds number, and $W = N^2d^3r_w/s$, Weber number.

The absorption number (Y) divided by power number (N_e) gives the simulation equation:

$$E = f_4 (h/d, b/d, \alpha, l'/d, n, V/d^3, F, R, W) \quad (10.24)$$

From experimental data two correlations, one between E and P^*, the non-dimensional parameter governing the theoretical power per unit volume (Figure 10.6a), and the other, between N_e and P^* (Figure 10.6b) were obtained for use of prediction of aeration performance. The correlation of oxygen transfer:

$$E = \{-21.58\, e^{(0.163P^*)} + 0.126\, (P^*)^{2.723} - 0.178 + 2.7\, P^* + 6.417 \ln(32.877\, P^*)\} \times 10^{-5} \quad (r^2 = 0.98) \quad (10.25)$$

showed that the whole simulation curve was divided into three zones. The linear equation, representing zone 1 ($P^* \leq 6.56$) as that of constant SAE, is:

$$10^5 E = 1.704\, P^* - 0.819 \quad (r^2 = 0.98) \quad (10.26)$$

whereas in zone 2 and zone 3 (P^* ³ 6.56 and beyond 10.68) effect of the sizes of the paddle wheel played a significant role.

The correlation of power consumption reflecting a combined effect of the relations between N and R, and N and F, expressed as:

$$N_e = 0.0157\, P^* - 0.595\, e^{(-0.715 P^*)} + 15.64\, P^{*(-0.0578)} - 12.854 \quad (r^2 = 0.98) \quad (10.27)$$

can be used for estimation of power consumption of a given diameter rotating at a particular speed.

Example 10.6

Find out the combination of paddle wheel speed (N) and diameter (d) such that maximum SAE is achieved based on an SOTR of 10 kg O_2/h. Assume that maximum SAE can be achieved at N^3d^2 = 2.744.

Solution

SOTR of 10 kg O_2/h requires power input = 10/1.656 = 6.04 kW

Now from eq. (10.23) N_e = 0.83, and from eq. (10.27), P^* = 13.068

Solving $P/(\rho N^3d^5) = 6.04 \times 1000/1000 \times N^3d^2 \times d^3 = 6.04/2.744 \times d^3 = 0.83$,

d = 1.383 m ≅ 1.39 m.

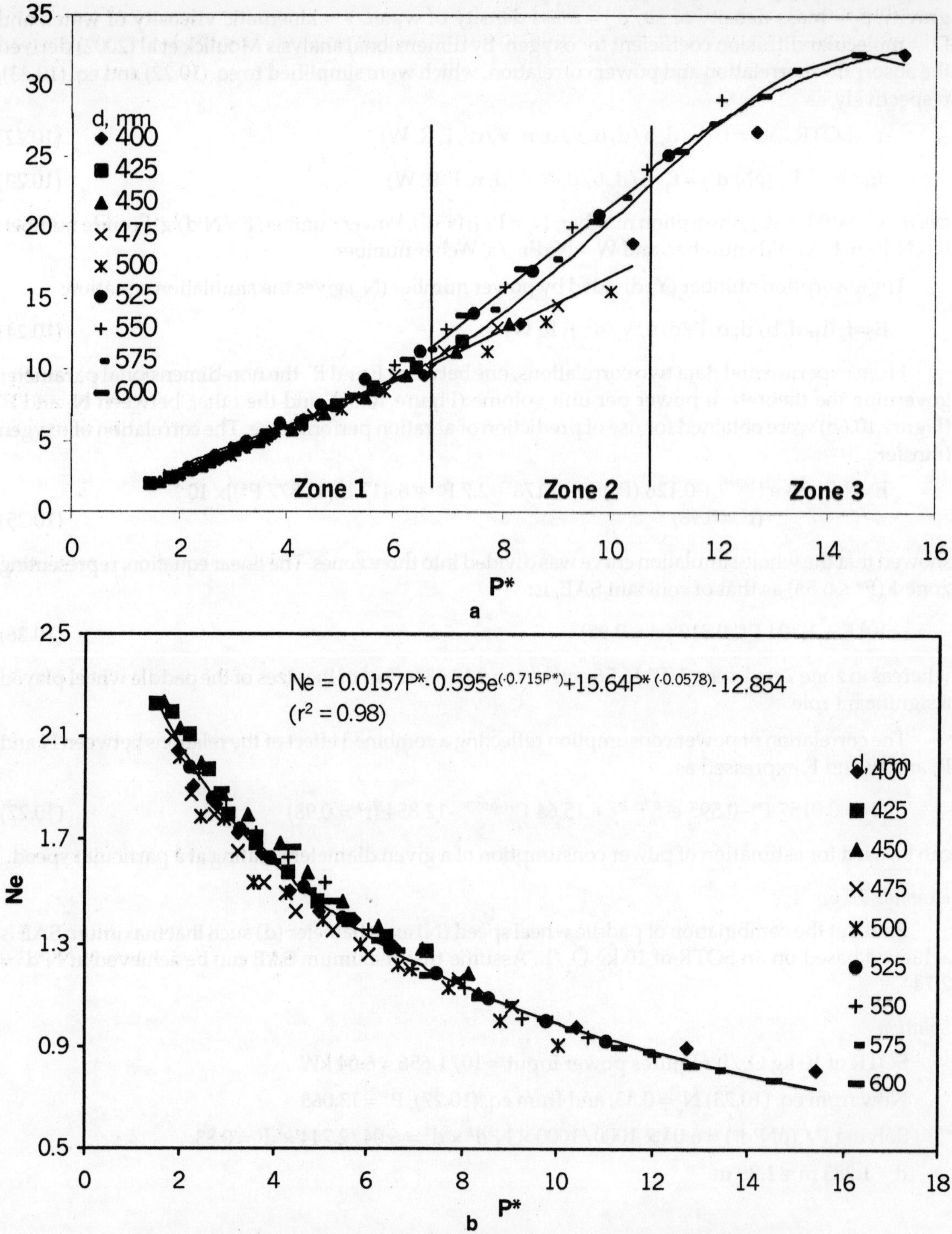

Figure 10.6: Simulation of Oxygen Transfer in Geometric Similar Conditions (a) Correlation between E and P*, (b) Correlation between Ne and P* (After Moulick *et al.*, 2005)

Thus required paddle speed = $(2.744/1.383^2)^{1/3}$ = 1,128 rps = 67.7, say 70 rpm. With these values of paddle speed and diameter, the optimum geometric parameters are obtained from the optimum geometric ratios (Moulick *et al.*, 2002):

$h/d = 0.125$, *i.e.* $h = 0.174$ m; $b/d = 0.325$, *i.e.* $b = 0.45$ m; $l'/d = 0.0625$, *i.e.* $l' = 0.087$ m; and $n = 6$, and $\alpha = 45°$.

11

Feed Management

11.1 Feeding Options

In general, there are four basic levels of input for feeding of the aquaculture species:

1. No fertilizer or feed input, where the rearing system is totally dependent on live food organisms naturally growing within the water body,
2. Fertilization to increase the growth of live food organisms,
3. Supplementary diet feeding, and
4. Complete diet feeding

On the basis of these levels, choice of feeding strategy that can be followed in aquaculture practice (Tacon, 1988) is shown in Figure 11.1. Subsistence farming is to produce fish or shrimp at the village level for home consumption, using locally available resources at minimum cost, usually in a single pond of 100 m^2, owned or operated by one or more family units. The farming activity is usually on a part-time basis, and requires little or no daily management. The most appropriate feeding strategy would be a low cost semi-intensive one using a combination of (*ii*) and (*iii*), *i.e.* pond fertilization with organic manures and supplementary diet feeding with agricultural by-products.

Commercial farming is for cash sale at maximum profit. It can be an extensive, semi-intensive or intensive culture, in lagoons, ponds, cages, concrete tanks or raceways. The feeding strategies vary with geographical location, the production unit used, and the species farmed. All feeding options of (*i*) to (*iv*) are used within commercial enterprises.

11.1.1 Pond Fertilization and Supplemental Feeding

Principles and method of pond fertilization have already been discussed in the previous section in Chapter 8. On salient features of supplementary feeding system an outline has been given in Section 6.5.2. In the following pages supplementary feeding practices have been dealt with.

Aquaculture Feeding Options

Subsistence Farming

i. No fertilizer or feed input

ii. Fertilization

iii. Supplementary diet feeding

Commercial Farming

i. No fertilizer or feed input

ii. Fertilization

iii. Supplementary diet feeding

iv. Complete diet feeding

Figure 11.1: Feeding Options (After Tacon, 1988)

Advantages of supplementary diet feeding combined with pond fertilization are: (*a*) use of higher fish and shrimp stocking densities, (*b*) faster growth of the cultured species, and (*c*) higher yields of crops. Selection of feed ingredients for use by rural or subsistence farmers must be based on (*a*) cost, *i.e.* feed ingredients at little or no cost, (*b*) availability–locally and throughout the year, (*c*) nutritional value, and (*d*) handling and processing.

1. Feed Ingredients

A list of potential raw materials for use in supplementary feeding at the rural farming level is given below (Tacon, 1988):

Kitchen/cooking waste–uneaten food and household scraps

Spoiled/contaminated animal feeds

Crop waste–leaves, stems, roots, tubers, peeling and seeds

Aquatic macrophytes

Mill sweepings

Slaughterhouse waste–animal offal, blood and rumen contents

Fruit wastes

Oilseed hulls and residues–bran, low value expeller cakes

Cereal grain hulls and bran

Sugarcane wastes–bagasse (as compost), molasses and filter press cakes

Terrestrial invertebrates animals–earthworms, snails and insects

Aquatic animals–worms, crustaceans and unwanted fish

In Asia, bulk of the ingredients used as supplementary feeds consists of agricultural by-products, or by-products of the animal husbandry industry. Fish are not capable of efficiently utilizing most of these ingredients, in particular the grasses. The feed conversion ratios of some commonly used typical single-component feeds selected from the above list vary widely such as 1.3 for clam meat, 6–9 for fresh sea fish (trash), 6–7 for wheat bran, 2–4 for groundnut cake, 4–6 for cereals and 10–40 for various grasses. Feed conversion ratios of some mixed diets are: fish flour and rice flour, 2.5–3.0; fish flour, soybean cake and yeast, 1.7–2.8; silkworm pupae, pressed barley, lemna and gammarus, 2.6–4 (Tacon, 1988). In mixed diet with several ingredients rice bran is the most widely used ingredient. Fish meal, trash fish and fish wastes are the commonest marine proteins used in fish feeds. In shrimp feeds several marine proteins are used–krill, shrimp meal, shrimp head meal, squid meal, clam and mussel meat.

A large number of feed ingredients in use in India are given in Table 11.1. Many ingredients are, however, reported to be of poor quality and not suitable for especially shrimp.

Table 11.1: Major Feed Ingredients Available in India (1992)

Ingredients	*Quantity (million ton)*	*Ingredients*	*Quantity (million ton)*
Rice bran	2.18	Fish meal	0.11
Deoiled rice bran	n. a	Shrimp waste	0.04
Wheat bran	1.20	Squilla	0.04
Groundnut oil cake	1.48	Clam meat	negligible
Sunflower cake	0.18	Mussel meat	Negligible
Soybean meal	0.84	Acetes shrimp	0.02
Rape seed cake	0.77	Poultry offal	0.06
Sal seed cake	0.098	Silkworm pupae	0.02
Mustard cake	0.098	Fish oil	0.01
Maize	9.10		
Sorghum	11.90		

Source: Dept. of Economics and Statistics, Dept. of Agriculture and Cooperatives, New Delhi, 1995.

2. Supplementary Feeds

We have already discussed in Chapter 6 the strategy of a supplementary diet formulation based on the standing crop or total biomass and the availability of natural food organisms in the culture system. Very little information is available on the types and extent of usage of different supplementary feeds for aquaculture. Various feeds range from single ingredients and simple mixtures to quite complex formulated ones. In Andhra Pradesh, India, there has been tremendous growth of polyculture practice of Indian major carp and a survey of 89 farms shows that 7 feed types are used by the farmers (Table 11.2).

In 1990s most of the global aquaculture output is produced using fertilizers and not with the use of feeds. It can be said that approximately 80 percent global aquaculture production is not using commercial feeds based on Asia's contribution of 85 percent of the world production (Akiyama, 1991). However, an estimate in world aquaculture shows that in 2000 more than 50 percent will be cultured through the use of feeds and a significant proportion will be reared through the use of dry and moist farm-made feeds ranging from simple mixture to relatively compounded pellet feeds (New and Csavas, 1995). In Asia, use of compounded pellet feed as supplementary diet has become increasingly popular, particularly in the shrimp industry. Some representative feed formulae adopted by Indian farmers are shown in Table 11.3.

Table 11.2: Supplementary Feed Ingredients Used in Semi-intensive Carp Polyculture in Andhra Pradesh, India

Type	*Farmers (%)*	*Average Total Input (kg/ha/year)*	*Other Ingredients Added*
Deoiled rice bran	8	20,340	Nil
Deoiled rice bran, cotton seed meal	4	25,280	Ground nut cake
Deoiled rice bran, deoiled ground nut cake	3	24,830	Cotton seed/Ground nut cake/ Sunflower meal
Deoiled rice ran, Ground nut cake	75	27,650	Cotton seed meal/ Sunflower meal/ Soybean meal/Sorghum/ Millets
Deoiled rice bran, rice bran	<1	22,260	Ground nut cake/ Cotton seed meal
Deoiled rice bran, Sunflower meal	1	46,800	Cotton seed meal/ Ground nut cake
Rice bran, deoiled bran	8	26,790	Ground nut cake/ Deoiled cake/ Cotton seed meal/Sunflower meal/ Sorghum/ Millet

Source: After De Silva, 1995.

Table 11.3: Farm-made Feed Formulae in Use in India (%)

Ingredient	*West Bengal*	*Andhra Pradesh 1*	*Andhra Pradesh 2*	*Tamil Nadu*
Fish meal	30	21	25	20
Shrimp waste	–	20	–	–
Squid meat	–	–	–	20
Meat/bone meal	–	20	–	–
Fish oil	–	1	–	–
Soybean cake	–	20	25	28
Ground nut cake	40	–	25	–
Sesame cake	–	–	–	5
Soya oil	–	1	–	15 (as corn oil)
Wheat bran	15	–	–	–
Rice bran	15	10	15 (as wheat flour)	12
Tapioca	–	6	5 (as millet)	–
Vitamin/mineral premix	–	1	5	–
Total	100	100	100	100
Form	Dry powder	n.a.	Balls	Pellets
FCR	2.5	n.a.	5	2.0–2.8
Yield, kg/ha/crop	400	n.a.	700	1000–1300

Source: After Nandeesha, 1995 (n. a. = data not available).

(De Silva, 1995). However, there is a great potential resource limitation of fish meal as focused by Wijkstrom and New (1989) and the actual fish meal requirements would be doubled by the year 2000. Recently there is an emphasis on supplementary ingredients *per se* but on use as substitute for fish meal. These ingredients are now commonly referred to as non-conventional feedstuffs, instead of as supplementary feed or ingredients. Comprehensive studies on such feed ingredients show that most oilseed meals and grains can substitute fish meal up to about 35 per cent of dietary protein and achieve more than 80 percent of the growth with fish meal as the primary source of protein in the control diet (De Silva, 1995).

In supplementary diet feeding use of a nutritionally balanced diet appears to be questionable because of resource wastage and economic reasons. The compound feeds for semi-intensive aquaculture need not be nutritionally complete and protein level in the diet could be reduced for economic reasons (De Silva, 1995). However, difficulties have been pointed out in formulating separate supplementary diets for every stage of growth condition, as the composition of supplementary diet changes with constantly changing conditions of natural productivity and standing crop of fish in the pond (Pillay, 1990).

Efficiency of a particular supplementary feed depends largely upon the presentation of the diet, which are available in physical forms ranging from powders, dough balls and moist pallets to machine-dried pellets. A single ingredient or simple mixture of two or more ingredients can be prepared through simple techniques of grinding and mixing. The raw materials may be all dry or moist, or partly dry. Simple mixtures such as groundnut cake and rice bran are often mixed without using any equipment or machinery at the farm level. The raw ingredients should be available as finely as possible from the market and then mixed by hand or shovel, or by pounding in a bowl. Care should be taken to avoid contamination by chemicals and microorganisms and to weigh the materials according to the correct proportions of the ingredients (New, 1987).

A compound feed formed into a shape involves several techniques of processing, which have been discussed in the following section.

11.1.2 Complete Diet Feeding

1. Types of Feed

Compound or compounded feeds are complex mixtures of several ingredients and are intended for complete diet feeding purpose. They are often fed simply as a meal or mash, or in some form of moist product, or in dry pellet. Types of compound feeds based on moisture content are dry (about 10 percent moisture), moist (about 30-45 percent moisture), wet (> 50 percent moisture), and intermediate between these. Compound feeds are mostly grouped as dry or moist (New, 1987).

The physical form of a compound feed is as important as its nutrient content. Feeds must be prepared in such a manner as to be available, acceptable, ingested and digested. The various forms are meals, crumbs, flakes, pellets and microcapsules.

Meals

Feeds containing a single or two ingredients are usually prepared as moist meals. For example, trash fish and rice bran or broken rice is minced to prepare a moist meal.

Pellets and Crumbs

Dry pelleted feeds can be processed as hard pellets (moisture content 13 percent), and soft pellets (moisture content about 18–20 percent). Further, sinking type and expanded or floating type pellets can be prepared depending on the manufacturing process. Crumbs are formed when pellets are broken into small pieces.

Flakes

Flakes are thin platelets produced by drum drying of a homogeneous wet suspension of liquid and dry materials on a heated drum. The flakes are fractionated into the desired size.

Microcapsules

Microcapsule is a miniature capsule in which dietary components of liquid and particulate materials are enclosed within. These feeds have broad applications in larval feeding.

2. Formulation

Formulation of a compound feed decides the inclusion levels of the various ingredients, once the choice of the feed type is taken. The ingredients are chosen on the basis of three major factors (New, 1987; Pillay, 1990):

1. Type and quantity of the ingredients for the species being cultured
2. Seasonal availability of ingredients at the local market and regularity of supply
3. Cost

A list of ingredients with an assessment of quality and composition is very essential before formulation work begins. Sufficient supply of the particular ingredients must be ensured in order to avoid frequent changes in formulation. Cost of ingredients constitutes a very high proportion of the total operating cost of the feed manufacture. In order to include each ingredient contributing a particular dietary component in the feed and to satisfy the dietary requirement of the species at the minimum cost, a 'least cost formulation' is worked out.

Least cost formulation is worked out most efficiently by optimization technique using a specifically designed computer program. The computer stores all the data of analysis and cost of raw ingredients, and nutritional requirement of the particular species. For the basic composition of a feed to be formulated, the main data required are the levels of crude protein, energy, specific amino acids, crude fiber and ash (Pillay, 1990). After the basic data has been established, its other compositional requirements can be examined and adjusted to arrive at the final, balanced formula.

Method of Computation

A simple formulation containing two or three ingredients can be achieved by square method or solving algebraic equation, as illustrated in the example below. Computation is done first with protein level, as protein is the most expensive component of a diet (New, 1987; Pillay, 1990).

Example 11.1

If the dietary protein requirement of a feed is 30 per cent, and the feed contains fish meal (55 per cent protein) and rice bran (12 per cent protein), calculate the inclusion levels of fish meal and rice bran in the feed.

Solution

Basis of calculation = 100 kg feed

Let x and y be fish meal and rice bran in kg/100 kg feed, respectively, then

$x + y = 100$ kg, and $0.55x + 0.12y = 30$

Solving the two equations, $x = 41.9$ kg and $y = 58.1$ kg.

Inclusion levels of fish meal and rice bran are 41.9 per cent and 58.1 per cent, respectively.

This is illustrated by the square or cross method below.

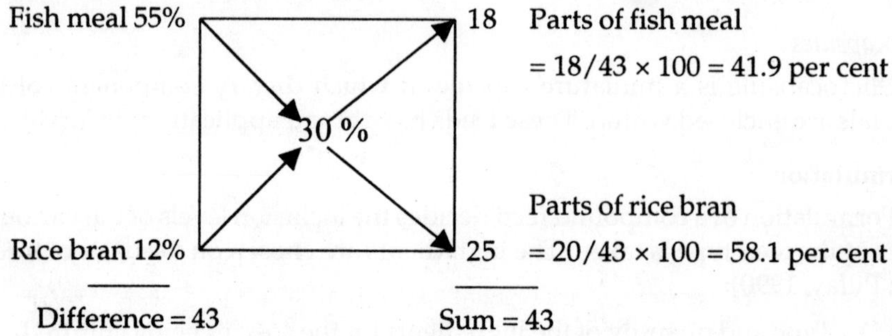

Example 11.2

In Example 11.1 the fish meal constitutes a high proportion and cost of the feed becomes high, if a good quality material is procured for formulation. In order to reduce the cost it is decided to substitute fish meal with soybean meal in the ratio of 1:2, and to balance the energy source of rice bran with wheat flour in the ratio of 1:1.

1. Estimate the inclusion levels of the ingredients and the lipid level of the compound feed.
2. What would be the total cost of the ingredients?

Table of 11.2.1 Data

Ingredients	*Protein (per cent)*	*Lipid (per cent)*	*Cost (Rs./kg)/1*
Fish meal (FM)	55	7	10
Soybean meal (defatted) (SM)	46	1.3	8
Rice bran (defatted) (RB)	12	2	5
Groundnut cake (extruded) (GN)	48	2	6
Wheat flour (WF)	14	1.3	7

/1 At 1995 price level

Solution

Average protein in one part FM and two parts SM = 55 + 92/3 = 49 per cent

Average energy in one part RB and one part WF = 12 + 14/2 = 13 per cent

Let x = protein source in kg/100 kg feed, and y = energy source in kg/100 kg feed.

Then, $x + y = 100$, and $0.49\,x + 0.13\,y = 30$.

Solving these equations $x = 47.2$ kg and $y = 52.8$ kg

Therefore, the components of the feed with inclusion level, contribution to protein and lipid, and cost are as shown in Table 11.2.2.

Table 11.2.2 Results

Component	Inclusion (%)	Contribution to		Cost (Rs./kg)
		Protein	Lipid	
FM	15.7	8.64	1.10	1.57
SBM	31.5	14.49	0.41	2.52
RB	26.4	3.17	0.53	1.32
WF	26.4	3.70	0.34	1.85
Total	100.0	30.00	2.38	7.26

Example 11.3

Suppose a formulation using all five ingredients given in Table 11.2.1 in Example 11.2 is to be worked out to meet the 30 per cent protein and 2.5 per cent lipid specifications. If fish meal inclusion level is taken as 16 per cent, work out the new formulation. To arrive at a least cost formula with five ingredients, which of the dietary specifications is to be adjusted and why?

Solution

Step 1. Four combinations of two ingredients, one with high protein and one with low protein, and one energy ingredient, have been worked out.

With FM = 16 per cent and other ingredients = 84 per cent, contribution to protein = 30–8.8 = 21.2 per cent, and FM portion of the diet = 21.2 × 0.84 = 25.2 per cent

Results obtained by square or cross method are given in Table 11.3.1

Table 11.3.1 Results of Example 11.3

Combination	Actual (%)	Final Diet (%)	Contribution to		Cost (Rs/kg)
			Protein	Lipid	
a) FM	-	16.0	8.8	1.12	1.60
GN	36.7	30.8	14.8	0.62	2.20
RB	63.3	53.2	6.4	1.06	3.17
	100.0	100.0	30.0	2.80	6.97
b) FM	-	16.0	8.8	1.12	1.60
GN	32.9	27.6	13.2	0.26	1.97
WF	67.1	56.4	7.9	0.10	4.70
	100.0	100.0	29.9	1.48	8.27
c) FM	-	16.0	8.8	1.12	1.60
SB	38.8	32.6	15.0	0.20	3.10
RB	61.2	51.4	6.2	0.12	3.06
	100.0	100.0	30.0	1.44	7.66
d) FM	-	16.0	8.8	1.12	1.60
SB	35.0	29.4	13.5	0.38	2.80
WF	65.0	54.6	7.6	0.71	4.55
	100	100.0	29.9	2.21	8.95

Clearly diet (a) meets the specifications best and minimum cost.

Step 2. Since all the five ingredients are available, a new formulation is to be worked out but with adjustment of either protein or lipid ingredients. Since increase in protein level would increase the dietary cost, the modified formula can increase the lipid level by selecting (d).

From (a) and (d), the new formula is worked out by using all the five ingredients with 30 per cent protein specification but 5 per cent lipid level. Combining (a) and (d), we get 2.80 + 2.21 = 5.01 per cent lipid.

Making a lipid balance with the desired level from all the five ingredients, *i.e.* lipid = 5.01–1.12 = 3.89, and solving by cross method:

Diet a) 2.80-1.12 = 1.68 → 3.53 3.53/5.74 x 100 = 61.5

3.87

Diet d) 1.48-1.12 = 0.36 → 2.21 2.21/5.74 x 100 = 38.5

The final formula is given in Table 11.3.2. The new formula meets the specifications with the five ingredients and also complies with the least cost criterion.

Table 11.3.2: Final Formula

Component	*Inclusion Level (%)*	*Final (%)*	*Contribution to Protein*	*Contribution to Lipid*	*Cost (Rs/kg)*
FM	16 per cent × 100 per cent = 16.00	16.0	8.8	1.12	1.60
GN	30.8 per cent × 61.5 per cent = 18.94	19.0	9.1	0.38	1.14
RB	53.2 per cent × 61.5 per cent = 32.72	32.7	3.9	0.65	1.64
SB	29.4 per cent × 38.5 per cent = 11.32	11.3	5.2	0.15	0.90
WF	54.6 per cent × 38.5 per cent = 21.02	21.0	2.9	0.27	1.47
		100.0	29.9	2.57	6.75

Usually a complete feed formulation containing more than two or three ingredients and even more complicated formulations are to be worked out. Information such as proportions of different protein sources and energy sources, besides the nutrient composition of the formulation is required. An elaborate calculation involving several simultaneous equations can be performed with a computer. The procedure involves optimization so that ingredients providing the necessary protein and energy levels may be changed according to availability, price and quality. Based on nutritional requirements of a particular species, *e.g.* protein, fat, carbohydrate, ash, fiber and energy (dependent variables), a linear programming model can be formulated by adjusting the decision variables so as to minimize the cost. Some critical nutrients such as methionine and lysine are specified as minimum, range, ratio or exact. Levels of minerals and vitamins need not be specified. Prices should be current and at the local market so that a realistic and effective least cost feed can be achieved (Pillay, 1990).

Example 11.4

A least cost formula for a marine shrimp feed is to be worked out from the ingredients given in Table 11.4.1 with proximate composition and coast (at 1993 price). The formula was developed by 30 per cent replacement of the imported fish meal with full fat soy flour. The dietary nutrient levels and energy were taken from the practical data available for the particular species (Tacon, 1993).

Table 11.4.1: Data of Example 11.4

Ingredient	*Proximate Composition (% dry basis)*					*Energy kcal/100 g*	*Cost (Rs/kg)*
	Protein b_1	*Fat b_2*	*Ash b_3*	*Carbohydrate*	*Fiber b_4*	*b_5*	
Fish meal X_1	76.18	10.82	12.64	0.36	0.21	403.54	26.00
Shrimp head meal X_2	46.49	5.30	36.69	11.52	8.39	279.74	2.50
Mussel meat X_3	47.42	16.61	27.86	8.11	4.72	371.61	22.00
Shrimp meal X_4	51.26	1.79	40.20	6.75	5.17	248.15	24.00
Wheat flour X_5	11.78	0.97	0.71	86.54	0.42	402.01	5.50
Full fat soy flour X_6	49.11	24.48	4.79	21.62	3.46	503.24	16.50
Deoiled rice bran X_7	17.18	0.51	18.81	63.50	15.45	327.31	2.00

Source: Gokulakrishnan and Bandyopadhyay, 1995.

Solution

The linear programming model with the objective function is to minimize

$$f(X) = C_1X_1 + C_2X_2 + \ldots + C_nX_n$$

subject to constraints $a_{11}X_1 + a_{12}X_2 + \ldots + a_{1n}X_n \geq b_1$

$$a_{21}X_1 + a_{21}X_2 + \ldots + a_{2n}X_n \geq b_2$$

$$\ldots$$

$$a_{n1}X_1 + a_{n2}X_2 + \ldots + a_{nn}X_n \geq b_n$$

where, $X_1, X_2, \ldots$ are the amounts of ingredients; $C_1, C_2, \ldots$ are the cost per unit weight of the ingredients; and $a_{11}, a_{12}, \ldots$ are the nutritional contents of the ingredients given in Table 11.4.1.

The recommended dietary intake levels (b's), given in the table are the constraints as set below: $44.44 \leq b_1 \geq 38.99$, $11.11 \leq b_2 \geq 6.66$, $17.78 \leq b_3$, $3.35 \leq b_4$, and $411.11 \leq b_5 \geq 316.67$

The other constraints are: $93.0 \leq X_1 + \ldots + X_7 \geq 92.9$, $X_1 \leq 18.62$, $X_5 + X_6 + X_7 \geq 40$, $X_2 \leq 10$, and $33.33 \leq X \geq 6$, all values are in per cent.

The maximum inclusion level of all the ingredients was kept at 93 per cent. The maximum inclusion level of imported fish meal was restricted to 18.62 per cent. The vegetable protein ingredients were kept at a minimum of 40 per cent in order to restrict the cost of the ingredients. The inclusion level of shrimp head meal was limited to 10 per cent due to high fiber content and uncertain availability. The minimum and maximum inclusion levels of any ingredient were fixed within 6 and 33.33 per cent considering the minimum and maximum values of most of the ingredients used for practical diet preparation for marine shrimps.

The designed formula arrived at by solving 25 equations is given in Table 11.4.2

Table 11.4.2: Final Formula

Ingredient	*Inclusion (per cent)*	*Cost (Rs/kg)*
Fish meal	18.61	4.84
Shrimp head meal	10.00	0.25
Mussel meat	6.00	1.32
Shrimp meal	6.00	1.44
Wheat flour	24.24	1.33
Full fat soy flour	20.20	3.33
Deoiled rice bran	7.95	0.16
Total	93.00	12.67

3. Preparation

Feed processing methods differ according to the type and form of the feed. Except for preparation of flakes and microcapsules, most of the processes follow a general procedure such as grinding the dry ingredients, mixing them in a mixer, blending the mixture with liquid ingredients such as oil, and preparing dough with sufficient quantity of warm water. Finally the dough is extruded through a die of a meat grinder or pellet mill or extruder. The extruded pellets are dried, broken into crumbles/granules and stored before use.

Processing involves a variety of operations:

1. Grinding or reducing the particle size for raw material preparation and pre-blending,
2. Separation by screening to select the desired particle size,
3. Mixing or blending into a uniform mixture,
4. Pelleting or forming pellets of a desired particle size and density by steam compaction or extrusion cooking,
5. Cooling/drying to reduce the moisture content to safe storage limit,
6. Packaging of the finished product, and
7. Storage.

Grinding is one of the very important operations, also called "milling" when reducing the grains into flours and "mincing" when reducing the size of large fish and crustacean meats into small pieces. Grinding improves feed digestibility, acceptability, mixing properties and pelletability. Mixing operation is performed to achieve uniformity of composition in the whole feed formulation. Cooking of starchy components such as rice bran and wheat flour during pelleting or extrusion helps in achieving partial gelatinization of raw starch, and results in binding and water stabilizing effect on all the ingredients. Drying of pellets is an essential operation, but should be preceded by cooling of hot pellets.

For successful application of feed the engineering properties that are to achieved during processing are:

1. Particle size and particle homogeneity,
2. Particle density and settling rate,

3. Water stability (solubility and differential leaching),
4. Storage stability of nutrients, and
5. Resistance to particle attrition during handling and storage.

Particle sizes vary from finely ground granules for first feeding, through crumbles and small pellets, up to large pellets for adult feeding. Particle density decides the type of sinking, floating or slow settling pellets in order to become available in the water column where fish swims. One of the major considerations of process technology is to get water stable pellets with minimum disintegration of particles and leaching of soluble nutrients, and maximum conversion ratio. Storage characteristics such as lipid oxidation, insect damage and dust formation are other vital properties of the feed.

11.2 Feeding Methods

11.2.1 Feeding Rate and Frequency

1. Feeding Rate

Feeding Rate is a vital factor in efficient aquaculture. While underfeeding can result in loss of production, overfeeding will cause wastage of costly feed. Unconsumed feed is additionally a potential cause of water quality deterioration, resulting in loss of animals or requiring expensive corrective measures for water treatment. Thus both overfeeding and underfeeding can lead to serious economic consequences and viability of the farm.

In general, feeding rate is calculated from the quantity of the feed to be given to the animals. The quantity of feed is normally based on a percentage of the biomass present (total weight of animals). For example, if a pond contains 20,000 animals weighing 15 g on average and the recommended feeding rate is 5 percent of biomass per day, the amount of feed to be given daily is: 20000 x 15 x 5 / 100 = 15000 g or 15 kg per day.

The percentage of biomass to be fed, however, should depend on the growth of the animal, *i.e.* average animal weight, and the number of animals in the production unit, *i.e.* survival. Average weight can be calculated directly from weighing samples or by measuring the length of the animals. Length measurements are a more accurate means of monitoring growth rate, if accurate length/weight relationships of the species have been pre-determined.

For average animal size accurate data must be obtained regularly such as through weekly, biweekly or at the worst, monthly sampling of the rearing pond. Sample is to be taken in several parts of the pond by seine, cast net or lift net. For a pond size of 5000 m^2, stocked at 10/m^2, minimum five samples comprising of 50 animals should be obtained at each sampling time and average weight calculated from the total weight measured.

Survival assessment is often extremely difficult to achieve in ponds. It can be calculated, during stock transfer from one pond to another, by taking the total weight of the stock and dividing by the average animal weight obtained from samples. In practice most farm managers apply an arbitrary survival factor based on the number of days since stocking. This factor is derived from the farm records, *i.e.* knowledge of previous culture cycles, modified by observation of actual mortalities, water quality, disease problems, etc. (New, 1987).

Example 11.5

Assume, for a pond stocked with 1,00,000 animals, 50 per cent mortality occur between stocking and harvesting in a 16 week cycle, and 20 per cent of the losses occur within the first 4 weeks with a further 10 per cent loss occurring during every 4 week period after that. Show how the assessment of biomass at 2 weeks' intervals would result and calculate the daily amount of feed to be applied for 15 g size beginning with the week 13 (Data adapted from New, 1987).

Solution

Assessment of biomass in weeks 3 and 4 = 1,00,000 × 90/100 = 90,000, in weeks 5 and 6 = 80,000, ...in weeks 13 and 14 = 60,000.

Thus, if the average weight of animals, at beginning of week 13 = 15 g, and percentage of biomass to be applied = 5, the daily amount of feed

= 15 × 60,000 × 5/100 = 45,000 g/day *i.e.* 45 kg/day.

2. Feeding Frequency

Besides feeding rate, feeding frequency (number of feeds per day) is an important criterion to achieve maximum benefit when the food is ingested entirely by the species particularly during complete feeding. The essential requirements of fixing feeding frequency are:

1. The period the feed remains in water are kept to a minimum, *i.e.* the feed should be consumed before disintegration and solubilization starts.
2. The feed is presented to the species in the correct physical form (*i.e.* size) and in such a manner over the working day so as to get a maximum feeding response.

Figure 11.2 shows the effect of feeding rate on growth and FCR in common carp post-larvae and fry culture, where

$$\text{Specific growth rate} = \frac{\log_e \textit{final body weight} - \log_e \textit{initial body weight}}{\textit{Total growth period (day)}} x100$$

It shows that the optimal feeding rate range depends on the species age and diet (Bryant and Matty, 1981). Similar relationship between feeding frequency and growth has been found, for example, optimum growth and feeding efficiency require 'satiation' feeding, (*i.e. ad libitum* or feeding to appetite), only once every two days with trash fish diet for juvenile grouper, and continuous or very regular feeding throughout a 24-h cycle for post-larval common carp and the African catfish. A direct relationship was observed between growth rate (*e.g.* fish length) and feed particle size in juvenile Atlantic salmon. Optimum feed sizes generally vary as 0.4–0.6 of the mouth size in warm water fish (Tacon, 1988).

Correct feed presentation and feeding level are very important. Feeding practices according to the convenience of the feeding technician shows that majority of semi-intensive shrimp farmers still feed only once or twice daily, during the early morning and/or late afternoon, instead of distributing their feed at regular intervals during the night from dusk to dawn. Shrimps have nocturnal feeding behavior and there occurs rapid loss of water-soluble nutrients from rations of prolonged immersion in water. Loss of water-soluble vitamins to the extent of 95 percent has been found to occur through leaching from a conventional pellet after one-hour immersion period (Tacon, 1988).

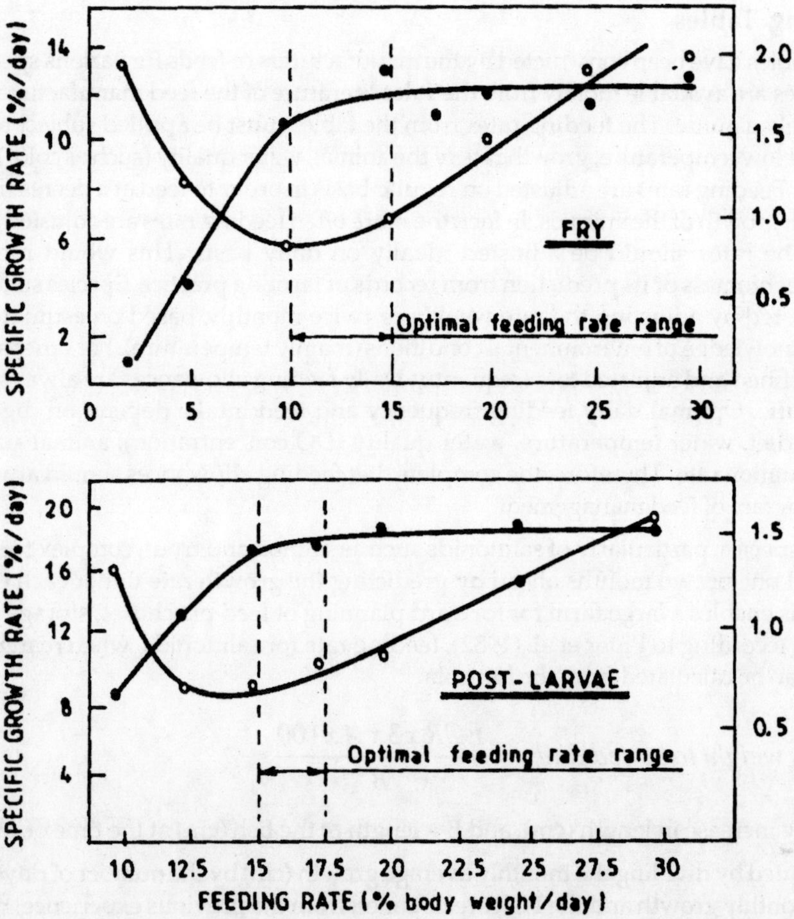

Figure 11.2: Effect of Feeding Rate on Growth (•) and FCR (o) of Common Carp Post-larvae and Fry Fed a Commercial Trout Fry Diet (After Bryant and Matty, 1981)

Some basic rules of feeding frequency, suggested by Piper et al (1982) are as follows:

1. For optimum growth and feed conversion, feed rate should ideally be only 1 percent of the body weight. For example, 5 feed/day means 5 percent of biomass/day.
2. Feeding frequency does not significantly affect survival rate after the initial growth stage of the animals.
3. Starvation and stunting of small fish are reduced by frequent feeding, thus resulting in better uniform growth of the group.
4. Irregular feeding results in feed wastage, poor FCR, water quality deterioration due to leaching out of water soluble nutrients.
5. Dry feeds should be distributed more frequently than moist feeds.
6. For fish 90 percent of the feed should be consumed within 15 minutes or less of the feeding time.

11.2.2 Feeding Tables

Feeding tables have been constructed by the manufacturers of feeds for various species. Examples of feeding tables are available mostly from the sales literature of the feed manufacturers. All feeding tables are merely a guide. The feeding rates from the tables must be applied subject to other factors such as times of low temperature, growth rate of the animal, water quality (such as color), and presence of old feed, etc. Feeding rates are adjusted on regular basis in order to feed at a decreasing percentage of biomass with growth of the animals. In fact, the more often feeding rates are adjusted; the better will be the FCR. The rates should be adjusted ideally on daily basis. This would require accurate measurement of biomass or its prediction from records of farming practice. Species such as salmonids and catfish are fed by adjusting the rate weekly or twice monthly, based on estimates of biomass, together with knowledge of environmental conditions (mainly temperature). For other species feeding rates are less refined and adjusted less frequently. Daily feeding allowances are always shown as diet and farm specific. Optimal daily feeding frequency and feed intake depend on digestible energy content of the diet, water temperature, water quality (DO concentration), animal size and gastro-intestinal evacuation rate. Therefore, the complete diet feeding allowances should always be treated as tentative program of feed management.

For some species, particularly of salmonids such as salmon and trout, complex feeding schedule can be planned one or two months ahead by predicting the growth rate under each environmental conditions. This enables a large farm for forward planning of feed purchases, storage requirements, cash flow, etc. According to Piper et al. (1982), feeding rate for salmonids, when reared at a constant temperature, can be calculated from the formula:

$$\% \, body \, weight \, to \, be \, fed \, daily = \frac{FCR \, x \, 3 \, x \, A \, x 100}{B}$$

where, A = daily increase in length (cm), and B = length of the fish (cm) at the time of calculation.

A is calculated by dividing the monthly average growth (cm) by the number of days in the month. The average monthly growth and the FCR are obtained from the previous experience, recorded in the farm data, of the animals of the same size, reared at the same temperature. Many years of experience with long-term record keeping in a sophisticated farm management system are required for this type of calculation. One feeding table suggested for salmonids is given in Table 11.4.

Example 11.6

For 100,000 fish in the pond, feeding rate was determined on the basis of 9.07 kg/1000 fish or 1.45 cm fish length. Average length increase during April = 0.0075 cm/day. FCR = 1.3. If the last feeding rate was adjusted on April 1, what would be the new feeding rate on April 15? (Data adapted from New, 1987)

Solution

A = 0.0075 cm. B = 1.45 + 15 × 0.0075 = 1.56 cm, *i.e.* 11.23 kg /1000 kg (obtained from table of length-weight relationship).

% Daily feed = 1.3 × 3 × 0.0075 × 100/1.56 = 1.88

Weight of feed = 100,000 × 11.23 × 1.88/000 × 100 = 18.8 kg

Table 11.4: Suggested Feeding Frequency for Salmonids

Species	*Fish size (g)*								
	0.3	0.45	0.61	0.91	1.82	3.60	6.10	15.10	>45.1
	No. of Feeding Times per Day								
Coho Salmon	9	8	7	6	5	3	3		
Rainbow Trout	8	8	6	6	5	4	4	3	2

Source: New, 1987.

Suggested feeding frequency for channel catfish is 8–10 feeds/day, reduced to 6/day at 2.5 cm length, further to 3/day at 7.5 cm length. For juveniles 2 feeds/day can be followed at 13–29° C 6–7 times, and at lower or higher temperatures, less frequent at 4–5 times/week can be followed (New, 1987).

For shrimps and prawns controversy exists on the optimum time and frequency. Some species burrow during the day and feed most actively at night. Others feed in the shallow parts of the pond but avoid these areas in daylight when temperatures are highest. Shrimps do not consume all of the feed presented at once, unlike most fish. In Taiwan, 4–6 per day in intensive farming are followed. It should be as frequently as possible, spreading the daily ration between these feeds (New, 1987).

Adjustment of feeding tables is shown in the following example (Tables 11.5 and 11.6).

Table 11.5: Experimental Feeding Guide for Carp

Temp.	*Pellet Size (mm)*	*% Body Weight (Biomass)/Day*					
		1.5	*1.5*	*2.7*	*4*	*5*	*5*
(° C)	Animal weight (g)	<5	5-20	20-50	50-100	100-300	300-1000
20-23		9	7	6	5	4	3
23-26		12	10	8	6	5	4
> 26		19	12	11	8	6	5

Source: Adapted from New, 1987.

In the above table the feeding rate is changed less frequently with the animal weight. However, more frequent adjustments can be made as shown for feeding rates at 20-23° C in Table11.6.

No "universal" feeding tables exist in case of supplementary feeding, but it varies with composition of the diet, natural food availability, water quality (mainly dissolved oxygen and temperature), and species (age, density and standing crop). The proportion of supplementary diet fed/unit body weight should be gradually increased over the course of the farming cycle. This is because natural food organisms play a gradually diminishing role in the overall nutritional budget as the standing crop increases within the pond with time. However, optimal feeding rates and frequency of feeding must be determined for the specific feed and farming (Tacon, 1988). Some examples of feeding rates and frequencies employed in semi-intensive practice are illustrated below in Tables 11.7

Table 11.6: Adjustments for Feeding as Shown in Table 11.5

	% of Biomass/Day	
Animal Size (g)	*Recommended*	*Adjusted Actual Rate*
5		7.0
10	7	6.7
15		6.3
20		6.0
30		5.7
40	6	5.3
50		5.0
Etc.		

Source: Adapted from New, 1987.

Table 11.7: Supplementary Feeding Regime for Common Carp Broodstock

Country	*Culture Condition*	*Type of Feed*	*Feeding Regime*
Bangladesh	Fertlized pond	Mustard oil cake: wheat bran (1:1), made into balls with water	3 per cent body weight, 10 per cent for spent fish
India	Manured pond, poly-culture system, stocking-1000-2000 kg/ha	Rice bran: oil cake (1:1)	1 per cent body weight/ day
Indonesia	Manured ponds, Stocking-2000kg/ha Fish weight, 300g–2kg, 1-4 kg	Rural farmers–Rice bran mixed with fresh vegetation, waste palm and groundnut oil.	–
		Hatcheries–pellet (20-25 per cent protein and 8 per cent fat, max.) containing rice bran, 50 per cent; fish meal, 25 per cent; leaf meal, 12 per cent; vitamin, mineral and antibiotic premix, 1 per cent	2-3 per cent body weight/day
Philippines	Ponds and concrete tanks	Rice bran and molasses, or Rice bran plus copra meal (1:1), or rice bran alone	5 per cent body weight/day
Thailand	Manured/fertilized Pond. Stocking, 20-30 m^2/2-4 kg fish size	Fish meal-based pelleted feed, 25 per cent protein	At various rates

Source: Data extracted from Jhingran and Pulin, 1985.

11.2.3 Feeding Devices

1. Hand–feeding or Manual Feeding

The most common method is feeding by broadcasting over water surface. Normally a boat is used for feeding by hand or shovel in large ponds and enclosures. Manual feeding is followed also by placing floating feeds such as vegetable items or pellets into a floating or fixed bamboo frame or by keeping sinking feeds on a submersible tray, or by tying bundles of vegetable materials and fixing to pond bottom (Tacon, 1988). Although labor-intensive, hand-feeding has the advantage that the operator can inspect his stock regularly and check any left-over feed. There are several mechanical aids to hand-feeding (New, 1987; Pillay, 1990):

Table 11.8: Supplementary Feeding Table for Common Carp in Fertilized Pond in Israel [1,2]

Fish Size (g)	*Carp Density/ha in Thousands*						
	2–4	*4–6*	*6–8*	*8–12*	*12–15*	*15–20*	*20–50*
20-50	–/ 1	–/ 2	1 / 1	1 /–	1 / 2	1 / 1	1 / 2
Up to 100	–/ 2	–/ 3	1 / 3	2 / 2	2 / 2	2 / 1	2 / 2
" 200	1 / 5	2 / 6	4 / 5	7 / 2	6 / 2	5 / 1	5 / 2
" 300	3 / 7	6 / 5	10 / 2	8 / 2	8 / 2	6 / 2	7 / 2
" 400	6 / 5	10 / 3	11 / 3	11 / 4	9 / 3	8 / 2	9 / 2
" 500	9 / 5	12 / 3	13 / 3	14 / 3	11 / 2	9 / 2	10 / 2
" 600	11 / 4	13 / 3	14 / 3	15 / 3	12 / 3	10 / 2	11 / 2
" 700	12 / 3	14 / 3	15 / 3	15 / 4	13 / 3	11 / 2	11 / 3
" 800	12 / 4	14 / 4	15 / 4	16 / 4	13 / 4	11 / 3	12 / 3
" 900	13 / 4	15 / 3	16 / 3	17 / 3	14 / 3	11 / 4	12 / 4
" 1000	14 / 3	15 / 4	16 / 4	17 / 4	15 / 3	12 / 3	13 / 4
" 1100	14 / 4	16 / 4	17 / 4	18 / 4	15 / 4	13 / 3	14 / 4
" 1200	14 / 4	16 / 4	17 / 4	18 / 4	15 / 4	13 / 3	14 / 4

Note: 1 First number in each column, g pellet (25 per cent protein)/fish/day; second number, g sorghum grains/fish/day

2. Feeding table refers to water temperatures above 24° C; for temperatures of 20-24 and 18-20° C, use 70 per cent and 50 per cent of the above quantities, respectively.

Source: Adapted from Tacon, 1988.

Table 11.9: Feeding Strategy for Freshwater Prawn *(M. rosenbergii)* in Fertilized Pond in Panama

Culture Condition	*Feed*	*Feeding Rate*
Nursery: 30 day cycle, ponds stocked with 100 PL (2-4 week old, 10-15 mm length/m^2	40 per cent protein diet	20 per cent body weight/day
Grow-out: 6 months cycle, stocking 3 juveniles/m^2	Artificial pellet (composition not given)	Daily feeding, g/1000 animals: 1st month–5 2nd month–25 3rd month–100 4th month–500 5th month–1850 6th month–3000 Feeding rate: 20 per cent body weight/day, after stocking, reducing to 6 per cent during the following 15 days, and later reduced from 5 to 3 per cent. Total ≤ 34 kg/ha/day

Source: Adapted from Tacon, 1988.

Table 11.10: Nursery Feeding Schedule for *P. monodon* in Malaysia

Culture Condition	*Feed*	*Feeding Rate*
PL 5-6 stocked at 4/m^2: First 7 days	100 per cent minced trash fish, spread on trays	20-40 per cent body weight/day
8th to 30th day	40 per cent rice bran and 60 per cent trash fish, formed into wet pellet	Fed once daily during the evening, 20 per cent
After 30th day	Same as above plus whole trash fish	10-15 per cent

Source: BOBP, 1984.

Table11.11: Feeding Scheme for Semi-intensive/Modified Extensive Shrimp Farming

a. After stocking of 1-30 days size

Age (Days)	*Feed Type and Size*	*Quantity per day No. of Shrimp*	
		50,000	*100,000*
1	Fine crumble, 0.42 mm	1	2
2–10	Protein, 42 per cent min. Fat, 5 per cent min.	1.2–2.8	2.4–5.6
11–30	Crumble, 0.89, or 1.41 mm Protein, 40-41 per cent; Fat, 5 per cent	3.1–8.8	6.2–17.6
31–50	Crumble, 1.41 mm or pellet, 1.8 × 3.5 mm Protein, 40–41 per cent; Fat, 5 per cent	9.05–13.8	18.1–27.6

b. For grow-out shrimps

Age (Day)	*Body Weight, Max., g*	*Feed, % Body Weight*[1/]	*Feed Type*
50–55	6–7	5.5–5.3	Pellet, 1.8x3.5 mm
55–60	7–8	5.3–5.1	Protein, 40-41 per cent;
60–65	8–9	5.1–4.9	Fat, 5 per cent min.
65–70	9–10	4.9–4.7	Pellet, 2.3 × 3.5 mm
70–77	10–12	4.7–4.5	Protein, 39 per cent min;
77–83	12–14	4.5–4.2	Fat, 5 per cent min.
83–90	14–16	4.2–4.0	
90–97	16–18	4.0–3.7	Pellet, 2.3 × 4.5 mm
97–104	18–21	3.7–3.4	Protein, 38 per cent min.
104–110	21–24	3.4–3.2	Fat, 5 per cent min.
110–117	24–27	3.2–3.0	
117–124	27–30	3.0–2.7	Pellet, 2.3 × 6.7 mm
124–131	30–33	2.7–2.5	Protein, 38 per cent min.
131–133	33–36	2.5–2.3	Fat, 5 per cent, min.

Note. Number of feeding per day = 4

Source: Sales literature of C.P. Group, Chennai, India.

1. Hand-operated blower inside a tank blowing the feed through a slanted pipe.
2. Blower mounted on a truck or towed by a tractor and propelled by a motor, transferring the feed by measurement also.
3. Hopper with a rotating arm at the bottom expressing the feed through a slit.
4. Hopper with a motor-driven screw at the bottom expelling the feed through a nozzle.
5. Mobile boat-mounted blowers, releasing differential amounts of feed through longitudinal slots in the bottom of the boat.

2. Automatic Feeders

Automatic feeders are devices operated mostly by electrical power, and fall into two categories–mechanically operated and compressed air feeders. Both are operated by timers to make it possible to dispense set amounts of feed at timed intervals. Many farmers prefer to dispense 70–80 percent of the feed by automatic device and the rest by hand so that feeding behavior of the stock can be checked (Pillay, 1990).

A compressed air feeder supplies feeds by ejecting through distribution pipe by timed release of a valve. Feed falls by gravity from a hopper into the distribution pipe mounted above it. The compressed air supply can be attached to a number of feeders. The mechanical feeders operate with the help of a number of devices (New, 1987). Movement by an electro-magnet of slugs over holes in a feed trough, or of two troughs one inside the other with coinciding holes supply feed by gravity. Hoppers motor-operated by discs, screw mechanism, blower, conveyor belt or a series of spikes on a revolving spindle are several devices attached to feeders. In modern farms use of computer-controlled automatic feeders regulate through appropriate programs based on variables like stock density, growth rate, conversion ratios, temperature, mortality, harvest time and so on (Pillay, 1990).

However, so far a power-free, inexpensive, automatic feeder is not available in many countries. An automatic feed dispenser, without using any electric power, was designed by Mal (1996). It is simple in construction and similar to that of the Hawaii-type (Baldwin, 1983), and can be fabricated using locally available, inexpensive materials. The feeder consists of a rotating bicycle wheel mounted on a slotted angle iron frame, feed boxes, trip arm, water tank and a float arrangement (Figure 11.3). The float gradually descends along with the slow rate of fall of the water level in the tank due to

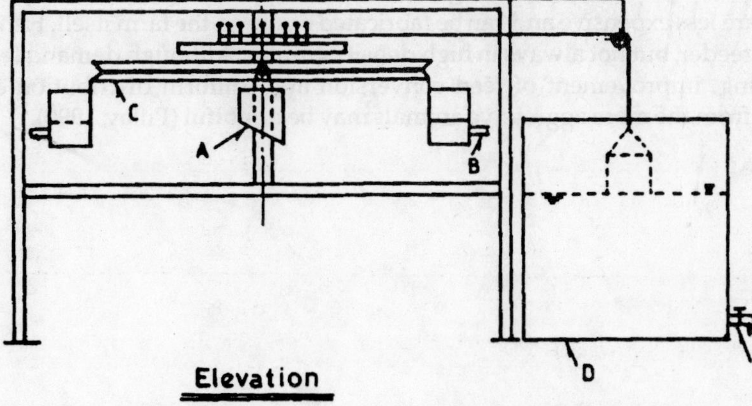

Figure 11.3: Elevation Drawing of the Hawaii-type Feed Dispenser
A: Feeder box; B: Trigger; C: Rotating wheel; D: Water reservoir; and E: Gate (After Mal, 1996)

discharge through the tank bottom. Discharge of feed is accentuated by rotation of the wheel and opening of the trap door by the trip arm. The timing of feed distribution depends upon the rate of descent of the float and the spacing of the boxes. At the initial setting when the tank is full, discharge through the opening is high and gradually reduces with the reduction in head.

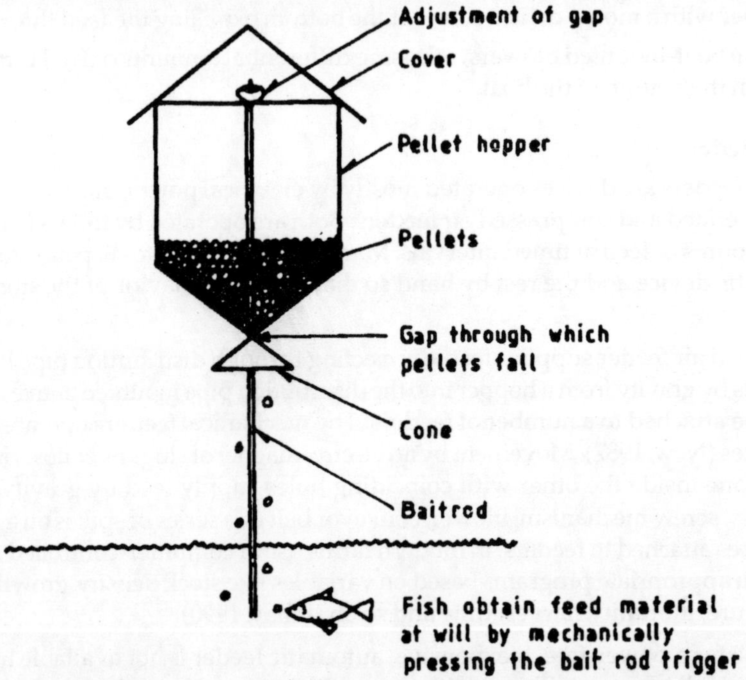

Figure 11.4: Demand Feeder with a Bait Rod (After New, 1987)

Another type of mechanical feeders, called demand feeders, operates by being activated by the animals themselves when they demand the feed. The feeder essentially is a hopper with an opening, through which bait in the form of a rod or plate moves to release a small amount of pellets. One such design of demand feeder is shown in Figure 11.4. The hopper can be refilled depending on its capacity. The feeders are less expensive and can be fabricated easily on the farm itself. Fish can learn to activate the demand feeder, but not always in high density culture. Although demand feeders manipulate *ad libitum* feeding, improvement of feed conversion and uniform distribution of feed impeded by competition from the more aggressive animals may be doubtful (Pillay, 1990).

12

Effluent Treatment Systems

Water quality conditions are maintained within the culture units by the methods of in-pond treatment designed for control and management of aquatic environment. On the other hand, water quality deterioration and problem of waste management occurs when large amounts of organic matter are discharged mainly from intensive culture units into the surrounding water bodies and create a serious pollution threat. Development of treatment systems for effluent from culture units thus becomes pertinent for maintenance of water quality standards set by the regulating agencies in different countries. Conversely, water circulation requirement for reuse of the effluent water in the closed system aquaculture is guided by the water quality criteria of the culture species and of the culture site. The most common water treatment procedures are followed for both systems of effluent discharge into the streams and water reuse in the culture units. The procedures have been referred to in Table 7.2 in Chapter 7.

12.1 Types of Waste Materials in Aquaculture Effluents

Wastes are discharged from farms as effluents and enter the receiving waters. They consist of solid and soluble types. Solids may be in suspended form or may settle on the pond bottom as sediment. Soluble wastes are usually derived from metabolites of the culture stock, or from decomposition and leaching of solid wastes. Dissolved waste products such as nitrogen and phosphorus compounds remain in the water, whereas organic wastes like carbon compounds reach the sediments and enrich the benthic system. As has been indicated earlier in Chapter 7, we have categorized four main waste components in the effluents: suspended solids, nutrients, biochemical oxygen demand (BOD) and pathogens.

12.1.1 Suspended Solids

Suspended solids in aquaculture effluent are an important and basic measure of effluent quality. Various types and sizes of solid particles are suspended in water–sediment particles such as clay and silt, organic materials such as detritus composed of plant and animal remains, waste food particles and fecal matter. In aquaculture, the type and size of solid particles depends primarily on quality of food used and feces produced, size and type of the animal species cultured, and amount of turbulence and resuspension in the culture system. Uneaten food and excreta are the main sources of solids in aquaculture. Fecal material typically comprises undigested dietary materials bound together in a mucus coat, frequently produced in long strings. These strings are normally denser than water, but sometime remain suspended due to mild turbulence created in water (Muir, 1982).

It was estimated that fecal production of salmonids, catfish, carps and shrimps ranged from 26–46 percent of the ingested feed, on a dry weight basis. Reported solids production values (uneaten and excreta) in fish culture systems range from 11 to 38 percent of the applied feed. The large variation can be attributed to differences in feed, species, management and differences in decay of organic matter within the culture unit (van Rijn, 1996).

The size of solid particles suspended in water range from 0.01 µm to 1 cm (Figure 12.1). The dispersed particles those are smaller than about 1 µm and larger than individual molecules (about 0.001 µm) exhibit colloidal properties. They can be distinguished only by means of an ultra microscope or an electron microscope, but usually they appear turbid even to the naked eye. Virus and bacteria fall in this range. Colloidal particles contribute little to the normal turbidity of water, but they are largely responsible for the color of natural waters. A colloid composed of liquid particles is called an 'emulsion'. Emulsions are prevalent in natural waters, but oils, fats and soaps are frequently found in the emulsified form in sewage and other liquid wastes. Mud particles, fine sand and sludge floc fall in the range of suspended solids.

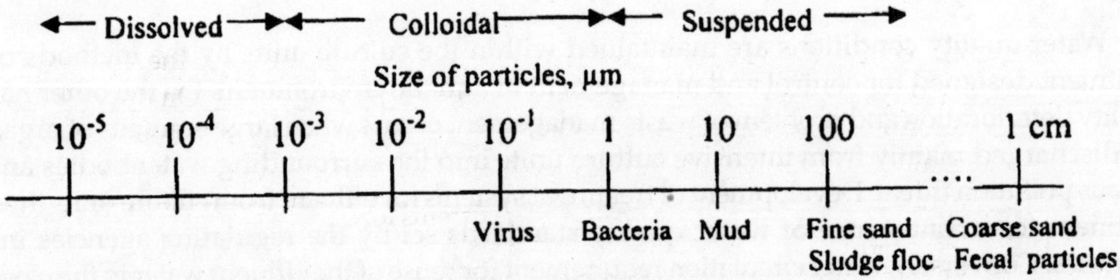

Figure 12.1: Classification and Size Range of Particles Found in Water and Wastewater

Size range of solids generated in aquaculture is about 1 cm (Muir, 1982). Studies on particle size distribution and specific gravity of suspended solids in fish culture effluents showed that size range was approximately 1–100 µm, and more than 95 percent of the suspended particles had a diameter less than 20 µm (Table 12.1)

Table 12.1: Particle Size Distribution and Specific Gravity of Suspended Solids in Fish Effluents

Culture System (Trout)	*Method of Treatment Used*	*% of Particles with*		*Mean Dia. (µm)*	*Specific Gravity*
		dia. < 10µm	*up to 20 µm*		
I	Screening and foam fractionation	85	95	11.7	1.19
II	Nitrification in a RBC	95	95	9.3	1.18
III	Sand filter and biological filter	60	95	13.5	–

Source: Chen *et al*., 1993.

Solid particles in the recirculating systems studied by Chen *et al.* (1993) were found to be slightly denser than water, *i.e.* 19 percent. Therefore the average specific gravity of the particulate matter was taken as 1.19. The specific gravity of solid particles largely depends on their organic nature, which originates from uneaten feed, excreta, or bacteria grown in the systems.

12.1.2 Nutrients and BOD

The two major nutrients produced as waste are nitrogen and phosphorus. Phosphates produced are in the particulate form, which may sediment out or may be later released into the surrounding water. Conversely, nitrogen is mostly in the dissolved form, with 26–32 percent of total nitrogen (TN) being in the particulate form (Bergheim et al., 1993; Cripps, 1995). In general, the inorganic and organic fractions of total nitrogen (TN) and total phosphorus (TP) in intensive aquaculture wastes differ, with about 60–90 percent of the TN and TP in the dissolved fraction and in the organic, particulate fraction, respectively (Costa-Pierce, 1996). However, in the low salinity areas such as in freshwater salmonid farm increased phosphorus concentrations were found as 60 percent soluble PO_4-P, 28 percent soluble polymerized and organic phosphorus and 12 percent particulate phosphorus (Pillay, 2004).

Although nitrogen and phosphorus are partly associated with the suspended particles in the effluent, various factors such as water temperature, feeding rate and feed composition affect the degree of association (Table 12.2). A direct relationship exists between the TP content of feed and the TP content of wastes discharged. Leaching of nutrients from the particulate into the solution is influenced by mechanical agitation or damage to the particles, and increased flow around the particles (Cripps and Kelly, 1996).

Table 12.2: Variation of Nutrient Percentage in Particulate Materials

Nutrient Concentration (%)		*Particle Size (µm)*
Total N	*Total P*	
26.6	30	> 0.45
7–32	47–84	> 60
20	25	> 0.45

Source: Cripps and Kelly, 1996.

Waste organic matters in effluent contribute to BOD as a result of their microbial breakdown within the culture water and also during discharge by using oxygen. Microbes directly use oxygen and produce carbon dioxide as end product. Depletion of oxygen in water assesses the environmental impact and the quantity of organic matter present in the effluent determines the treatment objectives. There are several limitations of BOD measurement in saline waters. So total organic carbon (TOC) and chemical oxygen demand (COD) are also used as water quality criteria. COD may be a more significant measure of pollutants (Cripps and Kelly, 1996). Table 12.3 shows the waste load of organic matter expressed as BOD and its production rate from 1 kg trout, fed on 1.1 percent dry feed /day, with respect to daily growth rate at 0.007 kg/day.

12.1.3 Pathogens

Pathogens naturally occur in the aquatic environment, and so are the normal inhabitants of all forms of culture practices. The indigenous pathogens increase populations and virulence as a result of environmental degradation from waste discharge. Enrichment of bottom sediments in land-based farms and in cage and pen culture in open water creates favorable conditions for increase in bacterial population. Increase in the number of microorganisms, particularly bacteria, is caused by the increased levels of dissolved organic matter. Filamentous fungi dominate the vegetation in water bodies heavily loaded with aquaculture wastes. Shellfish grown in polluted environment contaminated with human pathogens can act as carriers in new environments, as they can accumulate pathogens at much higher concentrations in these tissues than in polluted water.

Table 12.3: Waste Load of Nutrients and BOD

Component	*BOD*		*Nitrogen*		*Phosphorus*	
	g/kg	*g/kg.day*	*g N/kg*	*g N/kg.day*	*g P/kg*	*g P/kg.day*
Feed	2416	16.9	120	0.84	15	0.11
Feces	444	3.1	18.0	0.13	10.5	0.07
Excretion, NH_3	133	0.9	72.4	0.51	–	–
Respiration	992	6.9	–	0.51	–	–
Waste load (feces + excretion)	577	4.0		0.63		0.07

Source: Adapted from Pillay, 2004.

Although sewage fed farms show drastic reduction in coliform bacteria, reports show conflicting data on coliform bacteria. Fecal coliform population was either not present or drastically reduced in several farms in Norway, Poland and Scotland, whereas elevated concentrations of coliform and streptococci bacteria were found in effluents from fresh water farms in Finland and in hatchery effluents in USA (Pillay, 2004).

Introduction of exotic species or transplantation of culture stocks often significantly increases the diversity of pathogens in the environment. Salmonids, carp, tilapia and oysters have been very widely transplanted or introduced. Several reports show diseases spread by fish introductions. Cestode transmitted through grass carp from the Far East to Europe; furunculosis caused by an obligate pathogen through brown trout from Denmark to UK, and through salmon smolts from Scotland to Norway; massive mortality of salmon populations in Norway transmitted from Sweden through salmon molts; and infectious viral disease in Hawaiian shrimp culture carried through juveniles and adults from Panama are some of the examples (Pillay, 2004).

Any aquaculture farm is susceptible to disease outbreaks. Occurrence of disease depends on the complex interaction between the host species, disease agents and the environment. Favorable disease conditions occur in intensive and semi-intensive farming systems under high stocking densities with increased stress of stocks and feeding, and inadequate water exchange. In many cases possibility of disease transmission through contribution from bacteria has been raised without citing adequate evidence. Higher concentration of bacteria and increased bacterial composition in the effluent and disease transmission to the cultured fish through effluents from pond farms have been cited. Recent mass mortality caused by ulcerative syndrome occurred in many South East Asian countries. Although the disease has been transmitted through inflow water, exact causative agent was not known (Pillay, 2004). Disease outbreaks are comparatively less common in open water environments.

12.2 Treatability of Aquaculture Effluents

In comparison with domestic and industrial waste water treatment being conducted for more than a century, the field of aquaculture effluent treatment is new. Very few countries or region appear to have quantified data of effluent discharges or of total load released into the surrounding water bodies, and have or follow regulations for treatment system. Aquaculture farm effluent has been considered to be an unusual type not conducive to usual treatment and, therefore, presenting treatment problems. Reports indicate two problems of aquaculture effluents: (1) load and concentrations of the pollutants and (2) pollution potential of effluents (Cripps and Kelly, 1996; Pillay, 2004).

12.2.1 Load and Concentrations of Pollutants

Figure 12.2 shows that typical pollutant concentrations are low in comparison with common effluents like domestic and industrial waste water. The typical low concentrations of pollutants, from Figure 12.2, are suspended solids, 14 mg/l; total nitrogen, 1.4 mg/l; total phosphorus, 0.125 mg/l; and BOD, 8 mg/l (Cripps and Kelly, 1996). Report of waste water effluents from the shrimp farms, domestic sewage, and fish and other processing industries in India shows that concentrations of pollutants are much less (Table 12.4).

Table 12.4: Comparison of Wastewater Effluents from Marine Shrimp Farms, Fish and Other Processing Industries

Variable (mg/l)	*Wastewater Effluents*				
	Shrimp Farm	*Domestic Sewage*			*Processing Industries*
		Raw	*Primary Treated*	*Biologically Treated*	
BOD	4.0–10.2	300	200	30	2000–4000
TN	0.03–5.06	75	60	40	700–4530
TP	0.05–2.02	20	15	12	120–298
Solids	119–225	50	–	15	1880–7475

Source: Aquaculture Authority, 2002.

The loadings of nitrogen and phosphorus from intensive salmonid farms in Europe represent only a small fraction of the total discharges from other activities. Reported data in Figure 12.3 on nitrogen loads of aquaculture effluents and from other fields discharged in Danish seas show also that the extent and significance of the nitrogen loads is only about 0.2 percent from marine farms out of 150,000 tons/year from land based sources (Pillay, 2004). However, in waste water treatment pollutant concentrations rather than loadings are commonly used to describe the effluent content.

12.2.2 Pollution Potential of Effluents

In spite of the low concentrations of pollutants of aquaculture effluents in comparison with other sources, their mass flow rate is high and so the input to the environment is significant. In fact, the impact on the receiving water body is judged from the total load during a particular period of time such as kg of waste material per day. The major portions of solids in effluents, *i.e.* 'pulses' of solids occur during harvesting of semi-intensive and cage systems, and also in warmer temperatures and during increased feeding rates in the spring and summer. Solids production during tank cleaning

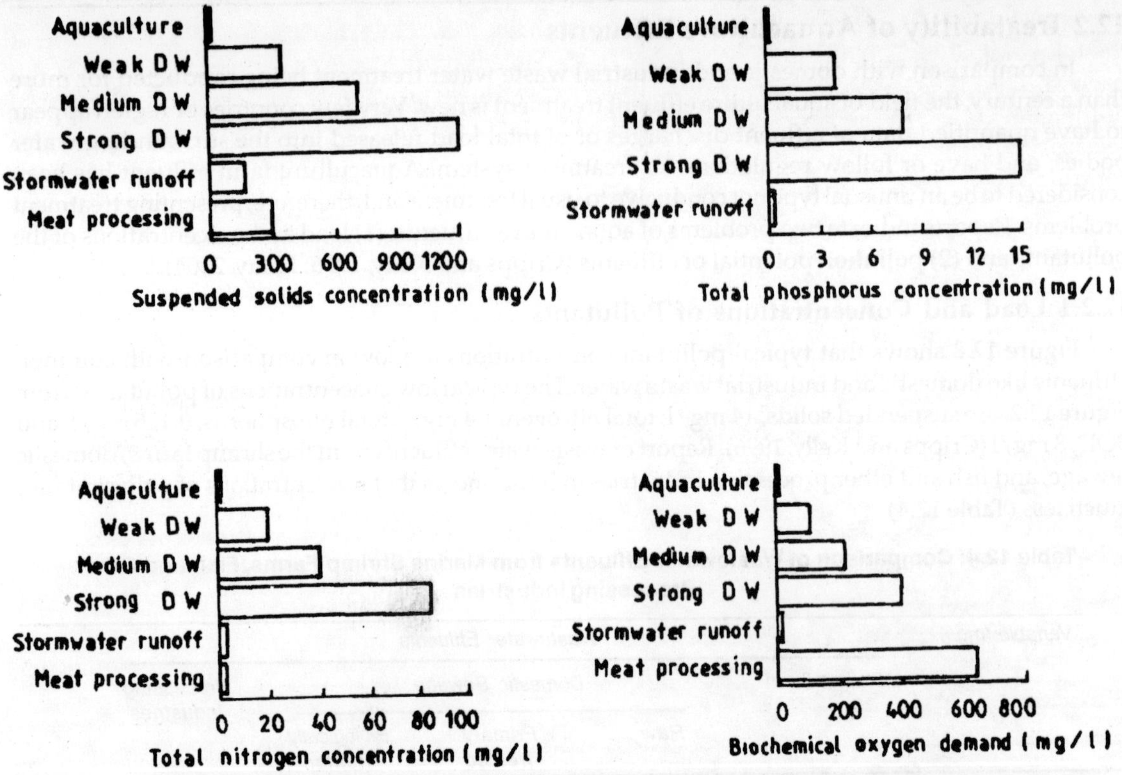

Figure 12.2: Comparison of Pollutant Concentrations of Aquaculture Effluents with Other Wastewaters. DW: Domestic wastewater (After Cripps and Kelly. 1996)

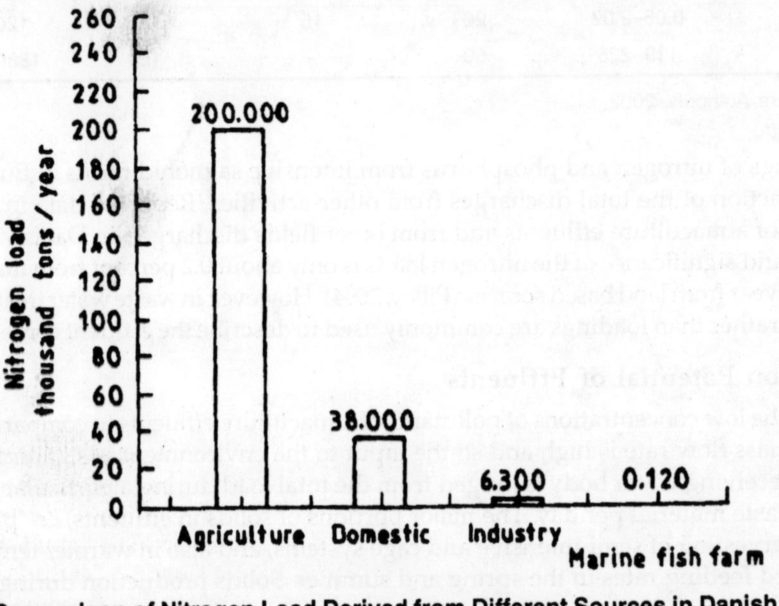

Figure 12.3: Comparison of Nitrogen Load Derived from Different Sources in Danish Seas in 1984 (After Pillay, 2004)

increases by 100–200 times the routine output (Bergheim *et al.*, 1982). In catfish ponds, 50 percent of the total suspended solids discharges in the last 5 percent of the pond water volume during drain harvest, although solids and nutrient concentration of effluent are comparable to or lower than those of the inlet water during normal operations (Schwartz and Boyd, 1994). Periods of highest waste release, *i.e.* quantification of the waste 'signal' from the farms is an important criterion for effluent control strategy. Table 12.5 shows that waste output problems are different in land-based farms.

Table 12.5: Waste Output Problems of Land-based Aquaculture Farms

Species	*Culture System*	*Water Flow*	*Waste Output Type*
Salmon, marine and fresh water	Tanks, batch type	Continuous	Daily pulsed peak
	Self cleaning tanks, batch type	Continuous	Continuous dilute
Trout, fresh water	Ponds/raceways, continuous	Continuous	Continuous dilute
Shrimp, marine water	Ponds, batch	Periodic	Occasional pulsed/ harvest peak
Carp, fresh water	Ponds, batch	Still water	Harvest peak
Catfish, fresh water	Tanks, continuous	Periodic	Harvest peak

Source: Cripps and Kelly, 1996.

Table 12.6: TN/TP Ratios of Various Wastewater Sources

Sources	*TN/TP Ratio*
Runoff from unfertilized fields	247.4
Runoff from rural croplands	60.9
Runoff from fertile soils	33.3
Ground water	28.5
Runoff from agricultural watershed	20.0
Seepage from cattle feedlot	8.9
Fertilizer	7.9
Urban stormwater drainage	5.8
Temperate zone cage culture	5.6
Human sewage, untreated	5.3
Intensive aquaculture (ponds, tanks, cages)	4.7
Septic tank effluent	2.7

Source: Costa-Pierce, 1996.

As important determinants of environmental impact, aquaculture waste quality and its temporal variability is evaluated by the TN/TP ratios. A list of TN/TP ratios of various sources given by Costa-Pierce (1996) shows that aquaculture effluents from intensive farms are highly rich in TP and are, therefore comparable in composition to untreated human sewage and urban storm-water runoff (Table 12.6). Although there are very few data in the tropics on waste production from intensive aquaculture, it has been reported that settleable solids and TP are considerably lower from tropical cages (*e.g.* TN/

TP = 0.001). Reasons may be selective enrichment of wastes with phosphorus and trace amounts of TN, volatilization of ammonia due to high pH in the water, and dominance of algal/bacterial food web. The study suggested the possibility of little environmental impact in case of low phosphorus content (0.8–1 percent) of the feed in tropical cage culture in eutrophic lakes or reservoirs (Costa-Pierce, 1996).

There is another criterion suggested for comparison for mass flow rates of pollutants in term of 'population equivalent' based on the population size and the output per person of BOD from domestic wastes (Pillay, 2004). However, use of the population equivalent values as an index of waste load in aquaculture has been questioned. Aquaculture farm effluent is not comparable to municipal waste water. While pollution from one kg trout is equal to that of 0.2–0.5 people, waste load of a fresh water farm producing 200–250 tons of trout per year in Norway is equivalent to untreated sewage from 1400–5000 persons (Bergheim *et al.*, 1982). A cage fish farm unit producing 50 tons/year discharges equivalent to a sewage treatment plant for about 7000 people, assuming a 90 percent reduction of phosphorus in the plant (Pillay, 2004).

12.2.3 Comparison of Effluents from Different Culture Systems

The quantity and quality of the total waste from any farm creating impact on the environment is directly related to the culture system followed. The main factors are stocking density, type and quantity of feeds provided, and water management procedures such as retention of water in the culture facilities. Routine farm operations such as cleaning of tanks, raceways and other facilities can affect the waste load in the discharges, as shown above.

1. Intensive Aquaculture Systems

In intensive land-based hatcheries and smolt farms using flow-through tanks and raceways, and sea-based floating cages, the waste discharges continuously load the water bodies. Waste loading data of land-based salmonid farms and floating cages in many European countries have been given in Chapter 6. The waste loads are reported to vary with fish stock, efficiency of production, feed quality and quantity, and feeding schedule. In spite of large variability, mean nutrient concentrations of TN and TP are similar among various intensive culture systems, as shown by the TN/TP ratio in Table 12.6.

Waste production data from intensive farming of other fish species and shrimps, as given in Chapter 6, show that 70–90 percent of the TN is released into the environment. In catfish ponds with proper sludge removal facilities, actual loadings correspond to about 9 g N and 1 g P per kg of fish produced, which is, however, much lower than that from single flow-through ponds. In shrimps ponds of Thailand, with FCR 2.0, estimated loads were 102±25 kg N and 46±14 kg P per ton of shrimp produced (Bergheim and Asgard, 1996).

2. Semi-intensive Aquaculture Systems

In semi-intensive fertilized ponds nutrient conversion efficiencies such as MCR are lower than in intensive ponds under direct feeding, as described in Chapter 6. As a result these systems would be less polluting than the intensive ones. For example, in some manure-fertilized ponds requirement of 170–491 g of N and 42–70 g of P per kg of fish showed that only 2 percent of the N and 6 percent of the P input were discharged into the surrounding water, and 83 percent of the N and 86 percent of the P were removed mainly through sediment deposition. Some additional phosphorus would be removed by flushing during water draining (Edwards, 1993). Figure 12.1 shows a comparison of nutrient budget between semi-intensive and intensive culture by partitioning of the nitrogen and phosphorus

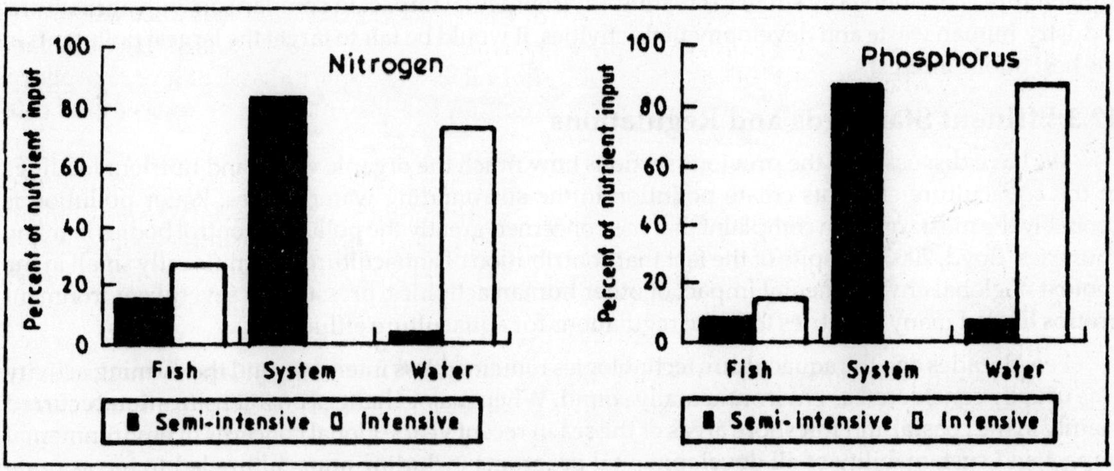

Figure 12.4: A Comparison of Nutrient Budgets Between Semi-intensive and Intensive Culture Systems. Semi-intensive system: Nile tilapia culture with organic manure and inorganic fertilizer, and only draining at the end of the culture cycle. Intensive culture: Cage or raceway system with complete feed and water exchange (After Edwards, 1993)

inputs assimilated in fish, removed in the system, and drained into the surrounding waters, as percentages of those in fertilizers and in feed, respectively. Subsequent nitrogen and phosphorus pollution of the water bodies was found to be 7–31 and 3–11 times more in intensive than in the semi-intensive systems (Edwards, 1993).

Quantitative data of the environmental impact from shrimp ponds with fertilizer and manure application are not available. However, these ponds as well as those under integrated aquaculture have demonstrated that they can be considered as an effective nutrient conversion system. Significant amounts of nitrogen accumulated in the pond sediments in integrated fish-cum-duck-alfalfa-rice system showed that the farm acts as a natural filter of nitrogen-loaded water (Bergheim and Asgard, 1996).

3. Extensive Aquaculture Systems

Extensive aquaculture systems are characterized by low stocking density and no input of feed or fertilizers, but high water consumption. Pond farms discharge water usually at the time of harvesting. In the majority of farms, discharge may be once a year or once in 2–3 years. If fertilizers are used, they are stopped several days before harvesting. Therefore, suspended solids and soluble nutrients are likely to be minimum and discharges are of little environmental significance.

According to Edwards (1993), aquaculture is much less widespread in Asia as it is widely supposed, and is an uncommon occupation throughout the world. Although the quality of aquaculture wastes is comparable to untreated sewage and urban runoff in terms of TN/TP ratio, magnitude of total discharge of farm waste is still lower, even on a larger scale. Most intensive farms can produce only localized effects of environmental impact. In this perspective, Costa-Pierce (1996) has raised the question of fairness of the regulating authorities and of their role in evolving a scientific method of regulation vis-à-vis a political process to halt the growth of aquaculture. While the contribution of aquaculture is undoubtedly small in creating environmental impact in comparison with agriculture, industry, human waste and developmental activities, it would be fair to target the largest pollutants in the best public interest.

12.3 Effluent Standards and Regulations

We have discussed in the previous sections how much the organic waste and nutrient loadings in the aquaculture effluents create pollution in the surrounding water bodies. Water pollution is probably the most common complaint that has concerned greatly the pollution control bodies in many countries (Boyd, 2003). In spite of the fact that contribution of aquaculture is significantly small in the context of global environmental impact of other human activities, pressure by several environment groups has led many countries to make regulations for aquaculture effluents.

Few decades ago the aquaculture technologies remained less intensive, and the farming activity was usually considered as environmentally sound. When major changes on intensification occurred mainly in the coastal and foreshore areas of the sea in recent years, global concerns of environmental impact and sustainability of all developmental programs including aquaculture led to incorporate ways of minimizing, mitigating or compensating negative impacts on the environment. Large number of restrictions is being imposed on aquaculture ventures in industrially developed and also in developing countries. Restrictions and regulations range from those in land and water use through water pollution to tourism and recreation. In many countries where aquaculture has developed as means of income generation and export earnings, some legal framework has been built up for control on the access to farming activities in the coastal areas. These are legal status of land and water use, the nature of land–coastal or inland, the nature of water–marine or freshwater; use of natural or chemical feed, wild or hatchery seeds; and so on (Edeson, 1996).

12.3.1 Effluent Standards

Water quality standards of the effluent and the criteria to meet the objectives are determined according to the particular use of the water resources. For environment protection of coastal areas, a management framework based on an approach termed as uniform emission standard (UES) has been adopted. The UES limits on the concentration of pollutant in the effluent applied to discharges of the substance in question or all processes of a particular type where it is produced. In marine pollution control harmful substances are grouped for specifying effluent treatment technology as 'black' which would not be, or as 'grey' which would be discharged subject to certain treatment requirements (Barg, 1997).

For discharge at the outfall, standards made by the government agencies normally contain specified limits of selected water parameters. Some examples of standards have been cited by Boyd (2003). One such standard recommended by the Global Aquaculture Alliance (GAA), an NGO in USA, gives initial limits and target (stricter) limits to be complied by the member organizations (Table 12.7). The standard may include restriction on discharge volume such as a maximum daily load of a variable such as BOD_5. There may be qualitative standard such as odor, foam or visible plume at the outfall. In the developed countries such as USA compliance of the standards, its monitoring and issue of permits to the farmers would be a situation completely different from those in developing countries where large number of small farms are operated and traditional effluent standards would be difficult to regulate by the government agencies.

In India, Aquaculture Authority under the Ministry of Agriculture, Government of India, has set norms of standards for the discharge of effluents at the final point (Table 12.8). Table 12.8 gives also a comparison between the standards of India and Thailand.

Table 12.7: Water Quality Standards for Shrimp Farm Effluents

Variable	Initial Standard	Target Value
pH	6.0–9.5	6.0–9.0
Total suspended solids (mg/l)	≤ 100	≤ 50
Total phosphorus (mg/l)	≤ 0.5	≤ 0.3
Total ammonia nitrogen (mg/l)	≤ 5	≤ 3
BOD_5 (mg/l)	≤ 50	≤ 30
DO (mg/l)	≤ 4	≤ 5

Source: Boyd, 2003.

Table 12.8: Standards of Effluent Discharge from Aquaculture Farms

Variable	Final Discharge Point in India at		In Thailand
	Coastal Marine Waters	Creek or Estuarine Waters/1	
pH	6.0–8.5	6.0–8.5	7.0–8.7
Suspended solids (mg/l), max.	100	100	100
DO (mg/l)	Not less than 3	Not less than 3	More than 4.0
Free ammonia (mg/l), max.	1.0	0.5	1.8–2.9
COD (mg/l), max.	100	75	–
BOD_5 (mg/l), max.	50	20	10
Dissolved Phosphate, as P (mg/l), max.	0.4	0.2	0.4
Total nitrogen (mg/l), max.	2.0	2.0	4.0

1. The same water courses are used as water source and discharge points.

Source: Aquaculture Authority, 1998, 2002.

12.3.2 Guidelines and Codes of Conduct

Several national and international organizations such as GESAMP (1991), FAO (1995), FAO/NACA (1995) and INFOFISH (1997) prepared guidelines with an objective to set out principles and standards to be followed for coastal environment management and ecological sustainability of aquaculture. The most important principle proposed in the guidelines is planning of aquaculture development in the coastal areas on the basis of resource availability and without any conflict with traditional uses as far as practicable.

Basically a code of conduct is a set of guiding principles addressed on how management and other operational activities should be carried on. Mostly aquaculture codes refer that proposed by FAO as the Code of Conduct for Responsible Fisheries (FAO, 1995, 1997). Although the most codes are meant for voluntary adoption without any legal binding, they usually demonstrate environmental stewardship for improvement of the industry (Boyd, 2003). Readers may consult the references cited for details of the guidelines and codes.

12.3.3 Codes of Practice

Codes of practice are supplements of the codes of conduct and are basically implemented as best management practices (BMPs) or good management practices (GMPs) for environmental management, as referred to by Boyd (2003). The practices are available for various specific operations in aquaculture. However, we will give lists of practices applicable only for pond or farm effluents and site characteristics for regulation of discharge by marine farmers in general.

1. Farm Effluents

In order to reduce nutrient loads in farm effluents the following management practices are recommended (Boyd, 2003):

1. Stocking and feeding rates should not exceed the carrying capacity of the water body.
2. Use fertilizer only to maintain phytoplankton bloom.
3. Use high quality feed, containing only the required quantity of nitrogen and phosphorus.
4. Avoid overfeeding and ensure as much consumption as possible.
5. Avoid or reduce water exchange as much as possible.
6. Minimize storm overflow during heavy rain by providing storage volume in the pond.
7. Artificial aeration should be applied, if necessary, to prevent chronic low DO concentration and to help self-purification of water by nitrification and other bio-processes.
8. Do not install deep water release structures in ponds in order to avoid discharge of dirty water from near pond bottom.
9. Use seine-harvest, if possible, instead of drain harvest.
10. Discharge pond water through a settling basin or vegetative ditch.

Minimum water exchange practice by reducing the pond water discharge rates from approximately 8–10 percent per day to less than 4 percent per day has shown significant improvements in pond management techniques in yields, growth rates and survivals (Jackson and Wang, 1997). The concept of 'zero effluent discharge', *i.e.* without effluent release from the pond, usually referred to as the daily pond management in a closed system, advocates that the pond may be topped up to counter evaporation effects, or refilled to the original level to avoid salinity increase. Commercial yields of *Penaeus monodon* without the need to discharge large volumes of pond water into the receiving waters could be achieved in north Queensland, Australia. Correct feed consumption and increased aeration also could reduce the need for water exchange (Robertson, 2000).

2. Site Characteristics for Discharge Regulations

In USA the EPA proposed rules for restrictions on the use of potential sites, mainly for marine farms. The rules require applications for approval of discharges from fish farm projects. In general, it is a permit system requiring clearances by the agencies of environment protection, navigation, commercial fisheries, and protection of sport fisheries and wildlife. For approval, the identification list includes:

1. Type and quantity of pollutant(s) to be used in the project and information on its conversion efficiency to product,
2. Species of organisms to be cultured and its potential increased yield,
3. Exhaustive list of water quality variables required for growth of the species,

4. Possible health effects, including diseases or parasites, potential escapes of non-indigenous species, and bio-concentration in the crop,
5. Pollutants produced by the species during culture and channeled into the effluent, and
6. Disposal method.

The proposal required to be enforced does not mean for all countries. However, the restrictions show the significance of site characteristics of environmental implications (Pillay, 2004).

Some other guidelines suggested are (Pillay, 2004):

1. Limit on annual production of 110 tons/km^2, with a minimum mean current velocity of 5 cm/s, in cage/pen culture. Depth requirements are 12 m for 10 tons; 15 m for 10-50 tons; and 20 m for more than 50 tons. With current velocity 25 cm/s, the depth requirements are 8, 12 and 18 m, respectively.
2. No permit on, or in the immediate vicinity of natural habitats like nursery areas, eel grass, kelp beds and so on.
3. Limit to the use of ground water for a certain period of the year to be determined as volume per sec. and volume per year. Use of surface water resources may be regulated for intensive aquaculture systems. But there are very few problems in less intensive cultures, except that for use of surface waters in irrigation during dry seasons.
4. Incorporate design to prevent release of sediments into receiving waters and avoid direct outlets from individual ponds.
5. No use of trash fish as feed.

12.3.4 General Regulations for Coastal Farm

Guidelines for the promotion of environment management of coastal aquaculture development given by GESAMP (1991) define certain factors affecting environmental compatibility of the farm. The factors, described in brief below, should be considered for sustainable development of coastal aquaculture.

1. Site factors such as availability of water–sea and fresh water; natural water movements like currents, tides for adequate water exchange and waste disposal; and various water quality parameters.
2. Species–their biological and eco-physiological characteristics.
3. Culture method–Factors include a) design and construction such as systems for water exchange and effluent discharge; b) operation for efficient farm management; and c) production level, if increased for future expansion and intensification.
4. Skill of the manpower and level of the aquaculture technologies.
5. Economic viability.

A comparison of economic and environmental suitability of shrimp farms in intertidal and supratidal zones for the culture systems is shown in Table 12.9. Some major guidelines for use of mangrove forests in Asia and Oceania have been cited by Pillay (2004): These are 'no alterations' with respect to: (*a*) the time and quality of freshwater runoff through mangrove communities; (*b*) tidal inundation frequencies or surface circulation patterns; and (*c*) physical structures, chemical properties and biological activities of the sediment substrates.

Table 12.9: Comparison of Shrimp Farms in Coastal Zones

Zone	*Tidal range*	*Soil condition*	*Construction*	*Initial pond productivity*	*Culture type*	*Impact*
			Intertidal zones			
Virgin mangrove forest	Submerged at LHWL	Peat/pyrite Reclaimed before use	Very difficult	Very low	Extensive	Destruction of mangrove forests
Shrub			Difficult		Extensive to semi-intensive	
Swamp	Submerged at MHWL	Pyrite/clay loam Reclaimed before use	Easy	Low		
Coconut plantation				Relatively low		
			Supratidal zones			
Rice field/dry land	Above HHWL	Clay loam to loam	Very easy	High	Intensive only	Landward sea water intrusion, salination of land and ground water
Coconut plantation		Sandy clay loam				
Grass and others		Sandy loam				

Source: Poernama, 1990.

In India, regulations of the Aquaculture Authority ban any shrimp culture within the coastal regulation zone as defined in the CRZ notification such as in all seas, bays, estuaries, creeks, rivers and backwaters. Adoption of environment-friendly improved traditional technology, resulting in optimum utilization of the resources and with production level of 1000 to 1500 kg/ ha/ crop, has been recommended. Salient features of these regulations are (Aquaculture Authority, 1998, 2002):

1. Farm design with individual culture units of within 5 ha, feeder channel system, and water tight sluice gates fitted with net filters.
2. All farms above 5 ha require keeping 10 percent of the area for waste stabilization ponds.
3. Maintain optimum soil and water conditions.
4. Remove pests and predators, but not by using any chemical piscicides.
5. Stock shrimp larvae from only hatchery reared and healthy seeds, in the range of 4–6 no./ m^2.
6. Supplemental feeding becomes essential after 4–6 weeks of rearing. Use only dry pellets of formulated feeds with minimum of 4 hour water stability.
7. Periodical water exchange shall be necessary to maintain optimal water quality conditions.

12.4 Effluent Treatment Practices

Aquaculture waste or effluent treatment technology was basically dependent on the methods of wastewater treatment followed in the environmental engineering field. Mainly sedimentation of suspended solids in aquaculture farm effluent was practiced in the beginning. Since the late 1970s growth of intensified aquaculture farming and subsequent imposition of regulations on environmental issues several treatment methods particularly for aquaculture effluents have emerged. Treatment systems based on unit operations and unit processes are usually grouped as primary, secondary and tertiary methods as in wastewater treatment of domestic and industrial effluents. However, properties of the aquaculture effluents being different from those of the domestic and industrial wastewaters, the design and operation of the equipment would require knowledge to match any separation operation in the treatment system (Cripps and Kelly, 1996). In the following pages, the treatment technologies in use in commercial farming will be outlined. Mostly they are primary treatment processes such as screening, filtration and sedimentation, and biological treatment in recirculating systems. Detailed design characteristics and analysis of the processes have been given in the following chapters.

12.4.1 Treatment Technologies in Use

1. Solids Removal from Pond Bottoms

In many land-based farms in Asia such as in stagnant and semi-stagnant ponds, much of the solids settle at the pond bottom after retention of water for about one year or more before harvesting. In case the final discharge of pond water is delayed after harvest, the re-suspended sediments would have settled. The sediments are normally removed during pond preparation, dried and used as fertilizer for vegetable crops.

For pond bottom conditioning in marine flow-through shrimp farms, pond bottom is flushed by a strong current of water and accumulated sludge and other undesirable materials are removed. In small ponds, removal by manual means or by use of suction pumps or dredges can be used. Hydraulic jets are commonly used in intensive pond bottoms. In tanks, intermittent tank cleaning is the most common and traditional method of waste removal. Wastes accumulated in days or weeks are re-suspended and flushed by increased inflow of water.

2. Solids Removal by Sedimentation Ponds

Sedimentation or settlement is one of the main methods of suspended solids removal and becomes probably the most common type of water treatment in aquaculture. In intensive flow-through pond farms where the water retention time is much less, sedimentation or settlement ponds have long been considered as a suitable method of effluent treatment. Even with high volume of effluent and low concentration of solids simple sedimentation has proved to be cost-effective in commercial farms. If the settlement ponds are situated downstream from the grow-out ponds, it would help reduce the amount of sediment leaving the farm, particularly during final drain harvest (Teichart-Coddington *et al.*, 1999).

The size of the settlement ponds varies from 10 to more than 50 percent of the total farm area. Settlement ponds with an equivalent area of 30 percent of the grow-out ponds could remove half of the total nitrogen discharged. Use of two settlement ponds, one drying and the other settling continuously, can help in their long term use (Robertson, 2000).

Sediment or particle traps are a common means of collecting and concentrating solids in tank-based culture systems. In-tank structures built to deflect the main flow and produce areas of increased residence time are common. Baffles located across the flow help accumulate solids in the settling areas of raceways. However, traps should be located in quiescent region to help settlement but without re-suspension (Cripps and Kelly, 1996).

3. Solids Removal by Filtration

Mechanical treatment by filtration was probably the only operation for removal of solids from aquaculture wastewater. Sand filtration, micro-straining and similar operations have been tried in commercial farms, and also in hatcheries. The most simple method is to screen the effluent by means of a stationary filtering mesh to remove the particulate materials.

Micro-screens are used more commonly at the farm inlets during water intake. Filter screens are used during transfer of water from reservoir pond into the grow-our pond to prevent the entry of extraneous organisms. Filters with a mesh size of 0.5 mm, in the form of a bag or box, are installed in all water inlets. A 3-tier filtration system consisting of a split bamboo cylindrical cage (1.5 m x 0.3 m diameter) loosely covered with a vellon net (40 mesh) and then with bolting silk net (80 mesh) fitted with inlet hume pipe was used for pond filling (Chakraborti *et al.*, 2002).

Media filters such as sand filters are mostly used in recirculating systems rather than in flow-through facilities. A filter box, perforated at the sides and bottom and filled with bags of sand, shells or pieces of broken coral, is installed in a closed-circulating shrimp farm. Water is treated by pumping through the middle of the filter media bags before refilling the grow-out ponds. A filter box capacity of 0.7–1.0 ton can reduce sediment and improve water quality at least 20–50 percent better than that in a untreated pond (Tookwinas, 2000).

4. Solids Removal in Cage Farms

Probably there are no adequate techniques of removing or treating wastes from shellfish or cage farms. Wastes usually settle below the cages and/or are carried away by the currents. So waste collection or removal from cages is an extremely difficult task. Several methods have been tried but with varying degree of success. These are collectors, closed cages, water column and sediment pumps. Collection of wastes by suspending funnel-like devices beneath the cages and pumping out the solid wastes at intervals resulted in removal of about 40–45 percent of the solid loadings, containing about only 15–20 percent of the TN and TP, from cages in some east European countries (Pillay, 2004).

Cripps and Kelly (1996) have referred a cage sludge collection system, which have collected 75–80 percent of the settleable solids during a 4 month period. A sludge sump is used at the bottom of a sludge trap for collection and daily pumping out of the solids. Sediment pumps may be a means of removing settled solids in heavily polluted areas.

5. Biological Treatment

Biological treatment of farm wastewater is usually employed to prevent accumulation of ammonia and nitrite in recirculating systems of aquaculture. The processes that are utilized to prevent water quality deterioration are similar to those which occur in static ponds under open system of aquaculture. Ammonia excreted by fish and liberated by decomposition of uneaten feed and fish feces, is converted through the biological processes of algal uptake, ammonification and nitrification, as discussed in nitrogen dynamics in Chapter 6. In extensive static fish ponds fish metabolites are converted by these processes, and oxygen depletion and accumulation of harmful compounds are prevented as long as the standing crop does not exceed the carrying capacity of the pond.

In intensive pond systems with recirculation facilities application of biological treatment, most commonly in practice, are: (*a*) culture unit coupled with an earthen pond or reservoir, (*b*) bio-filtration combined with sludge removal, and (*c*) bio-pond system with solids removal, nitrification and macrophytes-based nutrient removal (van Rijn, 1996). Figure 12.5 shows the treatment configurations for recirculating fish culture units.

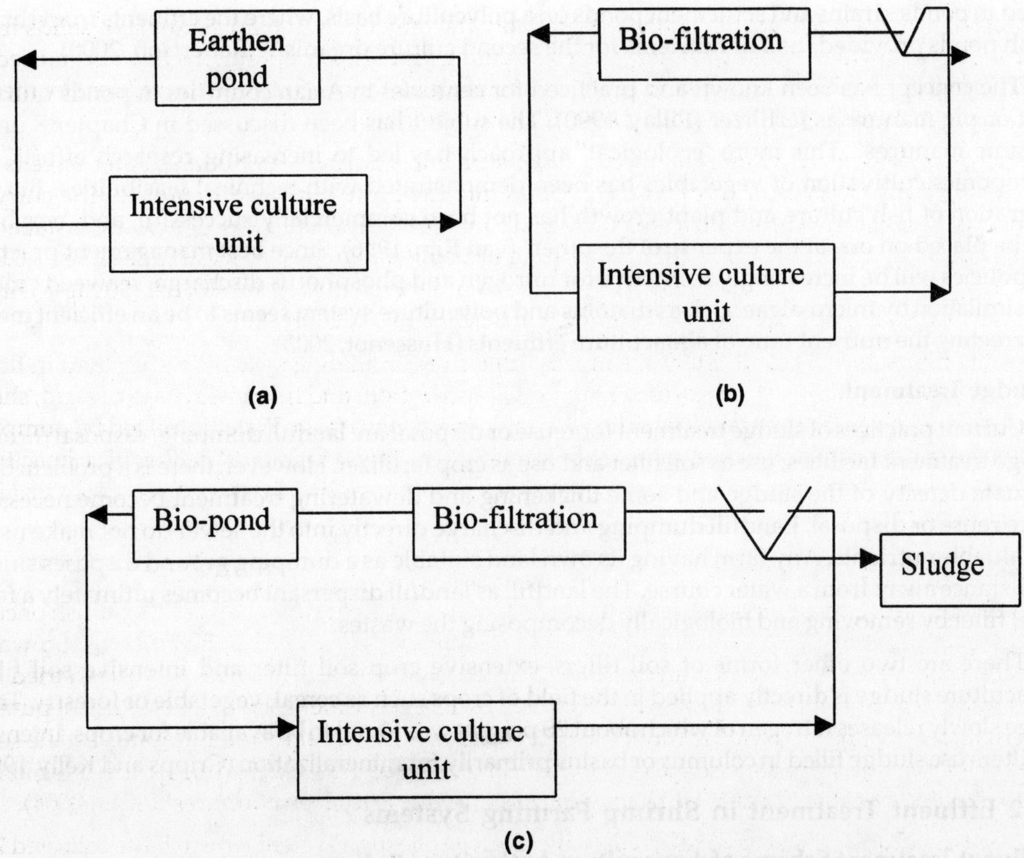

Figure 12.5: Treatment Configurations for Recirculating Culture Units (a) Earthen pond or reservoir; (b) Bio-filtration with sludge removal; (c) Bio-pond system with sludge removal (After van Rijn, 1996)

The earthen pond or reservoir is the simplest design in use in Taiwan, Israel, Singapore and several other countries, and utilizes the purification capacity of oxidation or stabilization ponds used to purify domestic wastewater. The main disadvantage is the unstable purification due to fluctuation of phytoplankton biomass. Treatment based on bio-filtration and sludge or solids removal is probably the most widely used in present day recirculating systems. Bio-filtration is carried out in a number of filter types and commercially available mechanical filters or sedimentation basins are used for solids removal.

In the bio-pond system plants, micro-algae or mollusks are utilized for nutrient assimilation. Nutrients of farm effluents are converted into plant or other biomass which can be removed and may often be utilized as valuable products. A phytoplankton production system converting fish pond effluent into nutrients for diatoms, which may also be used to feed bivalves (e, g. oysters, mussels), is essentially an integrated polyculture. These ponds, also termed as bio-remediation ponds, effectively act as biological filter and help reduce nutrients and suspended solids. Such systems have been trialed in ponds, drains and settlement ponds on a polyculture basis, where the effluents from shrimp or fish ponds provided food or nutrients for the second culture organism (Robertson, 2000).

The concept has been known and practiced for centuries in Asian countries in ponds utilizing duck or pig manure as fertilizer (Pillay, 1990). The subject has been discussed in Chapter 8 under 'Organic manures'. This more 'ecological' approach has led to increasing research efforts and hydroponics cultivation of vegetables has been demonstrated with technical feasibilities. But full integration of fish culture and plant growth has not been commercially successful, and 'emphasis may be placed on one at the expense of the other' (van Rijn, 1996). Since best management practices and policies will be increasingly restrictive for nitrogen and phosphorus discharge, seaweed culture or assimilation by micro-algae such as diatoms and polyculture system seems to be an efficient means of correcting the nutrient ratio of aquaculture effluents (Hussenot, 2003).

6. Sludge Treatment

Current practices of sludge treatment for reuse or disposal are landfill dumping, disposal through sewage treatment facilities, use as soil filter and use as crop fertilizer. However, there is a problem of an adequate density of the sludge and some thickening and dewatering treatment become necessary before reuse or disposal. Landfill dumping and discharge directly into the sewer do not make use of the valuable nutrients. Any farm having its own land suitable as a dumping ground disposes sludge at a distance away from a water course. The landfill as landfill dispersant becomes ultimately a form of soil filter by removing and biologically decomposing the wastes.

There are two other forms of soil filters–extensive crop soil filter and intensive soil filter. Aquaculture sludge is directly applied in the field of crops such as cereal, vegetable or forestry. Trout sludge slowly releases nitrogen of which about 20 percent was found to be available for crops. Intensive soil filters use sludge filled in columns or basins primarily for mineralization (Cripps and Kelly, 1996).

12.4.2 Effluent Treatment in Shrimp Farming Systems

1. Effluent Treatment Scheme of Aquaculture Authority of India

The shrimp farm effluent treatment recommended by the Authority is mandatory for farms of 5 ha area within the CRZ and farms of 10 ha or more outside the CRZ. The layout plan of the system should cover 10 percent of the farm area. Figure 12.6 shows the layout plan of a 5 ha marine shrimp farm effluent treatment. The area and construction of the various ponds will depend on the maximum production to the extent of 2 tons/ha/crop and water management system during the culture period (Aquaculture Authority, 2002).

2. Environment-Friendly Scheme for Intensive Shrimp Farming

The following facilities for water and wastewater management have been recommended for marine shrimp farms (Baliao, 2000):

1. One or two ponds as reservoir occupying at least 25 percent area of the grow-out ponds for storing water for at least one week before use in the grow-out ponds.
2. One tail pond with a control gate to hold the drained water and to allow most of the suspended solids settle before drainage.
3. Filters with a mesh size of 0.5 mm to be installed in all water inlets.
4. Sludge collector consisting of two 10 × 10 m net cages, made of mesh 22 knotless polynet, inside a net pen of 0.5 mm fine mesh at the center of each grow-out pond.

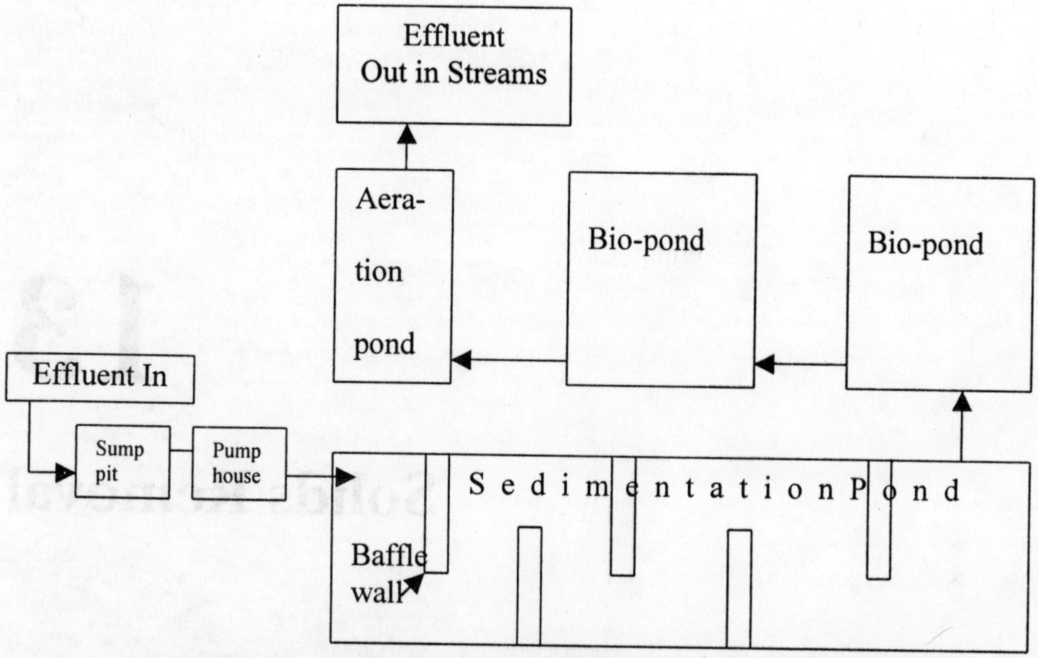

Figure 12.6: Layout Plan of a Shrimp Farm Effluent Treatment Scheme (Aquaculture Authority, 2002)

5. Stocking of saline-tolerant tilapia or milk fish in reservoir and net cages of the sludge collector as bio-manipulator which can help condition the water.

3. Closed-recirculating Shrimp Farming

A model shrimp farm in Thailand consists of two 35 × 90 m grow-out ponds, one 30 × 27 m treatment pond and a reservoir pond. The treatment pond uses a sand filter box (described earlier) to reduce sediment and improve water quality. The effluent from the grow-out pond is treated by passing it through the filter unit. For circulation an underwater jet and plastic sheets as waterway are used. The treatment pond grows filter feeding fish such as tilapia, aquatic plants and planktons. Water circulates through the grow-out pond and treatment pond in such a way that discharge of organic matter and nutrients are minimized or even completely avoided.

Pond bottom aeration system consisting of PVC pipes perforated with a row of tiny holes and evenly spaced at 10 m distance to cover the entire bottom area increases the oxygen supply through the entire water column in the grow-out pond (Tookwinas, 2000).

13

Solids Removal

The commonly used solids removal operations in primary treatment of wastewater are screening, sedimentation and filtration. Screening removes coarse and settleable solids by surface straining. By sedimentation settleable solids are removed and thickened as sludge. The terms sedimentation and settling are used interchangeably. Filtration removes fine residual suspended solids remaining after dissolved solids are removed by biological or chemical means.

Suspended particles are large enough to be removed from water by mechanical filtration. They settle in quiescent water to become part of the bottom deposits in ponds and lakes or to rise to the surface as scum if they are lighter than water. Colloidal particles settle so slowly that it is impractical to remove them by sedimentation. It is therefore necessary to build them into particles of suspended size, by flocculation, before they can be removed effectively from water. In aquaculture systems screening and filtration are followed normally for finer particles and sedimentation for coarse particles. Low settling velocity of fine particles (< 20 μm) makes gravitational settling methods impractical. However, as particles become smaller, their removal through mechanical filtration requires fine media or screen making removal more expensive due to pressure losses or frequent backwashing. Typical mechanical screen filters are designed to remove only particles larger than 80 μm, leaving smaller particles to accumulate. Table 13.1 shows a general guide to select a screening or filtration process for solids removal (Huguenin and Colt, 1989).

13.1 Screening

13.1.1 Types of Screens

Screening is the simplest means for separating particles from water. It is also the oldest and most widely used method in wastewater treatment. A screen is a device with opening, generally of uniform size, that is used to retain the solids in water. The screening element is made of parallel bars, rods or wires, grating, wire mesh, or perforated plate. The openings are usually circular or rectangular slots. The material may be steel, brass, nylon or other fabric. The mesh of screens is as small 40 μm, although screens with 70 μm or larger are preferred (van Rijn, 1996). These screens are commonly known as micro-screens. Although screens smaller than 1000 μm are referred as micro-screens, the term should be reserved for screen sizes less than 300–5000 μm (Huguenin and Colt, 1989).

Table 13.1: Ranges of Particle Size and Flow Rate for Filters and Screens

Particle Size (μm)	*Flow Rate (l/min)*			
	3.8 or less	*3.8–38*	*38–380*	*380–3800*
1 or less	Cartridge filters Diatomaceous earth (DE) filters	Cartridge filters DE filters	DE filters	DE filters
1–10	Cartridge filters	Cartridge filters Sand filters	Sand filters	Sand filters
10–75	Filter bags	Filter bags Sand filters	Filter bags Sand filters	Filter bags Sand filters
75–150	Filter bags Microscreens	Filter bags Microscreens Sand filters	Filter bags Sand filters	Microscreens
150–1000	Screen bags Microscreens	Screen bags Microscreens	Screen bags	Microscreens

Source: Huguenin and Colt, 1989.

Screens are of three types, static or stationary, rotary and vibratory. The simplest and the most basic type of screen is to place it across the fluid path and collect particles larger than the screen mesh on it (Figure 13.1). Coarse mesh screens or bar racks and mesh screens are commonly used. The screens are very cheap, and easy to clean, if clogging takes place. Manual cleaning by hand washing or scraping it with a brush, or mechanically by backwashing is practiced. However, cleaning labor costs are high for water carrying high concentrations of solids.

Rotary screens are screens attached to a rotating device and are partially submerged in the water. While one section of the screen is filtering the solids, back washing cleans the other portion above the surface of water. By this means the labor costs for cleaning are reduced, and problem of screen-clogging or shutting down the flow for cleaning is avoided. A number of rotary screen designs are available for use in aquaculture. Two basic types of rotary screens are axial flow and radial flow screens. In axial flow types, wastewater enters the disk type rotating screen and the filtered water exits axially (Figure 13.2a). By rotation the partially clogged screen comes in front of the backwash section, where high-pressure water spray dislodges the particles out of the screen. In radial flow type of screens, wastewater enters the rotating screen axially from one end, and the clean water exits radially through the screen. The screen is constructed of a drum or cylinder covered on the periphery by a screen mesh (Figure 13.2b). The screen cylinder is submerged about 40 percent in water at a time. The backwash spray, located on the outside of the screen at the top, periodically washes the debris into a fixed trough inside the drum.

The advantages of the rotary screens are that they are relatively inexpensive, easily operated and less labor-intensive. The major disadvantage of the axial flow screens is their capacity limitation, as

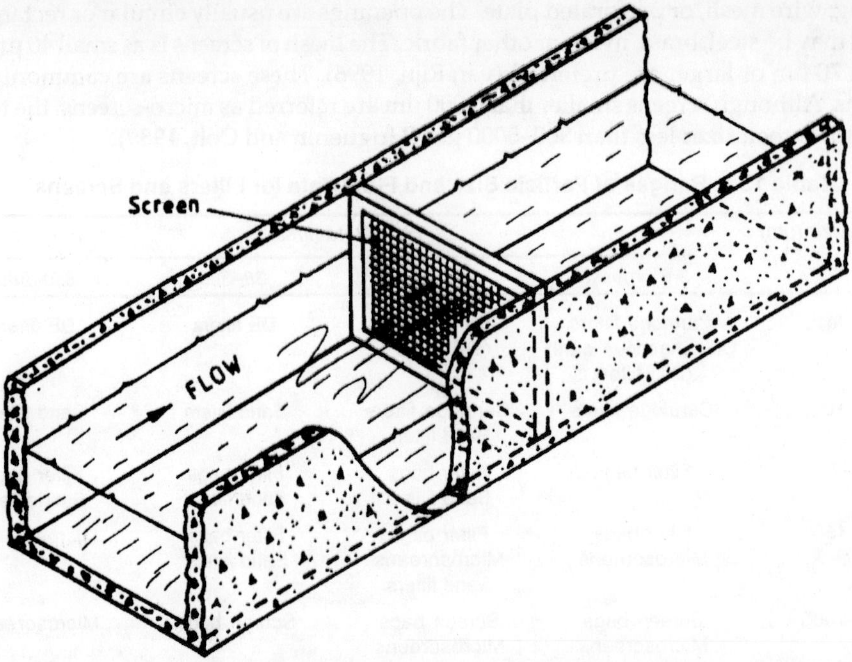

Figure 13.1: Stationary Screen (After Wheaton, 1977)

the influent water depth is limited by one-half of the screen diameter, which gives also the less surface area available in comparison with the radial flow types.

13.1.2 Typical Design Characteristics and Data

In wastewater treatment disk type rotary screen surfaces are available in sizes from 1.2 to 5.5 m in diameter, and drum screens in sizes 1 to 1.5 m diameter and 1.2 to 3.7 m in length. Stationary screens are rarely used for particles less than 1.5 mm size, or with high concentrations. Rotary screens are used for particles as small as 3 μm. They rotate at about 4 rev/min. For radial screens mesh sizes range from several centimeters to less than 1mm. The principal filtering fabrics of microscreens have openings of 23 or 35 μm (Metcalf and Eddy, 1979; Wheaton, 1977).

Important design factors of rotary screens are screen mesh size, head loss across the screen, amount and size of particles in the influent water, cross sectional area of the screen, and cleaning efficiency. All these factors decide the flow rate through the screen. Maximum allowable particle size in the screen effluent decides the screen mesh size. So removal efficiency depends on the screen size and therefore, care should be taken to select the screen size. In a test performance of a commercial axial flow rotating filter, a combination of 250 and 120 μm screens removed SS (16–94 per cent), TP (18–65 per cent) and TN (1–49 per cent), which were improved by using 60 μm on the downstream screen to SS (67–73 per cent), TP (43–74 per cent) and TN (38–67 per cent) (Cripps and Kelly, 1996).

Head loss depends on percentage of open area in the screen, amount and size of particles in the influent, efficiency of backwash, screen rotation speed, and flow rate. With increase of rotational speed up to a reasonable limit cleaning efficiency increases. This increase of speed can compensate for

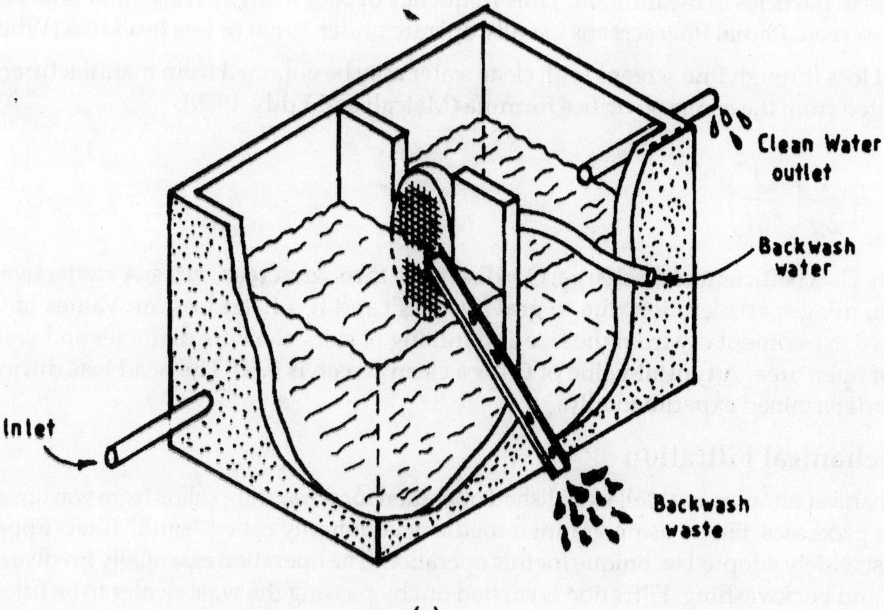

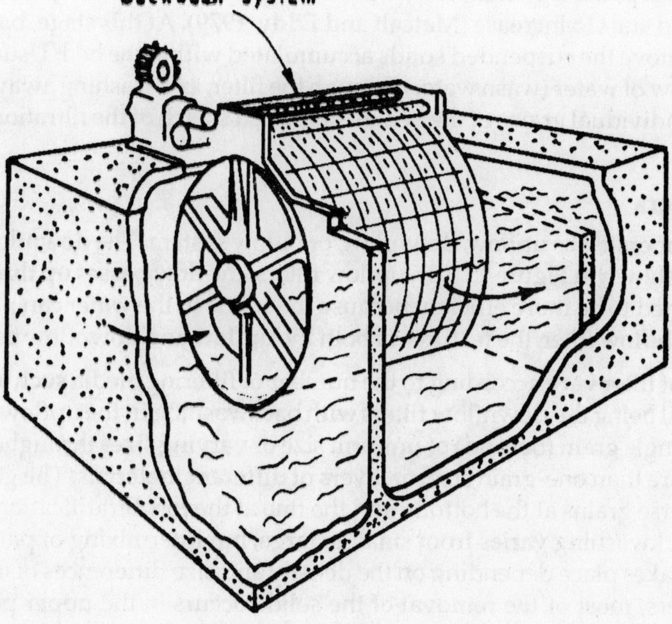

Figure 13.2: Rotary Screens (a) Axial flow and (b) Radial flow (After Wheaton, 1977)

an increase in particles in the influent. Thus frequency of backwash increases and limits the head loss across the screen. Radial flow screens usually operate under 30 cm or less head loss (Wheaton, 1977).

Head loss through fine screens with clear water may be obtained from manufacturers. It can also be calculated from the common orifice formula (Metcalf and Eddy, 1979):

$$h_L = \frac{1}{2g}\left(\frac{Q}{CA}\right)^2 \tag{13.1}$$

where C = coefficient of discharge; Q = discharge through screen, m^2/s; A = effective submerged open area, m^2; g = acceleration due to gravity, m/s^2; and h_L = head loss, m. Values of C and A are determined experimentally from the size and milling of slots, the wire diameter and weave, and the percent of open area. A typical value of C for a clean screen is 0.60. The head loss during operation should be determined experimentally.

13.2 Mechanical Filtration

Mechanical filtration is a well-established operation for separating solids from wastewater effluents of various processes. Filters using granular media or commonly called "sand" filters appear to be one of the most widely adopted technique for this operation. The operation essentially involves two phases: filtration and backwashing. Filtration is carried out by passing the wastewater to be filtered through a bed of granular materials. Within the bed, the suspended solids of the wastewater are removed by one or more of the mechanisms, such as straining, interception, impaction, sedimentation and adsorption. The filtration phase is terminated when the suspended solids in the filter effluent and the head loss across the bed start to increase (Metcalf and Eddy, 1979). At this stage, backwashing of the filter bed is done to remove the suspended solids accumulated within the bed. Usually this is carried out by reversing the flow of water (wash water) through the filter, and washing away the accumulated solids attached to the individual grains of the bed. A simplified sketch of the filtration system is shown in Figure 13.3.

13.2.1 Types of filters

The filter may be operated as upflow, downflow, or biflow system. The downflow filter, being the most common type, is shown in Figure 13.3. In upflow filters, the flow passes up through the bed. The upflow system is claimed to be more efficient because the solids in the water can penetrate a greater degree into the bed. In biflow filter, the features of both the upflow and downflow beds are combined.

In general, types of filter vary according to the number of filtering media such as single, dual and multi-medium beds, all being the downflow filters with backwashing in the upflow direction. Single-medium filters use a single grain (or sand) of uniform size or varying sizes throughout the bed. Dual-medium filters use more than one-grain size, or layers of different materials. The grains stratify after backwashing with coarse grains at the bottom, and the fine at the top. Stratification of grain sizes for each medium after backwashing varies from small to large, but intermixing of particles in dual and multi-medium filters takes place depending on the density and size differences of the various media. In single medium filters, most of the removal of the solids occurs in the upper portion of the bed, whereas in dual and multi-medium beds solids can penetrate deep into the bed to utilize more of the solids storage capacity within the filter.

The principal types of filters are classified again according to stratification as: (a) single-medium stratified, (b) single-medium or mixed-medium unstratified, (c) dual-medium stratified, and (d) multi-

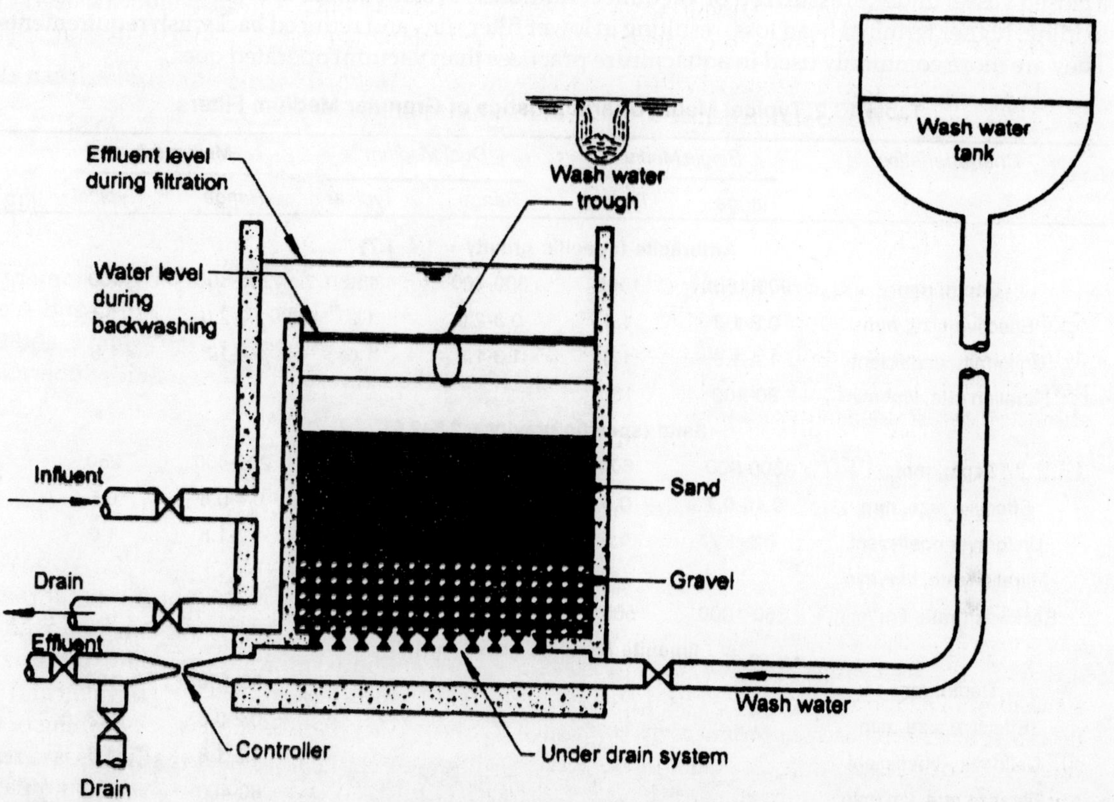

Figure 13.3: Gravity Flow Filter (After Metcalf and Eddy, 1979)

medium stratified. Typically, the beds are selected on the basis of the materials used in the bed such as their grain or pore size, shape, medium gradation, specific gravity, depth, hardness and solubility. The single-medium stratified filter beds are not used regularly in wastewater filtration, as they create an unfavourable head-loss buildup. The single-medium unstratified filters are disadvantageous, as they require uniform size of medium, high backwash velocities and costly backwashing facilities for deep beds. A single medium of varying sizes has proved to be an effective alternative to the single-medium filter with uniform size (Metcalf and Eddy, 1979).

The dual-medium filters in use are composed of combinations such as anthracite and sand, activated carbon and sand, and resin beds and sand. Multi-medium beds contain three materials out of these or with combinations of garnet or ilmenite also. The media are arranged in layers according to density and from coarse to fine in the direction of flow to increase the filtering capacity before clogging occurs. Typical media characteristics for commonly used single, dual and multi-medium filters are shown in Table 13.2.

Filters are also classified based on filtration driving force as gravity filters, and as rapid filters either under pressure or vacuum. Gravity filters operate in the same manner as downflow filter (Figure 13.3). These filters having low hydraulic loading are also called slow filters. Rapid filters are usually a closed vessel under pressurized or vacuum conditions. Pressure filters operate at higher hydraulic loading, higher terminal head loss, resulting in lower filter runs and reduced backwash requirements. They are more commonly used in aquaculture practices than vacuum operated ones.

Table 13.2: Typical Media Characteristics of Granular Medium Filters

Characteristics	*Single Medium*[a]		*Dual Medium*		*Multi-medium*	
	Range	*Typical*	*Range*	*Typical*	*Range*	*Typical*
			Anthracite (specific gravity = 1.4–1.7)			
Depth, mm	900-1800	1500	300-600	450	200-500	400
Effective size, mm	0.8-1.8	1.4	0.8-2.0	1.2	1.0-2.0	1.4
Uniformity coefficient	1.4-1.8	1.6	1.3-1.8	1.6	1.4-1.8	1.6
Filtration rate, l/m²/min	80-400	160				
			Sand (specific gravity = 2.5–2.6)			
Depth, mm	500-900	600	150-300	300	200-400	250
Effective size, mm	0.45-0.7	0.5	0.4-0.8	0.55	0.4-0.8	0.5
Uniformity coefficient	1.3–1.7	1.5	1.2-1.6	1.5	1.3-1.8	1.6
Filtration rate, l/m²/min			80-400	200		
Backwash rate, l/m²/min	360-1000	500[b]				
			Ilmenite (specific gravity = 4.8)			
Depth, mm					50-150	100
Effective size, mm					0.2-0.6	0.3
Uniformity coefficient					1.5-1.8	1.6
Filtration rate, l/m²/min					80-400	200

Source: Metcalf and Eddy, 1979 and Lawson, 1997.

[a] Separate single medium filters [b] For single medium sand filter only.

1. Gravity Filters

Figure 13.3 shows a gravity filter, where water containing solids enters at the top, flows down through the filter bed and is discharged through an under drain system. A common filter may consist of 3 to 5 layers of sand with particles of progressively larger sizes. For example, sand on the top layer, gravel on the second, and crushed rock of 1.25 cm and 2.5 cm size on the third and bottom layers, respectively, is one type of sand filter. These filters of 60–12 cm depth with flow rates of 82 l/m²/min are common in domestic sewage plants, but 30–90 cm depth filters appear to be sufficient (Wheaton, 1977).

2. Rapid Filters

Pressure sand filter is enclosed in a vessel, where the pressure is applied by a pump (Figure 13.4). Higher heads force the solids deeper into the bed, utilizing the filter depth better than the gravity filters. Major advantage is the smaller filter size for a given flow rate. Effluent quality is similar to that of a gravity filter running under similar conditions. The disadvantages are high head losses due to larger flows, and more frequent backwashing. Typical head losses range from 9.1 to 27.4 m (Huguenin and Colt, 1989). Capital cost is high due to pumping cost. In a vacuum filter higher head is achieved

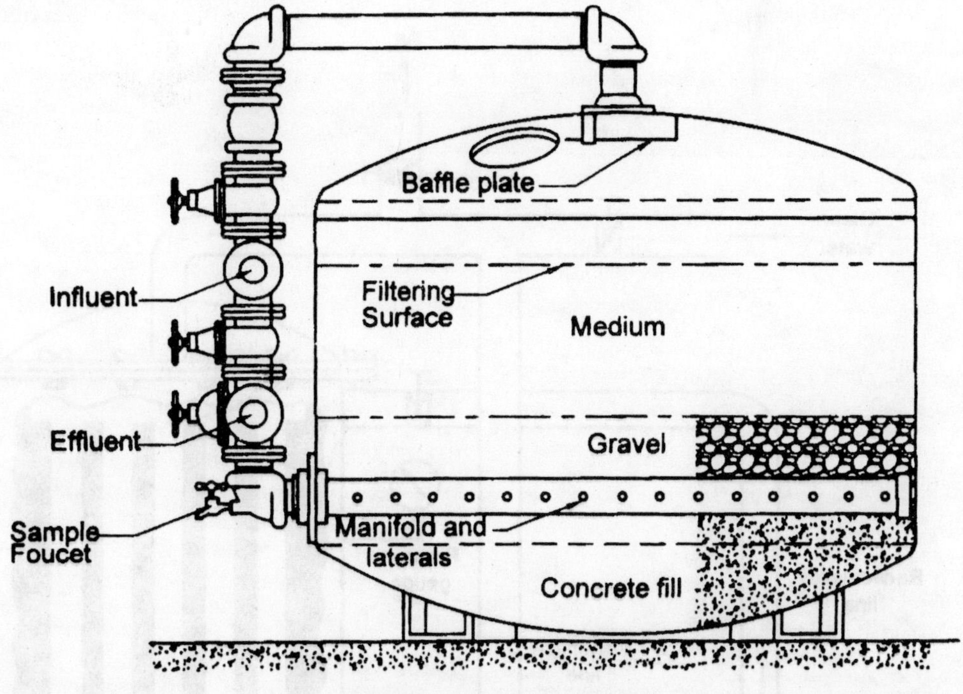

Figure 13.4: Pressure Filter (After Metcalf and Eddy, 1979)

by pumping the water out of the underdrains. Pressure type filters are more common in aquaculture applications than the vacuum operated ones.

3. Diatomaceous Earth Filter

Diatomaceous earth (DE) filters are basically a type of granular filter used in systems where high degree of water clarity and/or low bacterial populations is required. The granular material used is graded diatoms or diatomaceous earth, also known as *kieselguhr*, which is a mass of hydrated silica formed from skeletons of algae. It is a very porous and absorbent material, and there are a large number of grades of DE. Finer grade DE filter, usually a pressure filter, can remove the particles down to 0.1 µm.

The DE filter is composed of three basic elements: the central permeable core, fine wire mesh or fabric sleeve called septum, and the filter coat of DE. The core element made up of a rigid plastic material serves as a support for the septum, which fits over the core and together become a collection site for the DE. Figure 13.5 shows the filter element assembly and the whole filter system. The operation of the filter during the filter cycle starts by pumping clean water mixed with the earth through the system until all septum are uniformly covered with the earth to form a precoat. Water to be filtered is then mixed with more DE of a coarser grade (called the body feed), and pumped through the filter throughout the filter cycle. The new earth and solids from the water are trapped on the precoat and form new cake on the filtering surfaces. The clean water passes out of the filter through the core element. Pressure drop across the filter, or filter cake thickness decide the filter cycle. When the cake thickness increases, the distance between the filter elements decreases with excessive rise in pressure

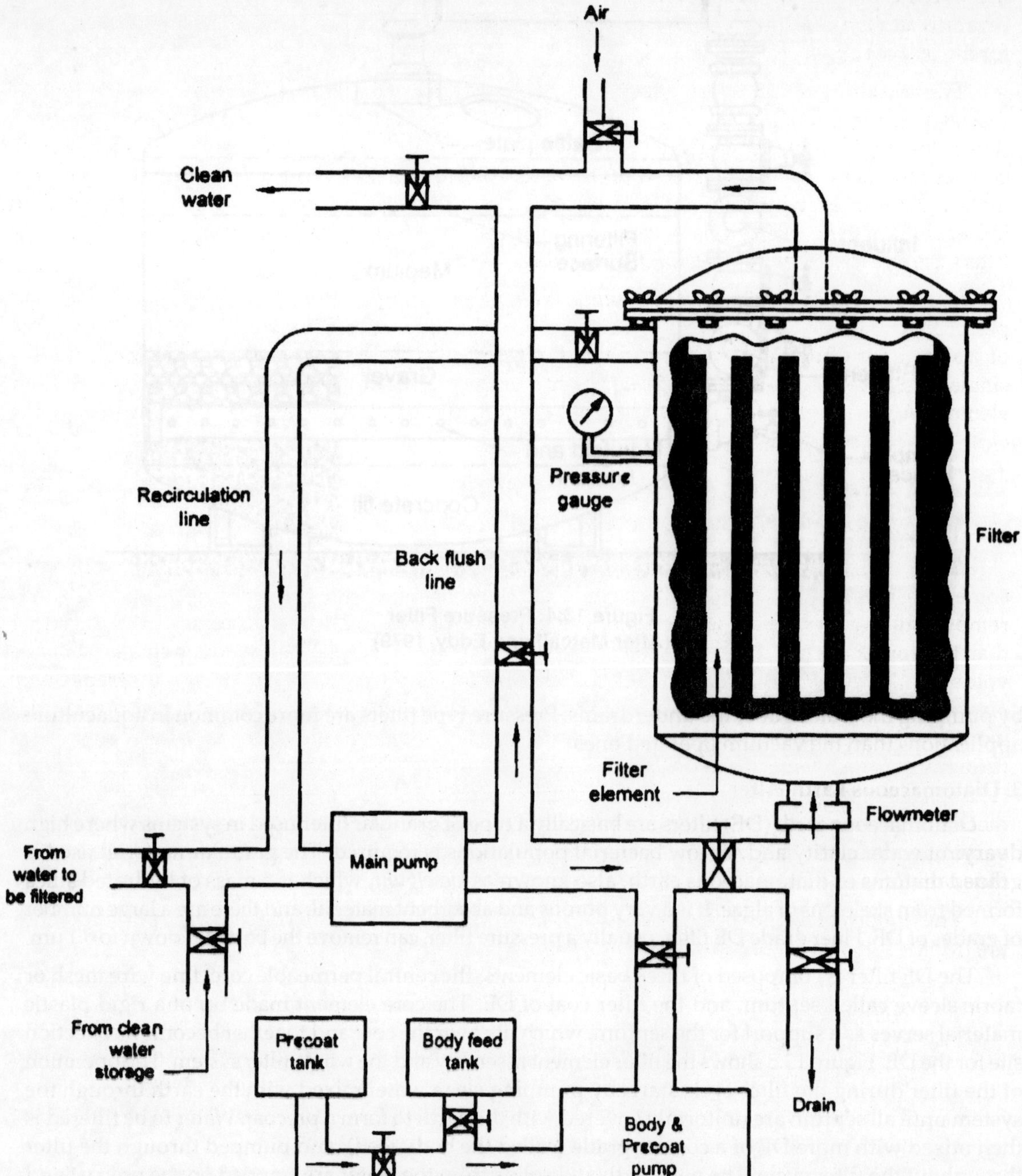

Figure 13.5: Diatomaceous Earth Filter (After Wheaton, 1977)

drop. At this point filter cycle is stopped and backwash cycle is started with clean water. Water flow is reversed and the used filter cake is forced off the septum during backwashing. The filter is then ready for the precoat cycle.

Filtration rate of DE filters depend on filter surface area, grade of DE used, and the type of wastewater being filtered. Although reasonable flow rates are in the range of 0.42 to 0.63 $l/min/m^2$, flow rates of 1.2 $l/min/m^2$ are recommended for getting high quality water. The initial and operating costs of DE filters are very high, and so the units with several thousand liters are recommended to make the system economically viable. Filters in the form of disposable cartridges, already coated with DE, are available for use especially in recirculating system. The initial and operating costs of cartridge filters are less than that of usual DE filter units.

Replacement cartridges are available from manufacturers in a number of sizes, materials and performances. They can remove particles as small as 0.01 µm and are used for filtrates of high degree of water quality. The filters can be used on-line, located in the downstream from a pump, and in a single to multiple units for large flows. A filter unit consists of a canister and a cartridge or filter element. An in-line pressure gauge is used to indicate high heads and cleaning or replacement of the cartridge (Lawson, 1997).

13.2.2 Filtration Process

1. Solids Removal Mechanisms

Suspended particles from wastewater passing through a granular filter bed can be removed by some form of straining action. It has been found that straining is the principal mechanism of solids removal during filtration of secondary effluents from biological treatment processes. Particles larger than the void space in the bed are strained out mechanically. But particles very much smaller than the voids are also effectively removed, when chance contact of the particles with the grains traps them. However, for smaller particles straining cannot be the only mechanism for solids removal by granular medium filtration. In fact, in most cases straining plays a relatively insignificant role in removal of turbidity of water. Removal of smaller particles in granular medium filters may take place in two steps: (1) transport of solid particles to the surface of the grains, and (2) removal of particles by attachment mechanisms. In most granular filters laminar flow conditions apply, the velocities within the filter bed varying from zero at the grain surface to a maximum at the center of the void. Transport mechanism thus can move solids out of the flow streamlines into lower velocity near the bed grains.

The important transport mechanisms, in general, for solids removal in granular medium filters are (Metcalf and Eddy, 1979 and Tebbutt, 1980):

1. *Straining*–Particles larger than pore space by mechanical means, and smaller particles by chance contact.
2. *Interception*–Particles moving along in the streamline may pass close enough to the grains and come in contact with the grain surface.
3. *Diffusion*–Very small particles such as colloidal particles (< 1 µm) are brought into the vicinity of the grain surface by random Brownian movements.
4. *Impaction*–Heavy particles not following the streamlines are impacted on the grain surfaces
5. *Sedimentation*–Particles move under gravity force and settle on to the grain surfaces.
6. *Hydrodynamic*–Heavy particles are subjected to lateral forces under a velocity gradient, and can move across streamlines.

The attachment mechanisms can be identified as:

1. *Adsorption*–When a particle is brought in contact with the grain surface, or with other particles, physical and/or chemical adsorption may take place to hold it there. Interaction between the particles and the bed grain takes place as a result of molecular attractive forces and electrokinetic forces.
2. *Adhesion*–Flocculant particles get attached to the grain surface as they come nearby the surface.
3. *Flocculation*–Particles join together under the influence of physico-chemical and molecular forces. The flocculants are then removed by one or more of the above mechanisms

2. Mathematical Analysis

The mathematical analysis of time rate of solid removal during granular media filtration has been developed by Tchobanoglous and Eliassen (1970). The analysis was based on the equation of continuity and worked out by using the basic concept of mass balance. Considering a filter cross section of area A, and thickness dx, measured in the direction of flow, the simplified statement of the suspended-solids mass balance equation becomes

$$-v\frac{\partial C}{\partial x} = \frac{\partial q}{\partial t} + \alpha\frac{\partial \overline{C}}{\partial t} \tag{13.2}$$

where, $\partial C/\partial x$ = change in suspended solids concentration in fluid stream with distance, mg/l.cm

$\partial \overline{C}/\partial t$ = change in average solids concentration in void space with time, $mg/cm^3.min$

$\partial q/\partial t$ = change in solids quantity deposited within the filter with time, $mg/cm^3.min$

$v = Q/A$ = filtration velocity, $l/cm^2.min$

Q = filtration rate, l/min

α = average voidage

Assuming the third term in eq. (13.2) insignificant, and the concentration-space curve as time-invariant, the equation can be written in the ordinary differential form as

$$-v\frac{dC}{dx} = \frac{dq}{dt} \tag{13.3}$$

For solution of eq. (13.3) a relationship between dC/dx and its variables governing the solids removal must be established. It has been found that dC/dx is proportional to some removal coefficient, which is again changing with degree of treatment or removal achieved in the filter. Therefore, dC/dx can be expressed as

$$\frac{dC}{dx} = \left[\frac{1}{(1+ax)^n}\right] r_o C \tag{13.4}$$

where, r_o = initial removal rate, cm^{-1}, and a, n = constants.

At n = 0, eq. (13.4) becomes

$$\frac{dC}{dx} = r_o C \tag{13.5}$$

which describes a first order or log removal function. At n = 1, the term within the bracket becomes progressively small with distance. This phenomenon indicates a functional relationship of the exponent n with particle size distribution in the filter effluent. The validity of eq. (13.4) was for filtration rates up to 400 l/m^2.min. The value of r_o can be computed at or near zero depth of filter. The values of a and n can be determined from the plot of

$$\left(\frac{Cr_o}{dC/dx}\right)^{1/n} = 1 + ax \tag{13.6}$$

A generalized removal rate equation was developed by considering the effects of the material deposited in the filter on the removal rate:

$$\frac{dC}{dx} = -\frac{1}{(1+ax)^n} r_o C \left(1 - \frac{q}{q_u}\right)^m \tag{13.7}$$

where, q = suspended solids quantity deposited in the filter, q_u = solids quantity that can be deposited ultimately in the filter, m = a constant related to floc strength.

At the start of the removal process by the filter, q = 0, *i.e.* $(1-q/q_u)^m = 1$. So eq. (13.7) becomes equivalent to eq. (13.4). When more solids are getting deposited in the upper layers leading to clogging condition, eq. (13.7) becomes zero, signifying no change of solids concentration with distance in the filter.

3. Computation of Head-loss

On the basis of hydraulics of clear water filtration through a granular bed several equations were developed, which can be utilized to compute the clear water head-loss. However, as clogging develops during filtration of wastewater, some of the void spaces reduces and voidage or porosity changes with increase in head-loss. So clear water equations need to be modified for estimation of head-loss. This approach makes the estimation extremely difficult.

Due to development of head-loss as a result of clogging, the total head-loss across a bed becomes the summation of clear water head-loss plus an additional head-loss due to deposits of materials within the bed.

$$h_t = h_o + \sum_{i=i}^{n} (h_i)_t \tag{13.8}$$

where, h_t = total head-loss at time t, m

h_o = total initial clear water head-loss, m

$(h_i)_t$ = head-loss in the i-th layer of the bed at time t, m, which is related to the quantity of solids deposited within the layer, *i.e.*

$$(h_i)_t = a(q_i)_t^b, \tag{13.9}$$

where, $(q_i)_t$ = quantity of solids deposited in the i-th layer at time t, mg/cm^3, and a, b = constants.

Total initial clear water head-loss is expressed in terms of Dercy's equation and filtration velocity:

$$h_o = \frac{vd}{K} = \frac{Qd}{AK} \tag{13.10}$$

where, d = depth of the filter, m; v = filtration velocity, m/s; Q = flow rate, m^3/s; and K = permeability constant, m^3/m^2.s

4. Filtration Process Variables

The major process variables considered in the design of filters for use in wastewater treatment are the characteristics of wastewater to be filtered, the filter media size, flow rate through the filter or filtration rate, filter bed voidage or porosity and depth, and allowable head-loss.

Wastewater Characteristics

The very important wastewater characteristics are concentration, particle size and distribution of solids, and also the floc strength, if flocculation process precedes filtration as a secondary settling facility. They affect the removal mechanism during the filtration process. However, these characteristics of wastewater generated in the secondary wastewater treatment processes vary from those generated in effluents of aquaculture farm. The types of solids generated in aquaculture and their particle size have been discussed previously.

Filter Media Size

Particle size of filter media affects the filtration process by changing the clear water head-loss and head-loss developed due to solids deposit in the filter. Selection of filter medium for aquaculture purpose is mostly based on the pilot plant studies or on field experience. As a guideline, particle size of sand should be 2–5 cm for culture effluents with high solids concentration and with no settling before the filter. With reasonable gravity settling before filtration, 2 cm or larger size particles can be recommended. Particle size is specified by standard screen analysis technique and with determination of 'effective size' and 'uniformity coefficient' (Wheaton, 1977).

13.3 Sedimentation of Solids

Sedimentation is the separation from wastewater, of suspended solids that are heavier than water. Gravitational settling of solids is the mechanism of separation by which density differences between the solid particles and particulate matter and water cause the particles to settle in a still or slowly moving water column. Sedimentation is one of the most commonly used operations in wastewater treatment for removal of grit, particulate matter and suspended flocs of both chemical and biological origin, and for concentration of solids in sludge thickening process. The purpose of sedimentation is to produce a clarified effluent from wastewaters, and also to produce sludge after solids concentration.

13.3.1 Types of Settling

Four types of settling phenomenon can take place depending on the concentration of particles in wastewater and their tendency to interact between them: discrete or free settling, flocculant settling, hindered or zone settling, and compression settling (Table 13.3).

Table 13.3 Types of settling encountered in wastewater treatment

Types of Settling	*Settling Phenomenon*	*Particles Removed*
Discrete or free (Type I)	Particles settle as discrete ones from a suspension of low concentration of solids, without significant interaction with other particles nearby.	Grit and sand particles
Flocculant (Type II)	Particles increase in size through coalescence or flocculation in a dilute suspension and settle at rapid rate.	Solids in primary and secondary settling process, chemical flocs
Hindered or zone (Type III)	Particles settle as a unit mass from suspensions of intermediate concentration. Inter-particle forces hider the settling of individual particles and a solid-liquid interface develops at the top of the unit mass of particles.	Solids in secondary settling facilities, biological flocs
Compression (Type IV)	Mass of particles in a suspension of high solids concentration forms a structure during settling and settles further only by compression of the structure. Particles are continuously being added by settling to increase the weight of the structure, and further compression takes place.	Solids thickening in a deep mass of sludge in secondary settling facilities

Source: Adapted from Metcalf and Eddy, 1979.

13.3.2 Types of Sedimentation Tanks or Basins

A large number of basin configurations and sizes are available for sedimentation purpose. Diagrams of typical configurations are shown in Figure 13.6. The principle of treatment by sedimentation is to place the wastewater under a relatively still condition so that the solids with a density higher than the water will tend to settle and the solids with a lower density will tend to float. After removing the settling and floating solids, the suspended solids content of the wastewater is reduced and clarified water is produced. Plain sedimentation takes place under natural forces only. Finely divided particles can be coagulated to increase settling velocity by addition of chemicals and/or other materials.

A commonly used sedimentation basin is divided into four sections: inlet, settling, outlet, and sludge areas. Influent water velocity gets reduced in the inlet area and reaches low velocity and laminar flow conditions for desired settling in the settling area. The effluent water, clear or with remaining unsettled solids, finally reaches a relatively high velocity in the outlet area for discharge. The sludge accumulates in the sludge area, and may be continuously withdrawn. The three principal designs following this principle are used in the sedimentation process: horizontal rectangular basin, circular basin, and plate and tube settlers. Characteristic features of the three designs are shown in Figure 13.7, and are compared in Table 13.4.

13.3.3 Mathematical Analysis of Settling

Discrete particle settling has been analyzed by the classical laws of Newton and Stokes. Considering that the spherical particles fall at a Reynolds Number < 0.3, the terminal velocity of settling has been expressed as:

$$v_c = \frac{\left[g(\rho_s - \rho)d^2\right]}{18\mu} \tag{13.11}$$

where, v_c = settling velocity, m/s; ρ_s = density of particle, kg/m^3; ρ = fluid, kg/m^3; μ = viscosity of fluid, Pa.s; d = diameter of particle, m; and g = acceleration due to gravity, m/s^2.

Table 13.4: Comparison of Settling Basins

Type	*Flow and Settling Device*	*Sludge Removal*
Rectangular basin	Flow in one end of the basin and out the other (horizontal flow). Bottom sloped towards either end	Automatic sludge removal device with drag chain for sludge scraping and surface skimming Sludge removal through a scum trough at the effluent side and hopper at the influent bottom
Circular basin	Flow, either into the centre and out from around the rim of the tank, or in the reverse direction (radial flow) Bottom sloped toward the centre	Rotating scraper at the tank centre removes sludge through a hopper
Plate and tube settlers	Bundles of small plain tubes or flat plates placed inclined in rectangular/circular tanks, used as settling devices Flow passes upward through the tubes, and solids settle by gravity and go out to the basin bottom	Sludge removed from the basin bottom

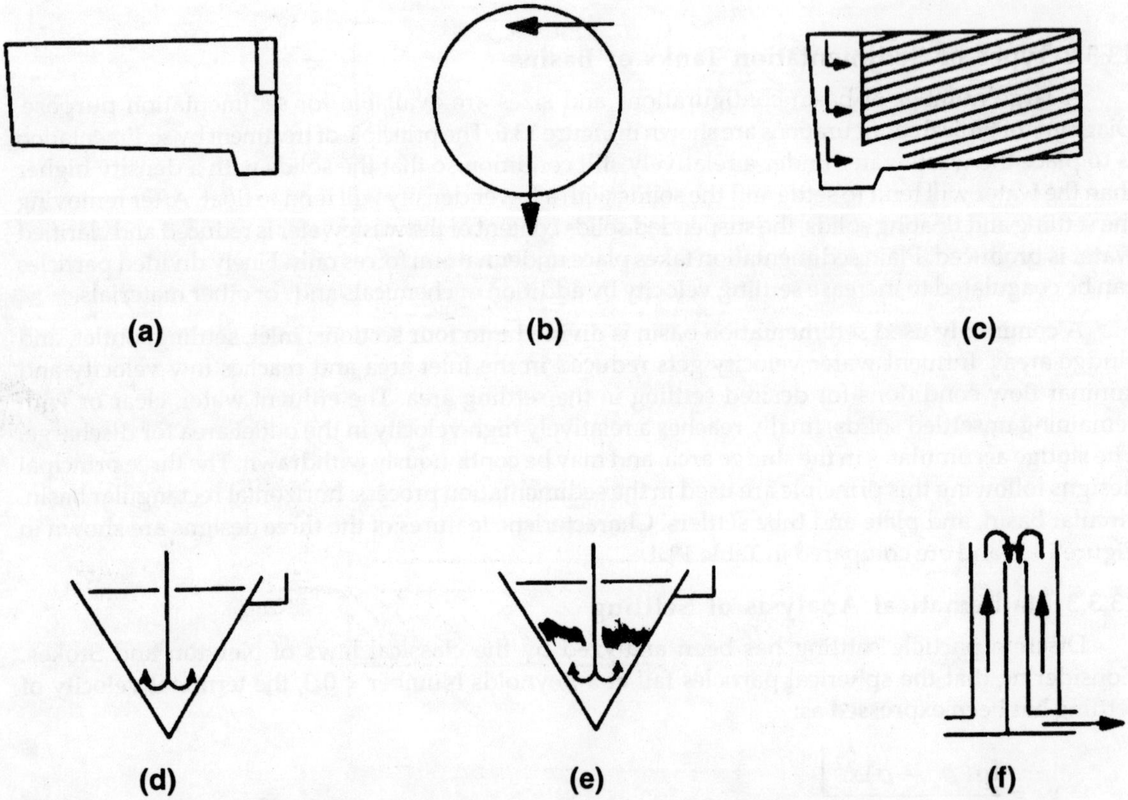

Figure 13.6: Sedimentation Basin Configurations
(a) Conventional horizontal flow, (b) Rotary flow, (c) Tilted plate interceptor, (d) Upflow settler, (e) Clarifier, (f) Settling tube (After Muir, 1982)

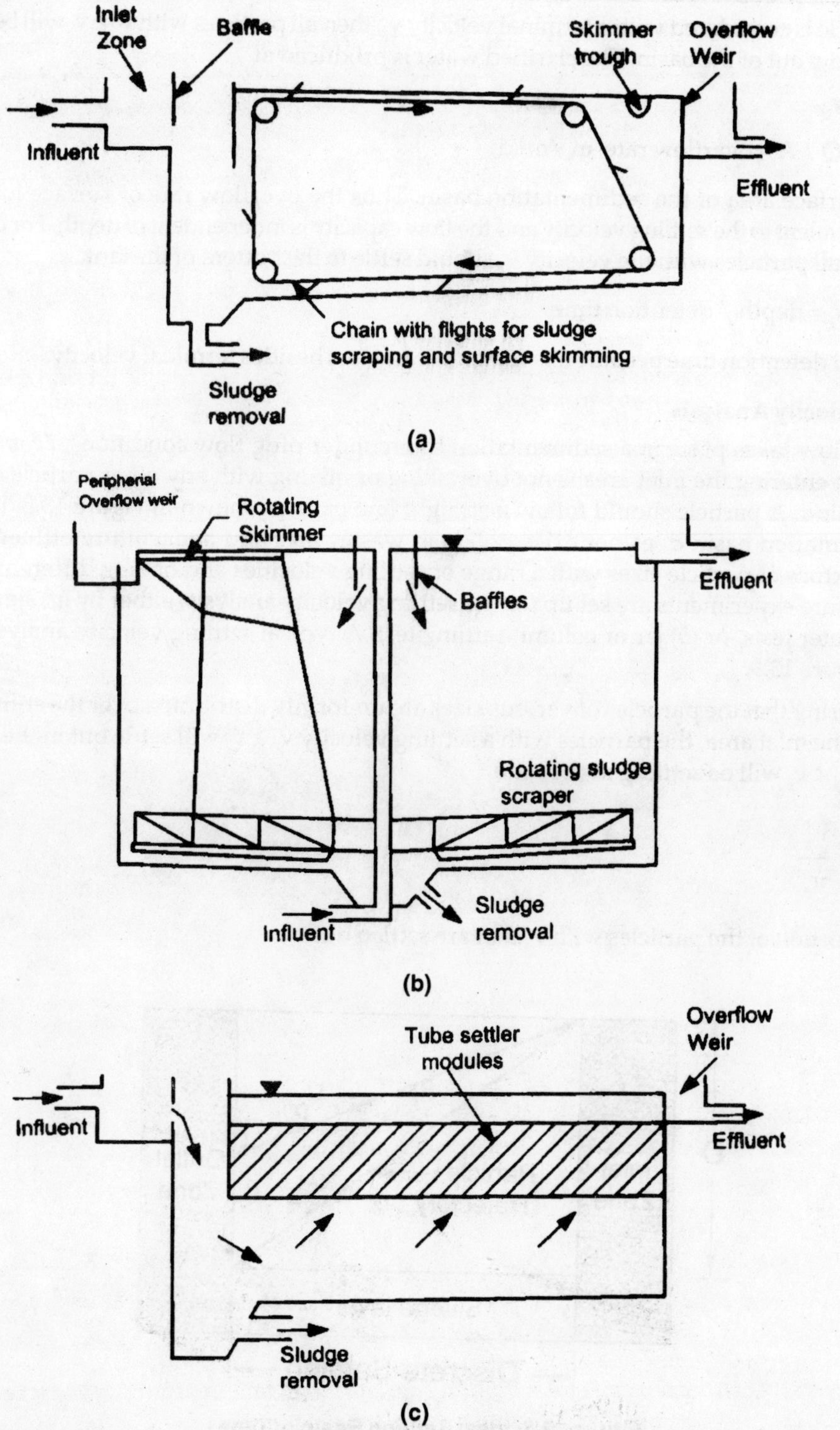

Figure 13.7: Sedimentation Basins (a) Rectangular, (b) Circular, (c) Tube settler

If a particle is considered with a terminal velocity v_c, then all particles with $v_p \geq v_c$ will be removed before they flow out of the basin. The clarified water is produced at

$$Q = A\,v_c, \tag{13.12}$$

$$\text{or } v_c = Q / A = \text{overflow rate, m}^3/\text{m}^2.\text{d}, \tag{13.13}$$

where, A = surface area of the sedimentation basin. Thus the overflow rate or surface loading rate becomes equivalent to the settling velocity and the flow capacity is independent of depth. For continuous flow system, all particles with the velocity v_c should settle to the bottom of the tank.

$$\text{Since } v_c = \text{depth / detention time}, \tag{13.14}$$

the depth and detention time become the design parameters besides terminal velocity.

1. Settling Velocity Analysis

Ideally flow takes place in a sedimentation basin under plug flow conditions, *i.e.* a particle or fluid element entering the inlet area is not overtaking or mixing with any other particle or element ahead or behind. A particle should follow a straight line path, as shown in Figure 13.8. In practice, ideal sedimentation basin does not exist. Solids in wastewater and aquaculture effluents usually contain a mixture of particle sizes with a range of settling velocities and having different detention times. Therefore experiments are set up for the settling velocity analysis, either by (*a*) sieve analysis and hydrometer tests, or (*b*) jar or column settling test. A typical settling velocity analysis curve is shown in Figure 13.9.

Considering that the particles of various sizes are uniformly distributed over the entire depth of the basin at the inlet area, the particles with a settling velocity $v_p \geq v_c$ will settle out in the basin, and those with $v_p < v_c$ will be settling in the ratio

$$X_r = \frac{v_p}{v_c} \tag{13.15}$$

where, X_r = portion of the particles with v_p that are settled out.

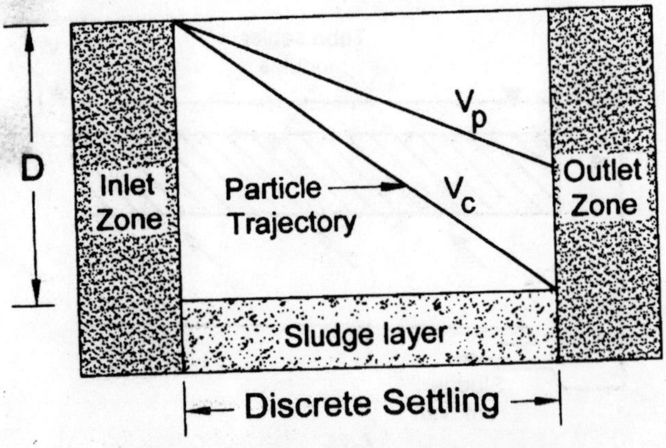

Figrue 13.8: Ideal Settling Basin of Type I

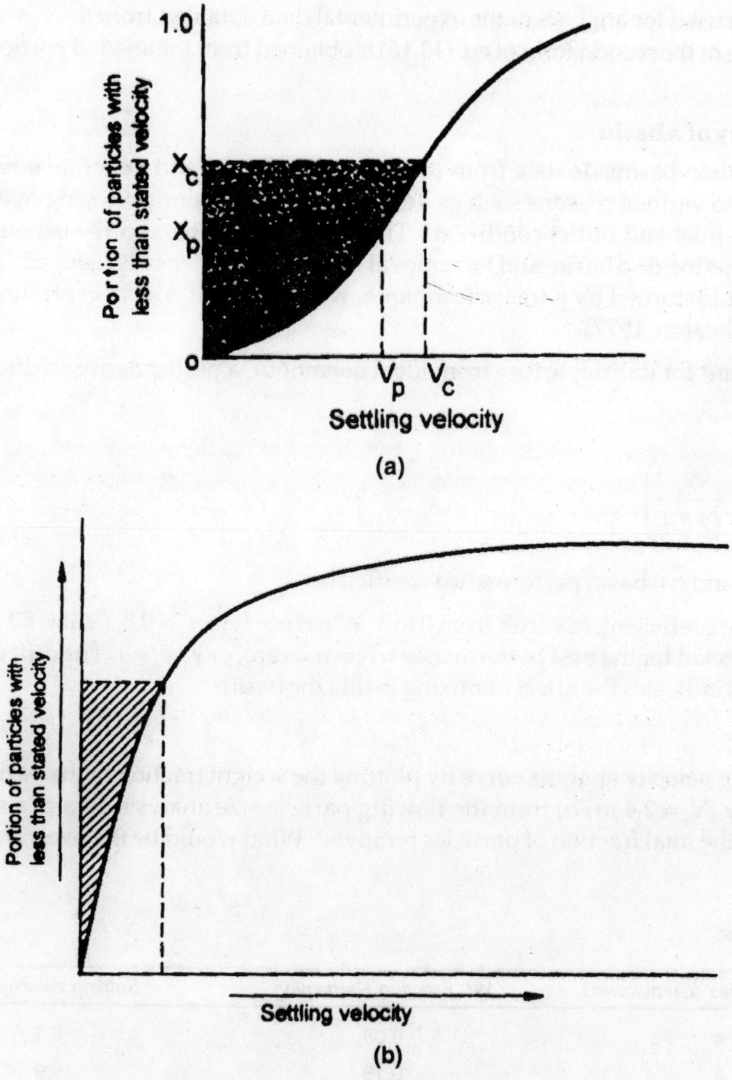

Figure 13.9: Settling Velocity Analysis Curve for Discrete Particles (After Wheaton,1977; Metcalf and Eddy, 1979)

Therefore, the total portion of particles settling out

$$= (1 - X_c) + \int_0^{X_c} \frac{v_p}{v_c}\, dx \qquad (13.16)$$

where, X_c = portion of particles with $v_p < v_c'$ $(1-X_c)$ = portion of particles with $v_p > v_c$, and $\int \frac{v_p}{v_c}\, dx$ = portion of particles settled out with $v_p < v_c$.

Eq. (13.16) can be used for analysis of the experimental data obtained from the size analysis or jar settling test. The value of the second term of eq. (13.16) is obtained from the shaded portion of the curve (Figure 13.9).

2. Removal Efficiency of a Basin

Open sedimentation basins deviate from plug flow conditions and generate internal currents inside the basin due to various reasons such as density variations, winds blowing over the surface, and variations in the inlet and outlet conditions. Thus short-circuiting and re-suspension of solids due to mixing develops inside a basin, and its removal or settling efficiency is reduced. The departure from ideal behaviour is studied by a tracer technique, which generates a concentration-time graph (Levenspiel, 1967; Wheaton, 1977).

In order to account for this departure from ideal behaviour, a performance coefficient has been defined as follows:

$$\frac{y}{y_o} = 1 - \left(1 + \frac{nv_c}{Q/A}\right)^{-1/n} \tag{13.17}$$

where, $y/y_o = (1 - X_c)$, and n = basin performance coefficient.

The performance coefficient, n, varies from 0 to 1. When n = 1, $y/y_o = 0.5$, *i.e.* it is 50 percent rather than 100 percent expected for the best performance when n = zero, or $y/y_o = 1$. The difference between these extreme values indicates the effect of mixing inside the basin.

Example 13.1

Draw the settling velocity analysis curve by plotting the weight fraction of the particles with less than settling velocity (V_c = 2.4 m/h) from the flowing particle size analysis data of a sedimentation basin, and calculate the total fraction of particles removed. What would be the overflow rate?

Solution

Data and results:

No. of Particles (Cumulative)	Wt. Fraction Remaining	Settling Velocity, m/h
4	0.08	0.4
5	0.18	0.8
9	0.36	1.2
14	0.64	1.6
8	0.80	2.0
5	0.90	2.4
2	0.94	2.6
2	0.96	2.8
1	1.00	3.4

Figure 13.1.1 shows the plot of settling velocity analysis. Area under the curve up to settling velocity = 2.4 m/h = 1.00. So,

Total portion of the particles settling out = (1–0.9) + 1/2.4 = 0.52, and overflow rate = 2.4 × 24 × 60 = 3456 m^3/m^2.d.

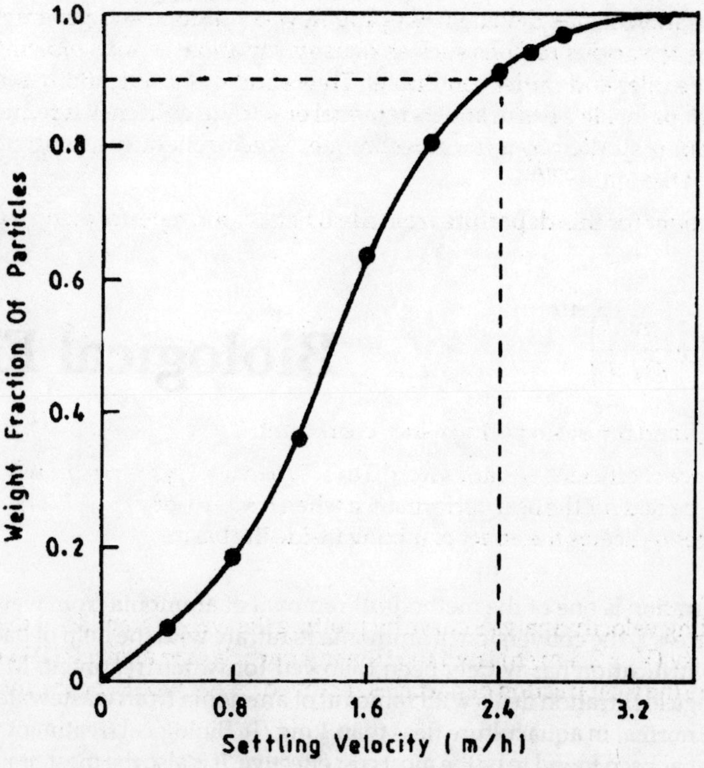

Figure 13.1.1: Settling Velocity Analysis of Example 13.1

14

Biological Filtration

Biological filtration is one of the methods of removal of ammonia from wastewaters, or culture effluents. The process is the conversion of ammonia to nitrate with the help of bacteria, and is known as nitrification. Nitrification has widely been followed for water treatment. Much of the literature available on biological filtration deals with removal of ammonia from wastewaters at concentrations higher than those normal in aquaculture (less than 1 mg/l). Biological treatment to remove ammonia from wastewaters has been found to be the most cost effective. It is also the most practical and economical method for aquaculture use (Muir, 1982). In water reuse culture system, fish metabolites and residues from uneaten food appear in effluent water as suspended solids, dissolved solids, and ammonia that must be removed to some degree before being recycled.

There are other methods of ammonia removal such as chemical oxidation with chlorine, physico-chemical processes like ion exchange, and biological methods such as bio-sedimentation. Nitrification has received considerable importance for its role in water treatment, wastewater treatment, and recently in aquaculture farm effluent treatment.

14.1 Principle of Ammonia Removal by Nitrification

14.1.1. Organisms

The principal organisms responsible for ammonia conversion are the autotrophic (independent of organic carbon sources) bacteria *Nitrosomonas* sp. and *Nitrobacter* sp. They oxidize inorganic compounds ammonia (NH_3) or nitrite (NO_2^-) for energy and obtain their carbon source from carbon dioxide. Some species of heterotrophs, which utilize organic compounds as their energy source, carbon source, and in most cases, as their nitrogen source, also reduce the ammonia concentration. In fact, the filter consists of a complex microbial flora that includes viruses, algae, fungi, protozoa and bacteria. In most systems, the microbial flora grows on some type of fixed media or support such as rocks or plastic media. The aquaculturist is mainly interested in many species of bacteria in the filter that grow predominantly in accordance with the environmental conditions. The final products of nitrification include nitrate, which is comparatively non-toxic to fish, and hydrogen ions, which decreases the filter effluent pH.

14.1.2. Reactions

Nitrification is a two-step reaction in which ammonium ion (NH_4^+) is oxidized to nitrite ion (NO_2^-) by *Nitrosomonas* and subsequently the nitrite ion is oxidized to nitrate (NO_3^-) ion by Nitrobacter. Ammonia is available in the forms of ammonium ion and free ammonia in any aquatic system because of the equilibrium relationship between them. The un-ionized form (NH_3) and nitrite are toxic to most aquatic organisms in concentrations below 1 mg/l. Nitrate is relatively nontoxic in concentrations up to about 400 mg/l.

The reactions are:

$$NH_4^+ + 1.5\,O_2 + 2\,HCO_3^- \rightarrow NO_2^- + 2\,HCO_3^- + H_2O \quad \Delta H = -270\ \text{kJ/mole} \tag{14.1}$$

$$NO_2^- + 0.5\,O_2 \rightarrow NO_3^- \quad \Delta H = -73\ \text{kJ/mole} \tag{14.2}$$

$$\text{Oxidation–overall: } NH_4^+ + 2\,O_2 + 2\,HCO_3^- \rightarrow NO_3^- + 2\,HCO_3^- + H_2O \tag{14.3}$$

The reactions require considerable oxygen, with the stoichiometric requirements of 3.43 kg/kg of NH_3–N for the first step, and 1.14 kg/kg of NO_2–N for the second step. The total requirement of 4.57 kg O_2, for the conversion of 1 kg NH_3-N to NO_3-N, is rather substantial amount. So it is necessary to supply at least this quantity of oxygen in a biological filter and oxygen availability becomes the limiting factor in a filter.

Nitrifying bacteria can synthesize simple chemical compounds into complex carbohydrates, fats, protein, nucleic acids, vitamins and other complex molecules that constitute the living cell. This indicates that the nitrifying bacteria have an elaborate capacity for synthesis (Gaudy and Gaudy, 1980). The cell synthesis reactions are shown below:

Synthesis *Nitrosomonas*:

$$13\,NH_4^+ + 23\,HCO_3^- \rightarrow 8\,H_2CO_3 + 10\,NO_2^- + 3\,C_5H_7NO_2 + 19\,H_2O \tag{14.4}$$

Synthesis *Nitrobacter*:

$$NH_4^+ + 10\,NO_2^- + 4\,H_2CO_3 \rightarrow 10\,NO_3^- + C_5H_7NO_2 + 3\,H_2O \tag{14.5}$$

The cell yields in these reactions are 0.15 mg $C_5H_7NO_2$ / mg NH_4^+-N, and 0.02 mg/mg NO_2^--N, respectively (Kaiser and Wheaton, 1984). As the bacteria die, they become particulate solids in the water. However, these are small amounts of cell mass per mass of substrate oxidized compared to heterotrophic bacteria. This is an advantage, as they cannot clog the filter bed. Excessive solids clog the filter, resulting in creation of anaerobic areas within the filter, high head loss through the filter, and other problems.

It is clear that one goal of filter design for aquaculture purpose is to take advantage of the conditions that favor a consistent growth and activity of the nitrifier within the biomass. The principal factors affecting the population, activity and nutrient supply for the nitrifier are shown in Table 14.1.

Table 14.1: Factors Influencing Nitrifier Biomass

Effects	*Factors*
Growth	Metals such as Mg, Fe, Cu, Ca, Zn, Mo, and PO_4; optimum K:N 1.35[a]
Inhibition	Thiourea, allylthiourea, 8-hydroxyquinoline, peptone, some lipid-soluble compounds, NO_2[a], NH_4[b], compounds with S and N on same C atom, metals
Activity	O_2 above 0.2 mg/l, 3.4 mg/l in bulk conc.; 5° C minimum, 35° C optimum; pH 5.5 minimum, optimum 6-9, adapt 6.0

a. *Nitrosomonas*, b. *Nitrobacter*

Source: Adapted from Muir, 1982.

Nitrifier growth is usually associated with formation of film or flocs of bacteria, or mixed-community materials present in the system. At the micro-level, nutrient transfer and bacterial growth take place in three successive steps: (*a*) transfer of nutrients from water to film or floc; (*b*) cell growth to provide and preserve the available surface area for contact and continuous transfer of nutrients and metabolites by absorption and adsorption; and (*c*) conversion of bacterial mass into removable form. A fourth step of stabilization of the bacterial mass into inert material may also occur (Muir, 1982). The rate of transfer through bacterial film or floc has been described by models based on plug flow or completely mixed systems of chemical reaction engineering. Modified Monod equations could describe nutrient removal and cell growth. Haug and McCarty (1971) found that the active film depth varied with higher NH_3-N concentrations and also with temperature, ranging from 700 μm at 25° C to 1300 μm at 5° C. Movement of nutrient over a range of 0.5–40 mg/l was proportional to the square root of its diffusion coefficient. An overall nutrient transport scheme of nitrifier film structure is shown in Figure 14.1 (Muir, 1982).

14.1.3. Environmental Factors Affecting Nitrification Rate

An understanding of the response of nitrification to environmental conditions is very vital compared to that on various reaction intermediates and enzymes involved in nitrifier growth and reproduction. The first step of biofilter design is examination of the effect of varying conditions of nitrification that consistently favor nitrifier growth and activity within the biomass. These are ammonia concentration, DO concentration, temperature changes, pH changes, and filter media types.

1. Ammonia Concentration

Nitrification reactions show that any of the four substrates ammonia, nitrite, carbonate or oxygen can become a limiting substrate, if the Monod's model of unicellular growth of bacteria is applicable. So describing the growth of *Nitrosomonas* and *Nitrobacter* by the Monod model, it can be written as

$$\mu = \mu_{\max} \frac{s_o}{K_s + s_o} \tag{14.6}$$

where, μ = growth rate of microorganisms, day^{-1}, K_s = saturation constant or substrate concentration, mg/l, at which $\mu = \frac{1}{2} \mu_{max}$, and s_o = initial concentration of the growth limiting substrate. It has been found by experiment that the maximum growth rate of *Nitrobacter* was considerably larger than that of *Nitrosomonas*, and K_s values for both the organisms were less than 1 mg-N/liter at temperatures below 20° C, and hence nitrite does not accumulate in large amounts in biological treatment systems under

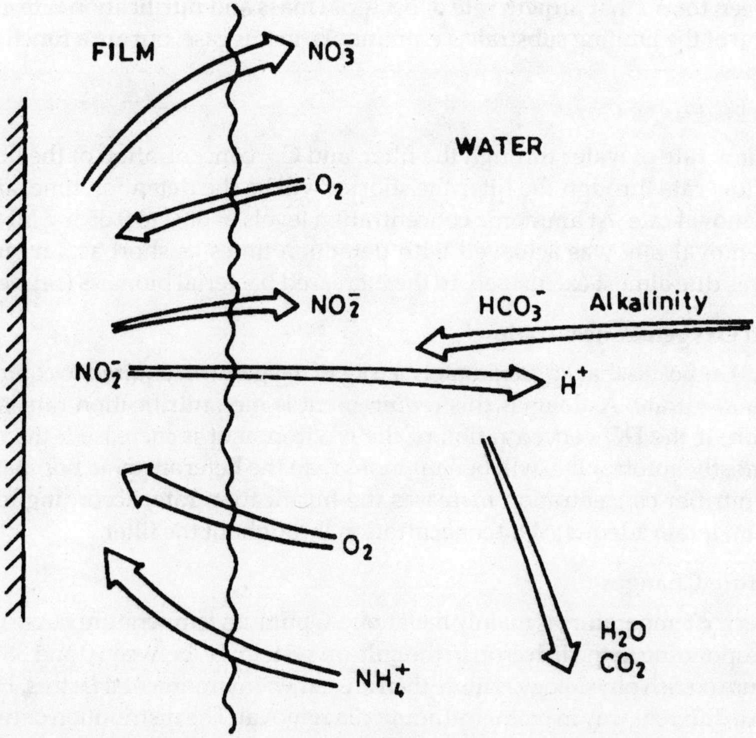

Figure 14.1: Overall Nutrient Transport Scheme of Nitrifier Film (After Muir, 1982)

steady state conditions (Kaiser and Wheaton, 1984). For this reason, ammonia conversion to nitrite, eq. (14.1), is the slower step of the two reactions and hence the rate limiting step, and can be used to model the nitrifier growth.

Eq. (14.6) can be modified to find out the substrate utilization rate as follows:

$$\frac{ds}{dt} = K\left(\frac{s_o.X}{K_s + s_o}\right) \tag{14.7}$$

where, ds/dt = rate of substrate utilization, X = bacterial biomass concentration in a system with limiting substrate concentration of s_o, and K = maximum observed reaction velocity when the substrate is in excess. This relationship clearly shows that the ammonia removal rate increases with increasing bacterial biomass. Thus, in order to promote the maximum nitrification rate, it is essential to build up and maintain a large population of nitrifiers. Conditioning or pre-activation by providing an artificial inorganic nitrogen source is usually followed as a start-up procedure of aquatic biofilters. As the aquaculture effluent contains low concentration of ammonia, food for growth is not sufficient, and stimulation of the nitrifier growth is necessary. Hence, maintenance of a large population of nitrifier is a common problem encountered in biofilters.

It has been found that growth rate of bacterial mass and nitrification rate are not function of the concentration of the limiting substrate, *i.e.* ammonia, in this case, but are a function of mass load of the substrate as

$$\text{Mass load} = Q \times C \quad (14.8)$$

where, Q = flow rate of water through the filter, and C = concentration of the limiting substrate. The higher the flow rate through the filter the shorter will be the detention time, and higher will be the ammonia removal rate. At ammonia concentration levels of 0.02 to 0.05 mg NH_3-N/l, 95–99 percent ammonia removal rate was achieved with detention times as short as 3 minutes. The increase in efficiency was due almost exclusively to the increased bacterial biomass (Brune and Gunther, 1981).

2. Dissolved Oxygen Concentration

Eq. (14.3) indicates that approximately 4.6 kg of oxygen are required to completely oxidize one kg of ammonia to nitrate. As long as this requirement is met, nitrification rate is independent of DO concentration. If the DO concentration of the environment is increased, the nitrifier populations, which are mostly autotrophic, will benefit more than the heterotrophic population. Thus, a gradual increase in nitrifier concentration increases the nitrification rate, according to eq. (14.7). It is thus essential to maintain adequate DO concentration throughout the filter.

3. Temperature Changes

The effect of temperature is mainly metabolic. Optimum temperature range for the nitrifier should be that corresponding to the fish crop in the culture system, *i.e.* between 0 and 38° C, because the effect of temperature on fish physiology is more than any other environmental factors. However, temperature effects act in a different way in biological ammonia removal. The distribution of two forms of ammonia, ammonia gas (NH_3-N) and ammonium ion (NH_4^+-N), in water is determined by temperature and pH. Therefore, when temperature rises in a nitrification filter, at constant pH, two things occur: (*i*) the autotrophic bacteria's metabolic rate increases, increasing the ammonia removal rate, and (*ii*) increase in the ammonia gas content, which escapes into the air. The distribution curve shows that at 30–35° C and pH 9.0 about 50 percent of the ammonia is in the ammonia gas form, resulting in the ready escape of the gas into the air. So it is the ammonia removal rate, not the nitrification rate, under these conditions. At lower pH values (6–7) a temperature change produces a much smaller change in the distribution of NH_3–NH_4^+, and the overall ammonia removal rate will be more closely approximate the nitrification rate.

4. pH Changes

The hydrogen ion concentration has a relatively strong effect on nitrification. As pH moves into the acid range, the rate of ammonia oxidation declines. Studies by Srna and Baggaley (1975) on waste water system showed that pH has a much less effect on *Nitrobacter* than on *Nitrosomonas*, and the former prefer lower pH values (6.6–6.7) than the latter (7.8). It has been suggested that multiple trickling filters be placed in series. The first filter would operate at higher pH values supporting more *Nitrosomonas* to oxidize the NH^+ ions. Since the oxidation of ammonia produces hydrogen ions (combined with carbonic acid–eq. 14.1), the second filter will be operating at a lower pH for oxidation of nitrite by *Nitrobacter* under a more favorable condition.

In fact the carbon dioxide produced and hydrogen ion generated cause the system pH to decrease steadily with time, if no buffering agent is added to water in a recycling system of aquaculture.

Alkalinity of 7.14 mg is destroyed for every mg of NH_3–N completely oxidized, and ntrifiers can adapt pH values as low as 5.5–6.0.

5. Effect of Minerals and Chemicals

The environmental and nutritional effects of some mineral ions were evaluated on the performance of biofilters in aquaculture systems. The results are shown in Table 14.2. The effects of specific inhibitors and poisons on the nitrifiers that are used as disease treatment agents are shown in Table 14.3.

Table 14.2: Effect of Mineral Ions on Nitrification and Bacteria

Mineral Ions	*Effect*
Calcium	High Ca content is very necessary for *Nitrosomonas* metabolism
Magnesium	High Mg concentration is essential for *Nitrobacter* activity. Concentration of 75 mg/l does not hinder the oxidation processes.
Sulphate	Does not obstruct nitrification of either bacteria.
Iron	Ferric iron concentration 10 mg/l hinders NO_2^--oxidation, but to some extent stimulates the NH_3 conversion to NO_2^-.
Potassium, Phosphate	Appear to have a positive effect on NH_3 oxidation by *Nitrosomonas*
Manganese, Chloride, Molybdenum	Have positive effect on oxidation on both NH_3 and NO_2^-
Cobalt, Copper, Zinc	Have negative effect on oxidation of NO_2^- to NO_3^-

Source: Adapted from Kaiser and Wheaton, 1984.

Table 14.3: Chemicals Toxic to Nitrifiers

Chemicals	*Contact*	*Effect*
Terramycin (Oxytetracycline)	Constant 10 mg/l 50 mg/l	Negligible No effect
Chloramphenicol	50 mg/l	No effect
Chloramine T	Constant 10 mg/l	Negligible
Erythromycin	50 mg/l	NO_3^- reduction, stopped nitrification
Malachite green	Constant 0.1 mg/l 3 h, alternate days	Slight disruption at 3 days Negligible
Formaldehyde	Constant 0.1 mg/l Constant 5 ml/l, 45 min	Slight disruption Disruption and recovery within 3 days
	10 mg/l, 3 times on alternate days	Negligible
Formaldehyde and Malachite green	10 mg/l + 0.1 mg/l, 3 times on alternate days	Negligible
Copper sulphate	1 mg/l, 3 times on alternate days	Negligible
Methylene blue	Single dose, 5 mg/l	Disruption over 16 days
Potassium permangaate	Single dose, 4 mg/l	Negligible

Source: Adapted from Muir, 1982; Kaiser and Wheaton, 1984.

6. Filter Media Types

It is clear that the larger the bacteria population in a biofilter, the higher the nitrification rate. This indicates that the most important aspect of filter media type is the large surface area of the media to accommodate the large bacterial biomass. However, the percentage void ratio decreases as the specific surface area increases, with the result that the filter is more easily clogged with the residues from the culture tank. A large variety of filter media is available for use in aquaculture, shown in Table 14.4. These are sand, gravel and rock materials such as limestone, and oyster shells, random plastic shapes and structured plastic modules, and miscellaneous waste materials such as net mesh, brushwood. Several filter media types were evaluated, and it was found that plastic beads appear to be the most promising media due to their high specific surface area and resistance to clogging (Kaiser and Wheaton, 1984).

Table 14.4: Types of Biofilter Media

Type		*Specific Surface Area (m^2/m^3)*	*% Void*	*'Block-ability'* [a]	*Weight (kg/m^3)*
Rock	½"	420	50	8.4	1300
	1"	210	50	4.2	"
	2"	105	50	2.1	"
Ceramic	½"	364	63	5.8	864
Raschig	1"	190	73	2.6	640
ring	2"	92	74	1.2	608
Carbon	½"	374	74	5.1	432
Raschig	1"	187	74	2.5	432
ring	2"	93	74	1.3	432
Berl saddle	½"	466	63	7.4	864
	1"	249	69	3.6	720
	2"	105	72	1.5	640
Pall ring	1"	217	93	2.3	528
(PVC)	2"	120	94	1.3	440
'Surfac'					
Standard		82	94	0.9	64
Close		187	94	2	48
'Flocor'	PVC	85	98	0.9	37
	RC	330	95	3.5	70
Norton ring 8.89 cm		92	97		–
Plastic bead		656	27		–

Source: Adapted from Muir, 1982. [a] 'Blockability'or the index of blockability is defined as surface area/percent void, indicating the relative ease with which a particular material will block as the film thickness.

Among the common media, dolomite (Ca Mg $(CO_3)_2$) has the highest buffering capacity followed by oyster shells, coral gravel and limestone, in the decreasing order (Spotte, 1970).

14.2 Filter Design

14.2.1 Filter Configuration

Biofilters usually consist of a porous solid medium on which the bacteria grow. The bacteria derive their nutrients, oxygen and other needs from water passing over the filter. A simple filter consists of a box or column filled with the media, through which water passes down, and runs out through the under drains in the bottom of the column. The filters are used in several configurations:

1. Submerged filters, where the filter media is kept completely under water level. There are three variations in the design, depending on the direction of water flow: downflow, upflow and side flow filters.
2. Trickling filter, in which the water is distributed on the top of the filter, and gets trickle through the bed in sufficient volume to keep a water film on the filter media but not to fill the pores in the bed. This allows natural aeration by air movement through the filter.
3. Rotating media filters incorporating principle of both fixed (*i.e.* submerged and trickling filters) and suspended-media operation. Two important designs are: rotating biodrum and biodisc, in which fixed media holding the film of bacteria are rotated through a suspended media bed or film held in a trough. The systems have gained importance in recycle or reuse of culture water.

1. Submerged Filters

Figure 14.2a shows the simplified sketch of submerged filter of downflow type. The filter consists of a vessel filled with a filter media, which support the growth of bacteria. Wastewater, distributed though a spray bar, flows by gravity, but the media remains completely submerged. Water flows out of the unit through a perforated bottom support.

Size of the filter media smaller than 19–25 mm is not usually recommended in order to avoid clogging of the bed. Clogging by solids in the deeper portion of the bed increases the head loss and develops short-circuiting. As a result anaerobic zones may develop forming toxic gases. This requires frequent backflushing, thereby increasing the operating expenses. Clogging is not serious in upflow filter designs, as water flows against gravity, and so energy costs for back-flushing are less. However, the most limitation of submerged filters is that the influent water must be aerated sufficiently before passed through the filter, and so oxygen-starving may become a problem for bacteria, if oxygen supply is limited.

2. Trickling Filters

The difference between the tricking and submerged filters is that the media is not submerged and the bacteria remain moist as the wastewater trickles down the bed in trickling filter. The basic functions are same in both the designs. Wastewater is sprayed over the surface of the bed with either fixed or rotating spray bars. The effluent water is collected below the filter support, and discharged through the under drain (Figure 14.2b). Air circulation is provided through the void spaces between the media particles. Hence the bacteria become rarely oxygen-starved, as the bacteria film grown on the media gets enough oxygen from the void spaces filled with air. Therefore, aeration and clogging problems are not as important as in submerged filters.

3. Rotating Media Filters

The two designs of biofilters, rotating biodisc and biodrum, are commonly referred to as rotating biological contactors (RBC). They are considered as the most promising designs of biofilters in

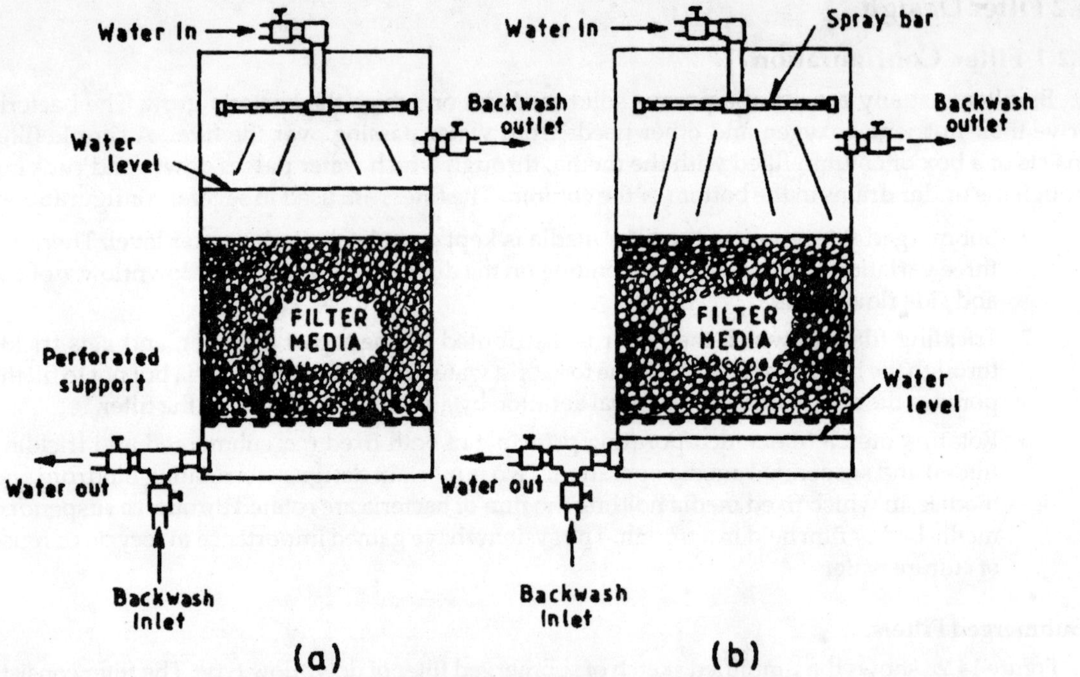

Figure 14.2: Biofilters (a) Submerged, and (b) Trickling (After Lawson, 1997)

aquaculture applications. The rotating biodisc is constructed from a series of flat or corrugated discs mounted on a horizontal shaft. Figure 14.3a shows a sectional view of a typical unit, where the disc system rotates in a trough or vessel through which the wastewater flows. The discs hold the fixed film of bacteria grown on them, and the water in the trough contains the suspended media or film. During rotation the film attached on the disc is kept moist by contacting the suspended media in the trough, and gets aerated. Suspended solids are normally separated and recycled within the trough below the discs. Solids separated from the discs, or running from the collecting trough, are normally more settleable. A two-stage biodisc system is shown in Figure 14.3b.

Discs are made of fiberglass or plastic sheet, which are roughened to facilitate holding of film. Spacing between the discs is restricted to not less than about 13 mm so that bridging of the bacteria film between disc and subsequent hindering of the water flow is avoided. About 40 percent of the disc surfaces are kept submerged at a time (Lawson, 1997). The disc system rotates at 2–3 rev/min, with peripheral speed not exceeding about 0.3 m/s, and so avoiding excessive shearing of the film.

The biodrum system developed by Holmberg (1980) employ rotating cylindrical cages filled with media. The media-filled drum rotates about a horizontal shaft through its center. Water flows through the unit in a manner similar to the disc system (Figure 14.3c). The drums provide more surface area and aeration effect, but with greater energy cost (Muir, 1982).

A typical arrangements of RBCs shows that wastewater can be fed parallel or perpendicular to the axis of shaft rotation (Figure 14.4). In the parallel flow system, water enters from one end of the unit and exits from the other. Here nitrification suffers in the anterior section of the unit due to more

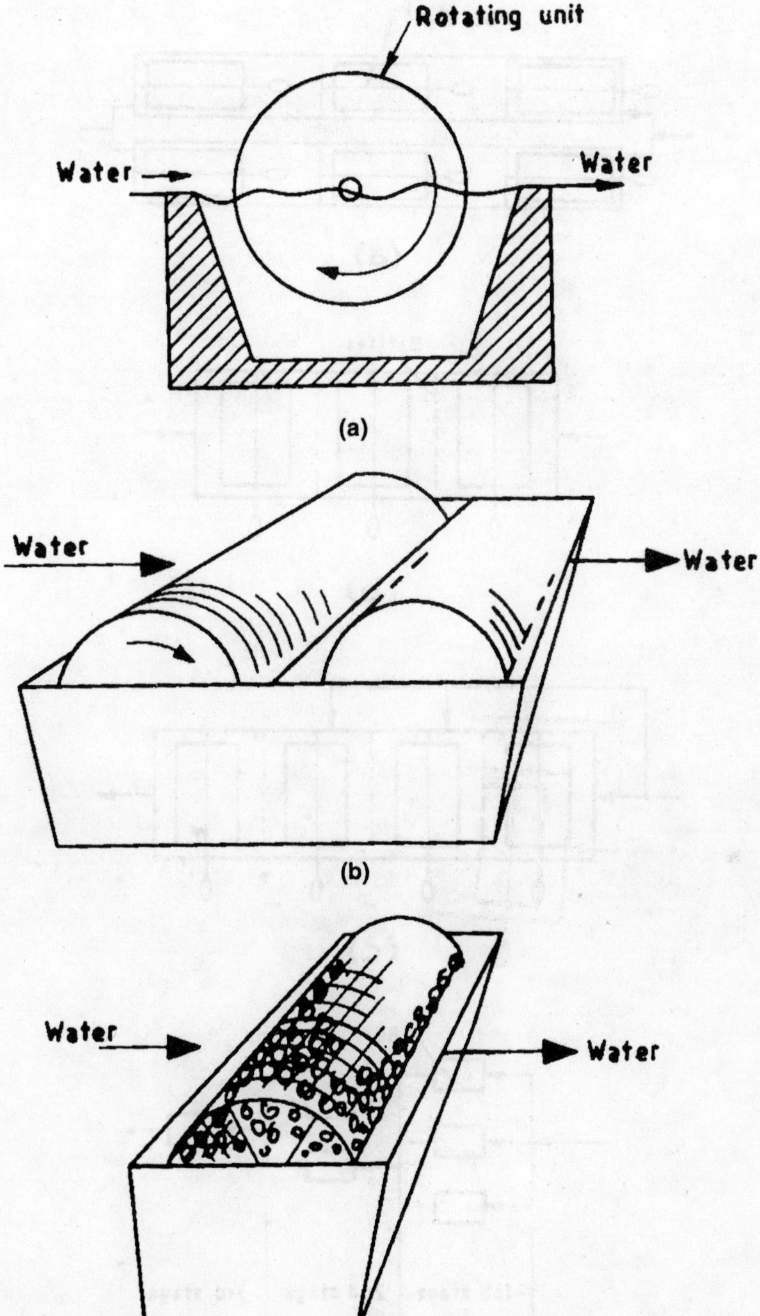

Figure 14.3: Rotating Media Filters (a) Sectional View, (b) Two-stage Biodisc, and (c) Single Stage Biodrum (After Muir, 1982)

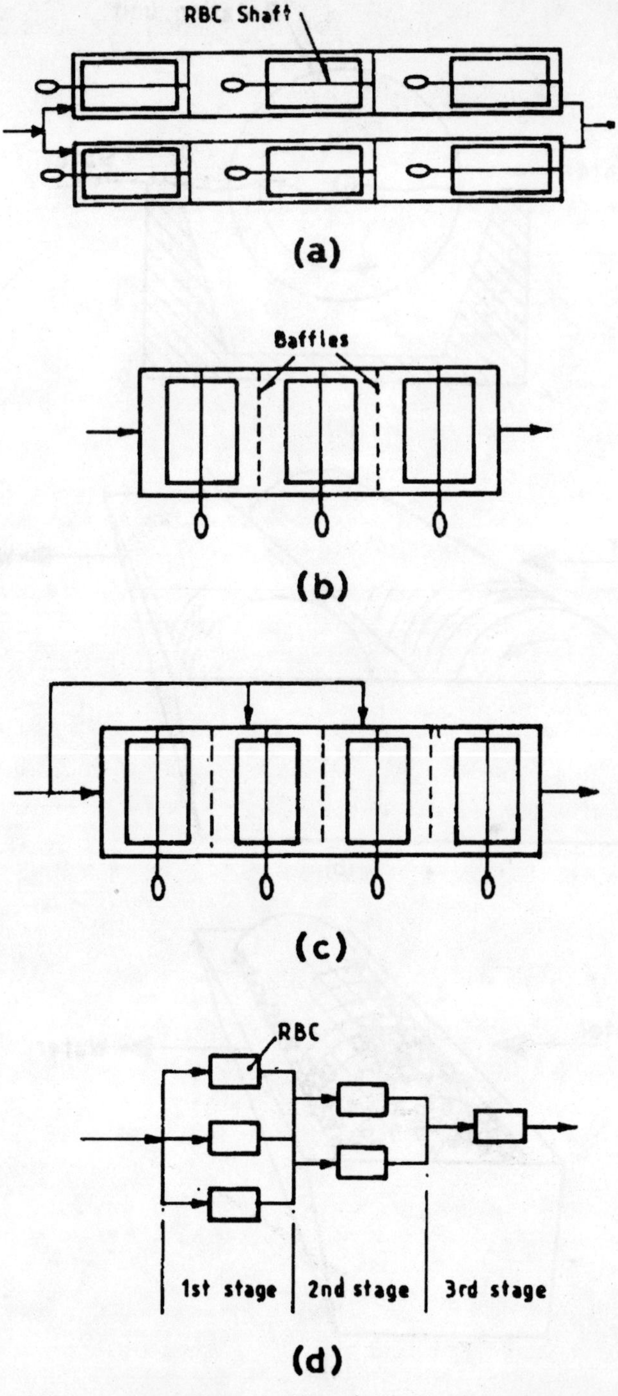

Figure 14.4: Type of Rotating Media Filter Arrangements (a) Parallel flow, (b) Perpendicular flow, (c) Step fed, and (d) Tapered fed (After Lawson, 1995)

favorable growth of heterotrophic bacteria. In comparison, water entering from the side of the unit in the perpendicular flow system favors growth of both heterotrophic and autotrophic bacteria evenly distributed, and so nitrification is more uniform. In large flow RBC units, flow may be introduced in series (step fed), or in parallel (tapered fed) system (Lawson, 1997).

14.2.2 Operating Parameters

The major operating parameters are filter configuration and flow distribution, available surface area, waste and hydraulic loading, duty cycle, besides environmental parameters (Muir, 1982). The environmental parameters have been discussed before. The filter media types providing the surface area for bacterial growth have also been described under environmental parameters. The other operating parameters are outlined below.

1. Flow Distribution

Rate of ammonia removal is directly related to surface area, if flow is evenly and perfectly distributed in submerged filters of both up and down flow types. In trickling filter the rate is more dependent on the thickness and distribution of bacterial film which holds up solid and liquid wastes, and directs the flow through the bed. The flow regime has been found to be nearer to the plug flow condition than to the perfect mixing condition. According to Levenspiel (1967) the reactor dispersion number (D/uL, where D = dispersion coefficient, u = flow velocity, and L = reactor length) ranges from plug flow (D/uL = 0) to perfect mixing (D/uL = ∞). Haug and McCarty (1971) found that dispersion number of a biofilter system without any bacterial film varied between 0.5 and 0.04 within detention time of 15 and 60 minutes, respectively. While normally developed filters have dispersion numbers of approximately 0.1, it increases to 0.2–0.4 during clogging.

2. Hydraulic Loading

Hydraulic loading is defined as the volume of wastewater flowing through the filter per unit of cross sectional area of filter bed per unit of time, expressed as $m^3/m^2/$ day, or liter/m^2/min, or as volume of flow per unit volume of filter material per day (m^3/m^3/day). Two operating parameters, maximum and minimum hydraulic loading rates, are decided based on filter and media types. In a submerged filter, minimum loading decides the lowest flow to satisfy the bacterial demand, whereas, in trickling filters and RBCs, the criteria is to keep the media supporting the bacterial film wet and to supply the bacteria all essential nutrients. The maximum loading is decided by the flow at which the film shears from the media surface, or by the excessive head loss through the filter bed. As shearing of film leads to loss of nitrification, very rapid flow velocity is avoided.

Muir (1982) reported several data showing the effect of hydraulic loading on nitrification. With simulated fish farm wastes ammonia removal rates were found to vary directly within the range of loading 0–100 m^3/m^3/day, removal efficiency remaining approximately constant. The other range of values, where removal efficiency was not affected, was 86.4–146.9 m^3/m^3/day. Therefore filter detention time is the primary factor affecting the ammonia removal. However, the ultimate factors are the available surface of actively growing bacterial film, and the effect of high flow velocity on the species, structure and performance of the film. In practice, the higher flow velocity is dependent on the pumping energy.

3. Duty Cycle

In a typical cycle of operation bacteria attach to and grow on the filter media and in the spore spaces between the media particles. As solids and the biomass accumulate, the filter becomes clogged slowly and the flow of water, oxygen and nutrients becomes impeded. In trickling filters with 2" media efficiency decreased sharply when loading reached 7.7 kg wet film/m^3 (Muir, 1982). Most filter systems are arranged to provide a backwash cycle in which air or water is allowed to scour excess material into the waste. Use of a backwash in a 3.5 cm rock filter in aquaculture showed that the initial designed flow of 0.12 cm/s was replaced by a 10-minutes flow of 0.46 cm/s for greater effect (Muir, 1982). Usually a backwash interval of one or two days is recommended.

14.2.3 Comparison of Existing Designs of Biofilters

A comparative advantages and disadvantages of principal designs of biofilter are presented in Table 14.5. Table 14.6 shows five different types of existing biofilter designs.

Table 14.5: Comparative Advantages and Disadvantages of Biofilters

Filter types	*Advantages*	*Disadvantages*
Trickling filter	Lighter and simpler design; good aeration; loss of accumulated biomass film	Lower loadings; poor dispersion, hold-up control; chances of fly problem; head-loss
Submerged downflow filter	Higher loadings; better dispersion, hold-up control; no head-loss; backwash in reverse flow	Usually heavier construction; poor aeration
Submerged upflow	Same as above, except backwash filter	Same as above, except backwash in same direction to flow
Recycled filter	Better efficiency at high waste loadings	More complex design and control

Source: Adapted from Muir, 1982.

Table 14.6: Comparison Between Different Designs of Biofilter in Water Reuse Systems

System	*Media Type*	*Fresh Water Required (l/min/kg fish)*	*Filter Surface Area (m^2/kg fish)*	*Hydraulic Loading (l/min. m^2)*	*Stocking Density (kg fish/m^3)*
Biodrum–Fish tank, settling tank, two bio-drums in series. No aerator. 94.56 per cent reuse	Plastic balls	0.0037	–	–	96–123.3 Rainbow trout
RBD–Fish tank with agitator, settling tank. 95 per cent reuse	Corrugated fiberglass	0.0141–0.008	0.5	0.378–0.389	40 Channel Catfish
Submerged upflow–Circulating raceways, settling basin, and aerator	Norton rings	0.065	0.68	3.26	31.95 Trout
Submerged upflow–Fish tank, sedimentation tank. Aerator in filter. 90 per cent reuse.	Koch rings	0.0694	0.0112	6.175	28.4 Trout
Trickling filter–Raceway, clarifier basin, no aerator. 95 per cent reuse	Limestone chunks	1.245	0.00777	160.1	42 Trout

Source: Adapted from Kaiser and Wheaton, 1984.

14.2.4. Filter Design Procedure

There are two design procedures available for biological filters outlined by Wheaton (1977). The procedures were based on data of cold water species and followed a lengthy approach. Lawson (1995) outlined a design procedure based on mass balance approach developed by Losordo (1991), Wheaton et al. (1991), and others. The procedure is applicable in recirculating tank-based systems of warm water aquaculture. The procedure is outlined below:

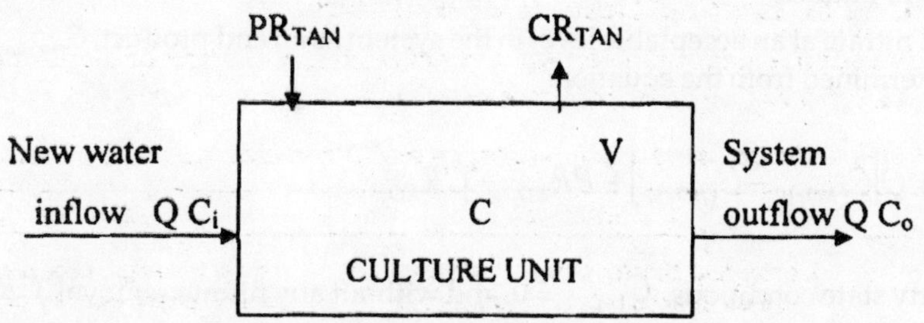

Figure 14.5: Recirculating Culture Unit
(TAN: Total ammonia–N, PR: Production rate, CR: Consumption rate, both in mg/min; Q: Flow rate through the system, l/min; V: System volume; C: Concentration, mg/l)

The system boundary for a recirculating culture unit (Figure 14.5) includes the biofilter, and input and output from the boundary. Assuming completely mixed tank conditions and biomass changing very insignificantly in a continuous production unit, the following mass balance equations can be written.

1. Ammonia Mass Balance

$$\frac{dC_{TAN}}{dt}V = Q\left[C_{(TAN)i} - C_{(TAN)o}\right] + PR_{TAN} - CR_{TAN} \tag{14.9}$$

PR_{TAN}, a function of fish metabolism rate and biological degradation of unconsumed feed, is expressed as:

$$PR_{TAN} = BM \times \text{per cent } BW \times K \tag{14.10}$$

where, BM = fish biomass, per cent BW = feeding rate, percent of body weight, and K = constant as g TAN production/kg feed = 30.

CR_{TAN}, consumption of TAN, or removal rate of filtration process in the system, is

$$CR_{TAN} = Q_f\, C_{TAN}\, E \tag{14.11}$$

where, Q_f = flow through the filter, l/min and E, the filter efficiency, defined as fraction of TAN removed in one pass through the filter, with an conservative estimate at about 25–50 percent.

Eqs. (14.9) and (14.11) are combined to give

$$Q_f = \frac{QC_{(TAN)i} - QC_{(TAN)o} + PR_{TAN}}{C_{TAN}E} \tag{14.12}$$

which gives the flow required to maintain steady state conditions.

Assuming $C_{(TAN)i} = 0$ in new water flow to the system, and $C_{(TAN)o} \approx C_{(TAN)}$ in a completely mixed system,

$$Q_f = \frac{PR_{TAN} - QC_{TAN}}{C_{TAN}E} \tag{14.13}$$

2. Nitrate–N Mass Balance

To maintain nitrate at an acceptable level in the system as an end product, $C_{(NO3)o}$, the flow of new water is to be determined from the equation:

$$\frac{dC_{NO_3}}{dt} = Q\left[C_{(NO_3)i} - C_{(NO_3)o}\right] + PR_{NO_3} - CR_{NO_3} \tag{14.14}$$

Under steady state conditions, $C_{(NO_3)_i} = 0$, and without any nitrate removal, $CR_{NO_3} = 0$,

$$Q = \frac{PR_{NO_3}}{C_{(NO_3)o}} \tag{14.15}$$

$$\text{Now, } PR_{NO_3} = CR_{TAN} \times 4.43 \tag{14.16}$$

where, 4.43 g nitrate is produced for each gram of ammonia nitrified. Assuming very small loss of TAN due to flow through the system, $QC_{(TAN)_o} \approx 0$, and we get $CR_{TAN} \approx PR_{TAN}$.

$$\therefore PR_{NO_3} = PR_{TAN} \times 4.43 \tag{14.17}$$

$$\text{and } Q = \frac{PR_{TAN} \times 4.43}{C_{(NO_3)o}} \tag{14.18}$$

which gives the required flow through the system in order to maintain steady state flow (Q_f) through the filter, eq. (14.13).

3. DO Mass Balance

Since DO is often a limiting factor in controlling the biofilter system, the DO mass balance equation is necessary.

$$\frac{dC_{DO}}{dt}V = Q\left[C_{(DO)i} - C_{(DO)_o}\right] + PR_{DO} - CR_{DO} \tag{14.19}$$

Here PR_{DO} depends solely on the aeration/oxygenation process incorporated in the total system.

$$\text{Now } CR_{DO} = CR_{Fish} + CR_{BOD} + CR_{N} \tag{14.20}$$

where,

(a) Respiration rate CR_{Fish} = BM × per cent BW × K (14.21)
where K = 200 g O_2 / kg feed (for salmonids),
= 214 g O_2 / kg feed (for channel catfish), and
= 300 g O_2 / kg feed (for warm water fish).

(b) CR_{BOD}, the carbonaceous oxygen demand due to biological degradation of fish wastes and unconsumed feed = 2.3 K_{BOD} × BM (14.22)

CR_{BOD} has been found to be about 2.3 times the K_{BOD}, which is the average unfiltered BOD_5 excretion rate of fish. For example, K_{BOD} = 2160 mg O_2 /kg fish/day for catfish fed 1 per cent BW of a 35 percent protein ration.

(c) Consumption rate due to nitrification (biofilter oxygen demand):

$$CR_N = 4.57\,[(PR_{TAN}) - Q\,C_{(TAN)o}] \quad (14.23)$$

where for every gram of TAN oxidized to nitrate, 4.57g of DO are consumed.

Rearranging eq. (14.19) under steady state conditions,

$$PR_{DO} = Q[\,C_{(DO)o} - C_{(DO)I}\,] + CR_{DO} \quad (14.24)$$

where $C_{(DO)o}$ = the desired DO to be reached at the system outlet.

Eq. (14.24) can be used to estimate the rate of oxygen addition required by the total system, or the required rate of DO input to the system by an aerator.

4. DO Mass Balance in Biofilter

This DO balance ensures that the filter will have adequate DO to function properly.

Under steady state condition

$$Q_{f^*} C_{(DO)_{fi}} - CR_{(BOD)_f} - CR_N - Q_f C_{(DO)fo} = 0 \quad (14.25)$$

where, biofilter flow rate to maintain DO

$$Q_{f^*} = \frac{CR_{(BOD)_f} + CR_N}{C_{(DO)fi} - C_{(DO)fo}} \quad (14.26)$$

where, $C_{(DO)fi}$ and $C_{(DO)fo}$ are DO concentrations in the flow in and out, respectively.

Q_{f^*} should be compared to Q_f obtained from eq. (14.13), and the higher of the two values should be used in the system. Alternatively, $C_{(DO)fi}$ could be increased, or $CR_{(BOD)f}$ reduced by settling or pre-filtration of the waste solids.

The steps for design calculation are summarized as follows:

1. Calculate the TAN production rate (PR_{TAN}) from eq. (14.10) with given data of BM, feeding rate and constant K.

2. Calculate the flow rate through the system (Q) from eq. (14.18). Use PR_{TAN} from step 1 and given data of $C_{(NO3)o}$.
3. Calculate the flow recycled through the filter from eq. (14.13). Use the given C_{TAN}, and filter efficiency.
4. Calculate the total DO consumed in the filter (CR_{DO}) from eq. (14.20). Assume $C_{(DO)fi}$ = 2 mg/l, minimum, and $C_{(DO)fo}$ = desired value to be maintained in the filter.
5. Calculate the oxygen production rate $(PR)_{DO}$ from eq. (14.24) to maintain $C_{(DO)o}$, the desired value of DO to be reached at the outlet.
6. Calculate the amount of TAN the biofilter can remove from the equation:

$$TAN_f = PR_{TAN} - Q \cdot C_{(TAN)o} \tag{14.27}$$

7. For biofilter design dimensions such as surface area and volume, use the practical data. For example, an upflow filter requires carrying capacity of the filter and the recommended hydraulic loading, and for RBC biodec, TAN removal capacity and surface area of the media.

Example 14.1

An upflow sand-filled biofilter is to be designed for use in a recirculating culture system carrying 1000 kg of biomass of a warm water species under the following conditions:

Data: Feeding rate = 1.5 percent of biomass daily

TAN produced = 25 g /kg feed

Oxygen consumed by respiration = 350 g /kg feed, and $C_{(DO)i}$ = 8.0 mg/l

System TAN concentration = 2.0 mg/l, and nitrate concentration = 300 mg/l

Biofilter efficiency = 35 percent, and filter BOD removal efficiency = 75 percent

1. Calculate the recycle flow rate through the filter.
2. If it is assumed that the DO concentration in the filter outlet is to be maintained at 2.5 mg/l, what would be the flow rate through the filter?
3. How much DO should be produced in the unit to maintain 7.5 mg/l of DO in the system?

Solution

1. $PR_{TAN} = 1000 \times 0.015 \times 25 = 375$ g/d = 15.6 g/h

$Q = 4.43 \times 15.6 / 350 \times 10^{-3} = 198$ l/h

$Q_f = 15{,}600–198 \times 2 / 2 \times 0.35 = 21{,}720$ l/h

$CR_{fish} = 1000 \times 0.015 \times 350 = 5250$ g O_2/d = 2,18,750 mg O_2/h

$CR_{BODf} = 2.3 \times 2160 \times 1000 \times (1\text{-}0.75) = 12{,}42{,}000$ mg O_2/d = 51,750 mg O_2/h

$CR_N = 4.57\ (15{,}600–198 \times 2) = 69{,}482$ mg O_2/h

∴ Total = 3,39,982 mg O_2/h

2. $Q_f^* = 51{,}750 + 69{,}482 / 7.5–2.5 = 24{,}246$ l/h = 404 l/min
3. $PR_{DO} = 198 \times 7.5 + 3{,}39{,}982–198 \times 8.0 = 0.34$ kg/h

For the biofilter dimension such as surface area and bed volume calculation, assume the carrying capacity with 1.2–2.4 mm diameter coarse sand as 541 g TAN/m^3 media surface /d, and maximum recommended hydraulic flux = 570 l/min/m^2.

Now the amount of TAN_f to be removed by the filter is

$$TAN_f = PR_{TAN} - Q\, C_{(TAN)o} = 15{,}600 - 198 \times 2 = 15{,}204 \text{ mg/h} = 365 \text{ g/d}$$

$\therefore$ Sand bed volume = 365/541 = 0.67 m^3 and cross-sectional area = 404/570 = 0.71 m^2

From practical data 102-cm diameter upflow sand filter having a cross sectional area of 0.81m^2 and a media depth of 38 cm will provide a media volume of 0.31m^3 (Lawson, 1997). So this design will require two of these filters.

15

Disinfection

The process of disinfection is applied to water in aquaculture enterprises for the purpose of (a) destroying or inactivating disease-producing organisms such as pathogenic bacteria and viruses, and (b) controlling predators. Mostly closed and semi-closed culture systems require more stringent control measures of disease prevention than open systems, because the former systems use increased animal densities and recirculation of a part of the water. However, all the systems require protection from predators of the cultured species.

Most disease-producing organisms causing mass mortalities in pond culture belong to bacteria, viruses or other unicellular organisms. These organisms produce either vegetative cells or spores. Spores are, in general, more difficult to destroy. The common harmful organisms are gill flukes of fish, polio virus, ich and infectious hepatitis virus. The action of disinfection on these organisms is to destroy them totally by killing, or partially by killing all or nearly all harmful organisms. The process of total destruction of organisms, known as sterilization may not be necessary, because it destroys food organisms or helpful bacteria such as nitrifier. Disinfection, therefore, refers to the selective destruction of disease-producing organisms (Wheaton, 1977).

15.1 Methods of Disinfection

Disinfection can be accomplished by the use of (1) chemical agents, (2) physical agents, and (3) mechanical means (Metcalf and Eddy, 1979).

Chemical Agents

Disinfection can be accomplished by several chemicals, mostly oxidizing agents, of which chlorine is the most universally used. Other oxidizing chemicals include iodine, bromine, phenolic compounds, potassium permanganate, ozone and hydrogen peroxide. Of these only potassium permanganate is commonly used in aquaculture enterprise.

Physical Agents

Heat and light are the most effective disinfectants. But heating by use of hot water is an expensive process due to its energy intensive method, and is normally not used in aquaculture. Light such as sunlight is also a good disinfectant, but is used in the form of ultraviolet radiation (special UV lamps) for small quantities of water in the industry.

Mechanical Means

Mechanical means such as screening, sedimentation and filters remove predators and bacteria to varying degree. Fine screens can remove 10–20 percent bacteria and plain sedimentation 25–75 percent in comparison to 98–99 percent by chlorination (Metcalf and Eddy, 1979).

15.1.1 Chlorination Process

1. Forms of Chlorine

The most common chlorine compounds used for chlorination are chlorine gas (Cl_2), and hypochlorites such as bleaching powder (Chloride of lime), and sodium or calcium hypochlorite ($NaOCl$ or $Ca(OCl)_2$).

The strengths of chlorinating chemicals are expressed as standard in terms of their 'available chlorine' content. It is also expressed for the doses of application and for the residuals in the water. This expression relates to the capacity rather than oxidizing or disinfecting power of any chemical. Available chlorine is determined as the equivalent amount of iodine liberated from an iodide in acid solution by a known amount of chlorine compound. The available chlorine content is compared with that of pure chlorine which is 100 percent (Palin, 1980).

2. Chemistry of Chlorination

When chlorine as gas is dissolved in pure water, two instantaneous reactions, hydrolysis and ionization, occur. Chlorine first hydrolyzes to form hydrochloric and hypochlorus acids:

$$Cl_2 + H_2O \leftrightarrow HCl + HOCl \qquad (15.1)$$

The hypochlorus acid then partly ionizes to give hydrogen ion and hypochlorite ion (OCl^-):

$$HOCl \leftrightarrow H^+ + OCl^- \qquad (15.2)$$

The ionization constant for the first reaction is

$$K_{Cl_2} = \frac{[HOCl][HCl]}{[Cl_2]} = 4.5 \times 10^{-4} \text{ at } 25^0 \text{ C} \qquad (15.3)$$

The ionization constant for the second reaction is

$$K_{HOCl} = \frac{[H^+][OCl^-]}{[HOCl]} = 3.7 \times 10^{-8} \text{ at } 25^0 \text{ C} \qquad (15.4)$$

(Metcalf and Eddy, 1979).

The relative proportions of the three forms chlorine, hypochlorus acid and hypochlorite ions depend on the water temperature and pH. The more important factor pH decides this proportion (Figure 15.1). The quantity of hypochlorus acid and hypochlorite ions present in water is called the "free available chlorine". The relative distribution of these two chemicals decides the extent of disinfection, as the killing efficiency of hypochlorus acid is about 40–80 times that of hypochlorite ions (Metcalf and Eddy, 1979).

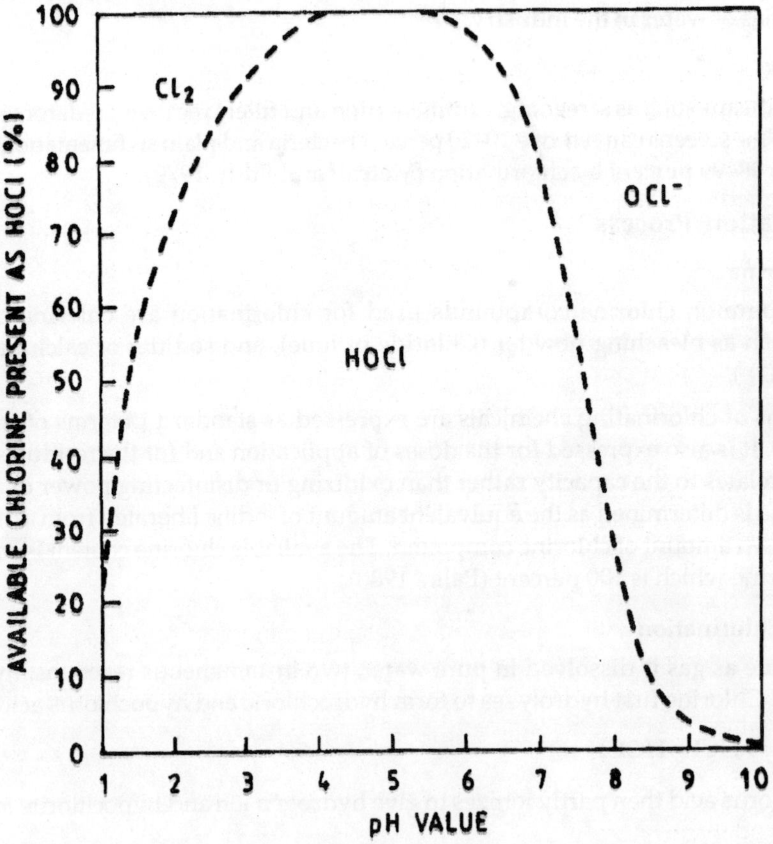

Figure 15.1: Effect of pH Value on Forms of Available Chlorine in Water (After Palin, 1980)

When hypochlorite salts are added in water, the reactions are as follows:

$$Ca(OCl)_2 + 2\,H_2O \rightarrow 2\,HOCl + Ca(OH)_2 \quad (15.5)$$

$$NaOCl + H_2O \rightarrow HOCl + NaOH \quad (15.6)$$

$$HOCl \rightarrow H^+ + OCl^- \quad (15.7)$$

Pond water contains significant amounts of nitrogen usually in the forms of ammonia or nitrate ions. Hypochlorus acid being a strong oxidizing agent, reacts readily with ammonia to produce chloramines and nitrogen chlorides as follows:

$$NH_3 + HOCl \rightarrow NH_2Cl + H_2O \tag{15.8}$$

Monochloramine

$$NH_2Cl + HOCl \rightarrow NHCl_2 + H_2O \tag{15.9}$$

Dichloramine

$$NHCl_2 + HOCl \rightarrow NCl_3 + H_2O \tag{15.10}$$

Nitrogen trichloride

The equilibrium constants for all these reactions depend on pH, temperature of water, and the initial ratio of chlorine to ammonia. Within the normal pH range of water the product consists entirely of monocloramine and the reaction is complete within one minute. The complete conversion of ammonia also requires about five times as much chlorine (Palin, 1980). The chlorine in the form of chloramines is called "combined available chlorine". The free available chlorine is the most toxic to the microorganisms and higher aquatic species. The chloramines also serve as disinfectant, but they are slow acting. The combined chlorine is not as easily removed as the free form, and is so often produced to have a more lasting effect (Wheaton, 1977).

If more chlorine is added than is required for the reactions (15.1), (15.2) and (15.8) to (15.10), continuing oxidations occur at a slower rate producing nitrous oxide (N_2O) and nitrogen, and the chlorine will be reduced to chloride ion. Possible reactions are:

$$NH_2Cl + NHCl_2 + HOCl \rightarrow N_2O + 4\,HCl \tag{15.11}$$

$$2\,NH_2Cl + HOCl \rightarrow N_2 + 3\,HCl + H_2O \tag{15.12}$$

$$NH_2Cl + NHCl_2 \rightarrow N_2 + 3\,HCl \tag{15.13}$$

The above stepwise chlorination reactions can be explained by referring to breakpoint chlorination curve (Figure 15.2a). Culture waters contain many organic and inorganic materials. Chlorine treatment of these waters produces chlorine residual. The chlorine dose–residual curve may be plotted for a given sample of water by treating a series of aliquots with increasing doses and determining the residual chlorine values after a predetermined period of contact. The curve can be divided into various stages:

1. From point 1 to 2 chlorine oxidizes many organic and inorganic matters such as nitrite (NO_2^-) and ferrous ions.
2. Chlorine residual increases with increase of chlorine dosage and chloramines form between the points 2 to 3 (reactions 15.8 to 15.10).
3. At point 3, free available chlorine residual builds up and starts oxidizing the chloramines (reactions 15.11 to 15.13).
4. The chloramines oxidation is completed at the breakpoint (point 4).
5. Continued addition of chlorine past the breakpoint will result in a directly proportional increase in the free available chlorine.

If the water contains only a small amount of ammonia the breakpoint curve will be obtained similar to that of Figure 15.2b.

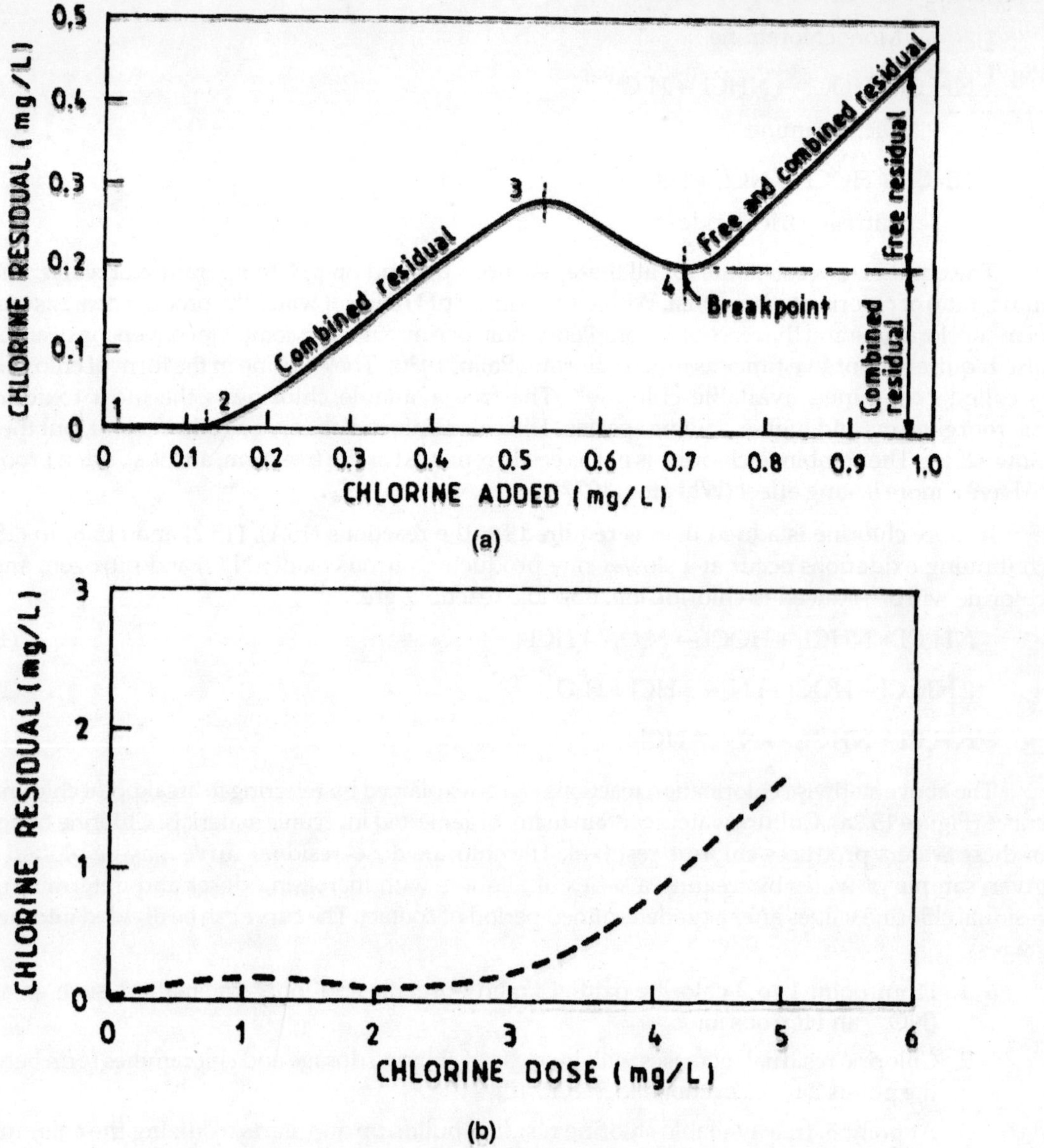

Figure 15.2: Breakpoint Chlorination Curves
(a) Generalized curve (After Metcalf and Eddy, 1979) and
(b) with ammonia nitrogen 0.06 mg/l (After Palin, 1980)

The main reason for chlorination beyond breakpoint is to obtain free chlorine residual so that disinfection can be assured. Chlorine demand of water is the difference between the applied dose and the residual chlorine. It is necessary, however, to specify whether the residual produced is in the free or combined form, and the time of contact.

Example 15.1

Determine the breakpoint dosage and the design dosage of chlorination to obtain a residual of 3 mg/l free available chlorine from the following data:

Chlorine dosage, mg/l	0.5	1.0	2.0	3.0	3.5	4.0	4.5	5.0	6.0	7.0	8.0
Residual chlorine, mg/l	0.4	0.8	1.5	1.9	2.0	1.5	0.6	0.6	1.6	2.6	3.7

Solution

From the plot of breakpoint chlorination in Figure 15.1.1 breakpoint dosage = 4.75 mg/l and design dosage = 7.3 mg/l.

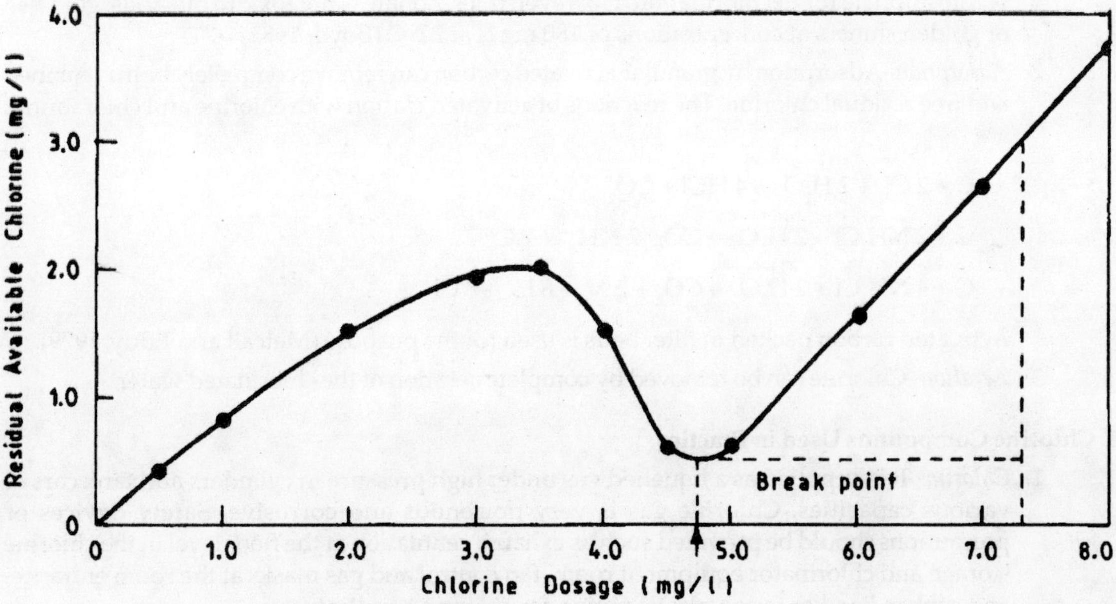

Figure 15.1.1: Breakpoint Chlorination Curve of Example 15.1

3. Disadvantages of Chlorination

1. *Odor*–Serious odor problems develop during breakpoint chlorination due to the formation of nitrogen trichloride and related compounds.
2. *Acid generation*–The hydrochloric acid formed by reactions (15.12) and (15.13), in practice, reacts with the alkalinity of water and pH of water falls. For each one mg/l of NH_3-N oxidized in the breakpoint chlorination process, in practice about 15 mg/l of alkalinity, as calcium carbonate will be required (Metcalf and Eddy, 1979).
3. *Toxicity of chlorine residuals*–It has been found that most aquatic organisms are highly sensitive to chlorine concentrations ranging from 0.09 to 0.295 mg/l (Wheaton, 1977). Because of the toxicity of chlorine and chloramines to aquatic animals, great care must be taken for use of chlorine as disinfectant of culture water. To minimize the effects of toxicity of chlorine residuals, it has been found necessary to allow sufficient time to disinfect the culture water and then to dechlorinate the water.

4. Chlorine Removal

Chlorine and chloramines can be removed from water by chemical treatment, aeration and adsorption techniques.

1. *Chemical process*–Reducing agents such as sodium thiosulphate and ferrous salts can be used for treatment of chlorinated water. Treatment with thiosulphate is the most effective method of chlorine removal. Reaction of chlorine with thiosulphate is

   $$Cl_2 + 2\,Na_2S_2O_3.\,5\,H_2O \rightarrow Na_2S_4O_6 + 2\,NaCl + 10\,H_2O$$

   The reaction shows that to remove one mg/l of Cl_2, 6.99 mg/l of sodium thiosulphate are required. Municipal water containing normally 0.5–2.0 mg/l of Cl_2 requires 3.4–14.0 mg/l of thiosulphate for dechlorination. However, thiosulphate is not toxic to bluegills, gold fish or golden shiners at concentrations of 180 mg/l at 22^0 C (Boyd, 1982)
2. *Adsorption*–Adsorption by granular activated carbon can remove completely both combined and free residual chlorine. The reactions of activated carbon with chlorine and chloramines are

   $$C + 2\,Cl_2 + 2\,H_2O \rightarrow 4\,HCl + CO_2$$

   $$C + 2\,NH_2Cl + 2\,H_2O \rightarrow CO_{2+}\,2\,NH_4^+ + 2\,Cl^-$$

   $$C + 4\,NH_4Cl + 2\,H_2O \rightarrow CO_2 + 2\,N_2 + 8\,H^+ + 8\,Cl^-$$

   Activated carbon packed in filter beds is used for the purpose (Metcalf and Eddy, 1979).
3. *Aeration*–Chlorine can be removed by complete aeration of the chlorinated water.

5. Chlorine Compounds Used in Practice

1. *Chlorine*–It is supplied as a liquefied gas under high pressure in cylinders and tank cars of various capacities. Chlorine gas is very poisonous and corrosive. Safety devices or precautions should be provided such as exhaust ventilation at the floor level in the chlorine storage and chlorinator equipment room, fan control and gas masks at the room entrance, and rubber-lined or tough plastic piping for chlorine handling.
2. *Sodium hypochlorite*–Solutions in strengths of 1.5–15 percent, stored in a cool location in corrosion resistant tanks.
3. *Calcium hypochlorite*–Available as a powder, granules, tablets, pellets in dry or wet forms. It is favored often over other compounds, because it is readily soluble in water and stable under proper storage conditions. Its strength is up to 70 percent by weight in fresh form.
4. *Solid hypochlorites*–Bleaching powder or chloride of lime is one of the best known disinfectant. The essential constituent is calcium oxychloride, which is decomposed in water to produce Calcium hypochlorite. Available chlorine strength is about 33 percent by weight in fresh form.

Probable dosages of chlorine for various types of wastewater are for untreated wastewater, 6–25 mg/l; primary sedimentation, 5–20 mg/l; chemical precipitation plant, 2–5 mg/; and activated sludge plant, 2–8 mg/l (Metcalf and Eddy, 1979).

15.1.2 Potassium Permanganate Treatment

Potassium permanganate, being a strong oxidizing agent, oxidizes the organic matter and reduced inorganic substances in water by being reduced from permanganate ion (MnO_4^-) to manganous ion as manganese dioxide. The oxidation-reduction reaction produces oxygen as follows:

$$4\,KMnO_4 + 2\,H_2O \rightarrow 4\,MnO_2 + 4\,KOH + 3\,O_2 \tag{15.14}$$

In practice, first manganese dioxide is formed according to the reaction:

$$MnO_4^- + 4\,H^+ + 3\,e^- \rightarrow MnO_2 + 2\,H_2O \tag{15.15}$$

If the water contains much organic matter and therefore, is deficient in dissolved oxygen, much of the permanganate is used immediately to oxidize the organic and inorganic matter. However, as long as any reduced substance remains to react with permanganate, the reaction producing oxygen will not proceed. Therefore, only the permanganate remaining in excess of the requirement for oxidation reactions will decompose to release oxygen. On the basis of stoichiometric reaction (15.14), the potassium permanganate needed to release one mg/l of dissolved oxygen by decomposition will be 632.16/96 = 6.58 mg/l.

The reactions involved in removing inorganic substances such as hydrogen sulphide and ferrous ions from water are as follows:

$$3\,H_2S + 4\,KMnO_4 \rightarrow 2\,K_2SO_4 + S + 3\,MnO + MnO_2 + 3\,H_2O \tag{15.16}$$

$$3\,Fe(HCO_3)_2 + KMnO_4 + 7\,H_2O \rightarrow MnO_2 + 3\,Fe(OH)_3 + KHCO_3 + 5\,H_2CO_3 \tag{15.17}$$

Ferrous hydroxide formed according to reaction (15.17) can absorb orthophosphate and remove it as insoluble precipitate.

Besides being a disinfectant potassium permanganate has many other uses in aquaculture such as parasite and disease treatment of fish, algal control, detoxification of piscicides, and alleviation of oxygen depletion in ponds. For detail discussion on the effects of these treatments, Boyd (1982) may be consulted.

15.2 Mechanisms and Kinetics of Disinfection

Several mechanisms have been suggested in the action of disinfection. All these are based on microbiological and/or biochemical action on the cellular systems, irrespective of disinfectant treatment (Pelczar and Reid, 1965).

1. Destruction of the cell wall by gross oxidation of surfaces and membranes leading to cell lysis and death.
2. Alteration of cell permeability of the membrane resulting in vital nutrients such as nitrogen and phosphorus to escape from within the cell.
3. Alteration of the colloidal nature of the protoplasm by coagulation of the proteins with heat, radiation, acids of alkalis.
4. Inactivation of the enzymes

The rate of destruction of microorganisms by most disinfectants described by the first order death rate equation

$$\frac{dN}{dt} = -kN \tag{15.18}$$

or its integrated form: $$\ln\frac{N_t}{N_o} = -kt \text{, or } t = \frac{2.3}{k}\log\frac{N_o}{N_t} \tag{15.19}$$

The rate constant, k, is dependent on contact time of disinfectant; temperature and pH of water; concentration of disinfectant; the nature, number and type of organisms; and the nature of suspending liquid where the destruction of microbial cells is being studied. For a given concentration of disinfectant, the longer contact time effects the greater destruction of bacteria (Figure 15.3). This effect has been expressed in logarithmic law, eq. (15.19).

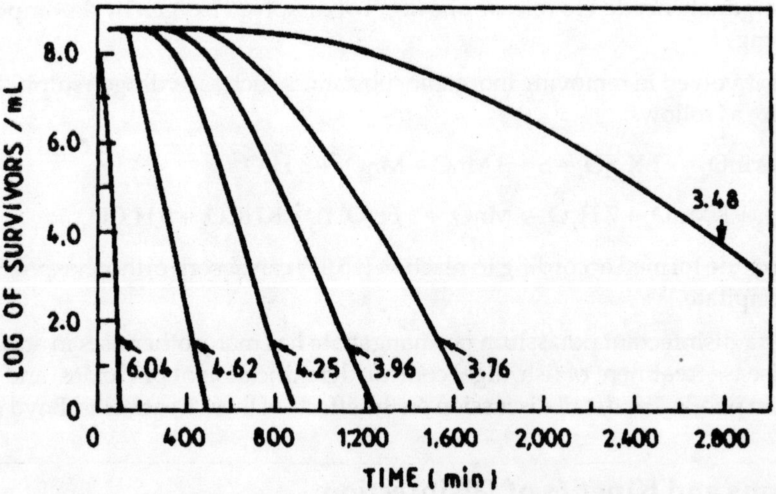

Figure 15.3: Effect of Contact Time and Concentration of Phenol as Disinfectant on *E. coli* at 35° C (After Metcalf and Eddy, 1979)

Contact time for a definite percentage of destruction of microorganisms, t, can be related to disinfectant concentration, C, empirically as follows:

$$t\,C^n = K_1 \tag{15.20}$$

where, K_1 = constant and n = constant. Rearranging and taking logarithm of eq. (15.20)

$$\ln t + n \ln C = \ln K_1 = K_2, \text{ or } \ln t = K_2 - n \ln C \tag{15.21}$$

Constants n and K_2 can be determined from the slope and intercept of the plot of eq. (15.21). When $n > 1$, disinfectant concentration influences strongly the disinfection process. When $n < 1$, the contact time is more important than disinfectant concentration. When n is near unity, both the contact time and concentration of disinfectant have equal effect on disinfection (Fair, *et al.*, 1968).

For a given percentage of bacterial destruction, eq. (15.19) can be reduced to

$$t = \frac{K_3}{k}, \text{where } K_3 = 2.3 \log \frac{N_o}{N_t} = cons\tan t \tag{15.22}$$

From eqs. (15.20) and (15.22), solving for k, we get

$$k = \frac{K_3}{K_1} C^n = K_4 C^n \tag{15.23}$$

where, K_1, K_3 and K_4 are all constants. Thus eq. (15.23) relates the rate constant to disinfectant concentration.

The effect of temperature on the rate constant, k, can be described by the Arrhenius relationship as follows:

$$k = Ae^{-E_a/RT} \tag{15.24}$$

where, E_a = activation energy, J/mol, R = gas constant, 8.314 J/mol.°K, T = absolute temperature, °K, and A = frequency factor.

Some typical values of the activation energy for aqueous chlorine compounds at various pH values are shown in Table 15.1.

Table 15.1 Activation energy of chlorination process

Chlorine Compound	*pH*	*E, kJ/mol*
Aqueous chlorine	7.0	34.3
	10.7	62.8
Chloramine	7.0	50.2
	9.5	83.7

Source: Adapted from Metcalf and Eddy, 1979.

References

Ahmad, T. and Boyd, C. E. 1988. Design and performance of paddle wheel aerators. *Aquacultural Eng.* 7, 39-62.

Akiyama, D.M. 1991. Future consideration for the aquaculture feed industry. In *Proceedings Aquaculture Feed Processing and Nutrition Workshop* held in Thailand and Indonesia, 19-25 September, 1991. D. M. Akiyama and R. K. H. Tan, eds. American Soybean Association, Singapore.

APHA. 1980. American Public Health Association, American Water Works Association, and Water Pollution Control Federation, *Standard Methods for the Examination of Water and Wastewater*, 15th Ed. American Public Health Association, Washington D. C.

Aquaculture Authority. 1998, 2002. *Guidelines for Adopting Improved Technology for Increasing Production and Productivity in Traditional and Improved Traditional Systems of Shrimp Farming*. Aquaculture Authority, Ministry of Agriculture, Government of India, New Delhi.

ASEAN (Association of Southeast Asian Nations). 1978. *Manual on Pond Culture of Penaeid Shrimp.* A project of the ASEAN with the assistance of the FAO/UNDP South China Fisheries Development and Coordinating Programme, Manila. ASEAN National Coordinating Agency of the Philippines, Ministry of Foreign affairs, Manila.

Baldwin, J. W. 1983. The design and operation of an automatic feed dispenser. *Aquaculture* 34 151-55.

Barg, U. C. 1997. Guidelines for the promotion of environmental management of coastal aquaculture development. *FAO Fisheries Technical Paper. No. 328.* Rome, FAO. 1992. Daya Pub. House, New Delhi.

Bay of Bengal Programme (BOBP). 1984. Coastal Aquaculture Project for Shrimp and Finfish in Ban Merbok, Kedah, Malaysia. A Report prepared for the project GCP/RAS/040/SWE development of small-scale fisheries in the Bay of Bengal. *FAO Field Document BOBP/REP/20,* December 1984, p. 31.

Bergheim, A. and Asgard, T. 1996. Waste production from aquaculture. Ch. 3, p. 50-80. In *Aquaculture and Water Resource Management.* B. J. Baird, M. C. M. Beveridge, L.A. Kelly, and J. F. Muir, eds. Blackwell Science, Oxford.

Bergheim, A., Sanni, S., Indrevik, G. and Holland, P. 1993. Sludge removal from salmonid tank effluent using rotating microsieves. *Aquacultural Engineering* 12, 97-109.

Bergheim, A., Siversten, A. and Selmer-Olsen, A. R. 1982. Estimated pollution loadings from Norwegian fish farms. I. Investigations 1978-79. *Aquaculture* 28 (3-4), 347-61.

Boyd, C. E. 1974. *Lime Requirement of Alabama Fish Ponds.* Auburn University Agricultural Experiment Station, Auburn, Alabama.

Boyd, C. E. 1979. *Water Quality in Warm Water Fish Ponds.* Auburn University Agricultural Experiment Station, Auburn, Alabama.

Boyd, C. E. 1982. *Water Quality Management for Pond Fish Culture.* Elsevier, Amsterdam.

Boyd, C. E. 1989. *Water Quality Management and Aeration in Shrimp Farming.* Fisheries and Allied Aquacultures Departmental Series No. 2. Alabama Agricultural Experiment Station, Auburn University, Auburn.

Boyd, C. E. 1991. Empirical modeling of phytoplankton growth and oxygen production in aquaculture ponds, p. 363-395. In *Aquaculture and Water Quality, Advances in World Aquaculture,* Vol. 3. D. E. Brune and J. R. Tomasso, eds. World Aquaculture Society, Baton Rouge, Louisiana.

Boyd, C. E. 1998. Pond water aeration systems. *Aquacultural Eng. 18,* 9-40.

Boyd, C. E. 2003. Guidelines for aquaculture effluent management at the farm-level. *Aquaculture* 226, 101-112.

Boyd, C. E. and Watten, B. J. 1989. Aeration systems in aquaculture. *CRC Critical Reviews in Aquatic Sciences,* Vol. 1, Issue 3, 425-472.

Boyd, C. E., Romaire, R. P. and Johnston, E. 1978. Predicting early morning dissolved oxygen concentrations in channel catfish ponds. *Trans. Amer. Fish. Soc. 107,* 484-492.

Brett, J. R. and Groves, T. D. D. 1979. Physiological energetics. In *Fish Physiology,* Vol. 8, W. S. Hoar, D. J. Randall and J. R. Brett, eds. Academic Press, New York., 279-352.

Brune, D. E. and Gunther, D. C. 1981. The design of a new high rate nitrification filter for aquaculture water reuse. *Journal of World Mariculture Society 12 (1)*, 20-31.

Bryant, P. I. and Matty, A. J. Adaptation of carp (Cyprinus carpio) larvae to artificial diets. 1. Optimum feeding rate and adaptation age for a commercial diet. *Aquaculture* 23, 275-286.

Chakraborti, R. K., Sundaray, J. K. and Ghosal, T. K. 2002. Production of *Penaeus monodon* in the tide fed ponds of Sunderbans. *Indian J. Fish*. 49(4), 419-426.

Chen, S., Timmons, M. B., Aneshansley, D.J. and Bisogni, J. J. Jr. 1993. Suspended solids characteristics from recirculating aquaculture systems and design implications. *Aquaculture* 112,143-155.

Clifford III, H. C. 1992. Marine shrimp pond management: A review. *Proceedings of the Special Session Shrimp Farming*. J. Wyban, ed. World Aquaculture Society, Baton Rouge, LA, USA.

Colt, J. and Orwicz, C. 1991. Aeration in intensive culture. In *Aquaculture and Water Quality, Advances in World Aquaculture*, Vol. 3. D. E. Brune and J. R. Tomasso, eds. World Aquaculture Society, Baton Rouge, Louisiana.

Colt, J. and Watten, B. 1988. Applications of pure oxygen in fish culture. *Aquacultural Eng 7*, 397-441.

Costa-Pierce, B. A. Environmental impacts of nutrients from aquaculture: Towards the evolution of sustainable aquaculture systems. Ch. 4, p. 81-113. In *Aquaculture and Water Resource Management*. B. J. Baird, M. C. M. Beveridge, L.A. Kelly, and J. F. Muir, eds. Blackwell Science, Oxford.

Cripps, S. J. 1995. Serial particle size fractionation and characterization of an aquaculture effluent. *Aquaculture* 133, 323-39.

Cripps, S. J. and Kelly, L. A. 1996. Reductions in wastes from aquaculture Ch. 7, p. 166-201. In *Aquaculture and Water Resource Management*. D. J. Baird, M. C. M. Beveridge and L. A. Kelly, eds. Blackwell Science, Oxford.

Department of Economic and Statistics, Department of Agriculture and Cooperative, New Delhi. 1992. Cited in Nandeesha, 1995.

De Silva, S. S.1995. Supplementary feeding in semi-intensive aquaculture systems. In Farm-made Aquafeeds. M. B. New, A. G. J. Tacon and I Csavas, eds. *FAO Fisheries Technical Paper No. 343*, Rome, FAO, 1994.

Edeson, W. R. 1996. The legal regime governing aquaculture. Ch 8, p.202-214. In *Aquaculture and Water Resource Management*. D. J. Baird, M. C. M. Beveridge and L. A. Kelly, eds. Blackwell Science, Oxford.

Edwards, P. 1982. Lecture notes on utilization of animal and plant wastes. Report of Consultancy at the Regional Lead Center in China for Integrated Fish Farming. Network of Aquaculture Centers in Asia (NACA). *FAO Field Document, No. NACA/WP/82/6*, Bangkok.

Edwards, P. 1993. Environmental issues in integrated agriculture-aquaculture and wastewater-fed fish culture systems. p. 139-70. In *Environment and Aquaculture in Developing Countries*. R. Pulin, H. Rosenthal and J. Maclean, eds. ICLARM, Manila, Philippines.

Fair, G. M., Geyer, J. C. and Okun, D. A. !968. *Water and Wastewater Engineering*, Vol. 2. Wiley, New York.

FAO. 1995. *Code of Conduct for Responsible Fisheries*. FAO, Rome, p. 41

FAO. 1997. Aquaculture Development. *FAO Technical Guidelines for Responsible Fisheries*, vol. 5, FAO, Rome, p.40.

FAO/NACA. 1996. Regional Study and Workshop on Environmental Assessment and Management of Aquaculture Development. *NACA Environment and Aquaculture Development Series No. 1*, FAO, Rome; NACA, Bangkok.

Gaudy, A. F. and Gaudy, E. T. 1980. *Microbiology for Environmental Scientists and Engineers*. McGraw-Hill, New York.

GESAMP (IMO/FAO/Unesco /WMO/WHO/IAEA/UN/UNEP Joint Group of Experts on the Scientific Aspects of Marine Pollution, 1991. Reducing Environmental Impacts of Coastal Aquaculture. *Rep. Stud.GESAMP*, 47, p. 35.

Gokulakrishna, P. and Bandyopadhyay, S. 1995. Formulation and characterization of some pelleted feeds for *Penaeus monodon*. *Fishery Technology* 32(1), 19-24.

Haug, R. and McCarty, P. 1971. *Nitrification with Submerged Filter*. Technical Report No. 149, Department of Civil Engineering, Stanford University, California.

Hickling, C. F. 1962. *Fish Culture*, Faber and Faber, London.

Hollerman, W. D. and Boyd, C. E. 1980. Nightly aeration to increase production of channel catfish. *Trans. Amer. Fish. Soc. 109*, 446-452.

Horvath, I. 1984. *Modelling in the technology of wastewater treatment*. Pergamon, New york.

Huguenin, J. E. and Colt, J. 1989. *Design and Operating Guide for Aquaculture Seawater Systems*. Elsevier, Amsterdam.

Hussenot, J. M. E. 2003. Emerging effluent management strategies in marine fish-culture farms located in European coastal wetlands. *Aquaculture* 226, 113-128.

INFOFISH. 1997. Sustainable Aquaculture. In *Aquaculture 96*. K. P. P. Nambiar and T. Singh, eds. Infofish, Kuala Lumpur.

Jackson, C and Wang, Y. 1998. Modelling growth rate of *Penaeus monodon* Fabricus in intensively managed ponds: effects of temperature. Pond age and stocking density. *Aquaculture Research* 29, 27–36.

Jhingran, V. G. and Pullin, R. S. V. 1985. A Hatchery Manual for the common, Chinese and Indian Major Carps. Manila, Philippines, ADR/ICLARM Publication, *ICLARM Studies and Reviews* 11, 91 p.

Kaiser, G. E. and Wheaton, F. W. 1984. *Nitrification Filters for Aquatic Culture Systems: State of the Art*. Scientific Article No. A3350, Contribution No. 6422 of the Maryland Agricultural Experiment Station, Department of Agricultural Engineering, University of Maryland, Maryland.

Lawson, T. B. 1997. *Fundamentals of Aquacultural Engineering*. Indian ed., CBS Publishers and Distributors, New Delhi.

Levenspiel, O. 1972. *Chemical Reaction Engineering*. 2nd ed. John Wiley and Sons, New York.

Losordo, T. M. 1991. An introduction to recirculating production systems design. *Engineering Aspects of Intensive Aquaculture, Proceedings from the Aquaculture Symposium*, Ithaca, New York, Cornell University, April 4-6, 1991.

Loyacano, H. A. 1974. Effects of aeration in earthen ponds on water quality and production of white catfish. *Aquaculture* 3, 261-271.

Mal, B. C. 1996. Performance of Hawaii-type automated fish feed dispenser. *Aquacultural Eng* 15 (2), 81-90.

Metcalf and Eddy, Inc. 1979. *Wastewater Engineering: Treatment, Disposal, Reuse*. Tata-McGraw Hill, New Delhi.

Moav, R., et al. 1977. Intensive polyculture of fish in fresh water ponds. 1. Substitution of expensive feeds by liquid cow manure. *Aquaculture 10*, 25-43.

Moulick, S., Bandyopadhyay, S. and Mal, B.C. 2005. Design characteristics of single hub paddle wheel aerator. *J. Environ. Eng*. 131(8), 1147-1154.

Moulick, S., Mal, B.C. and Bandyopadhyay, S. 2002. Prediction of aeration performance of paddle wheel aerators, *Aquacultural Eng*. 25, 217-237.

Muir, J. F. 1982. Recirculated water systems in aquaculture. In *Recent Advances in Aquaculture*. J. F. Muir and R. J. Roberts, eds. Croom Helm, London.

Nandeesha, M. C.1995. Aquafeeds and feeding strategies in India. In Farm-made Aquafeeds. M. B. New, A. C. J. Tacon and I. Csavas, eds. *FAO Fisheries Technical Paper No. 343*, Rome, FAO, 1994.

New, M.B. and Csavas Imre. 1995. Aquafeed in Asia. In Farm-made Aquafeeds. M. B. New, A. G. J. Tacon and I Csavas, eds. *FAO Fisheries Technical Paper No. 343*, Rome, FAO, 1994.

New, M. B. 1987. *Feed and Feeding of Fish and Shrimp*–A Manual on the Preparation and Presentation of Compound Feeds for Shrimp and Fish in Aquaculture. ADCP/REP/87/26. FAO, Rome.

Ognean, T. 1993. Aspects concerning scale-up criteria for surface aerators. *Water Res*. 27 (3), 477-484.

Palin, A. T.1980. Disinfection In: *Developments in Water Technology–2*, Ch. 5, W. M. Lews, ed. Applied Science, Essex, England.

Pelczar, M.J. Jr and Reid, R. D. 1965. *Microbiology*, 2nd ed, McGraw-Hill, New York.

Perkins, E. J.(1976. The evaluation of biological response by toxicity and water quality assessment. In *Marine Pollution*. R. Johnston, ed. Academic Press, London.

Pillay, T. V. R. 1990. Nutrition and feeds. In *Aquaculture–Principles and Practices*, Ch. 7. Fishing News Books, London.

Pillay, T. V. R. 2004. *Aquaculture and The Environment*, 2nd Ed., Blackwell Publishing, Oxford.

Piper et al. 1982. *Fish Hatchery Management*, Washington D. C. U. S. department of the Interior, Fish and Wildlife Service, pp. 517. Cited in New, 1987.

Poernama, A. 1990.Site selection for coastal shrimp ponds. p. 3-23. In Technical and Economic Aspects of Shrimp Farming. Proceedings of the AQUATECH '90 Conference, Kuala Lumpur, Malaysia, 11-14 June 1990, M. B. New, H. D. Saram and T. Singh, eds. Organized by Infofish in collaboration with the Fisheries Development Authority of Malaysia.

Robertson, C. 2000. *Sustainable Prawn Farming–Prawn farm effluent treatment and recirculation systems. A review of options in sustainable farm design and operation*. Q100021. Department of Primary Industries, Queensland, Australia.

Ross, D. A. 1970. *Introduction to Oceanography*. Appleton-Century-Crofts, New York.

Schmidtke, N. W. and Horvath, T. 1977. Scale-up methodology for surface aerated reactors. *Prog. Water Technol*. 9, 477-493

Schroeder, G. 1980. Fish farming in manure-loaded ponds. In *Integrated Agriculture-Aquaculture Farming Systems*. Pullin, R. S. V. and Shehadeh, Z. H, eds. ICLARM Conference Proceedings 4, 258. International Center for Living Aquatic Resources Management, Manila and the Southeast Asian Center for Graduate Study and Research in Agriculture College, Los Banos, Philippines.

Schwartz, M. and Boyd, C. E. 1994. Effluent quality during harvest of channel catfish from watershed ponds. *Progressive Fish-Culturist* 56, 25-32.

Shelton, J. L., Jr. and Boyd, C. E. 1983. Correction factors for calculating oxygen-transfer rates of pond aerators. *Trans. Amer. Fish. Soc. 112*, 120.

Smith, J. M. and Van Ness, H. C. 1975. *Introduction to Chemical Engineering Thermodynamics.* 3rd ed. McGraw-Hill, New York.

Sorensen, D. L., Knieb, W. A., Porcella, D. B. and Richardson, B. Z. 1980. Determining the lime requirement for the Blackbird Mine spoil. *J. Environ. Qual.* 9, 162-166.

Spotte, S.1970. *Fish and Invertebrate Culture: Water Management in Closed Systems.* Wiley-Interscience, New York.

Srna, R. F. and Baggaley, A. 1975. Kinetic response of perturbed marine nitrification systems. *A report on Grant No.2-35273, Office of Sea Grant, NOAA.* Dept. of Commerce, University of Delaware, Delaware.

Stickney, R. R. 1979. *Principles of Warmwater Aquaculture.* John Wiley, New York.

Tacon, A. G. J. 1988. The Nutrition and Feeding of Farmed Fish and Shrimp–A Training Manual 3. Feeding Methods. *FAO Field Document No. 7* GCP/RLA/075/ITA. FAO, Brazil.

Tacon, A. G. J. 1993. Feed ingredients for crustaceans–natural foods and processed feedstuffs. *FAO Fisheries Circular No. 866.* FAO, Rome.

Tchobanoglous, G. and Eliassen, R.1970. Filtration of treated sewage effluent. *J. Sanitary Eng. Div. ASCE*, Vol. 96, No. SA2.

Tebbutt, T. H. Y. 1980. Filtration. In: *Developments in Water Technology–2*, Ch, W. M. Lews, ed. Applied Science, Essex, England.

Teichert-Coddington, D. R., Rouse D. B., Potts, A. and Boyd, C. E. 1999. Treatment of harvest discharge from intensive shrimp pond by settling. *Aquacultural Eng* 19 (3), 147

Tisdale, S. I. And Nelson, W. L. 1956. *Soil Fertility and Fertilizers.* The MacMillan Co., New York.

Tomasso, J. R. and Brune, D. E. 1991. Aquaculture water quality–The emergence of an applied discipline. In *Aquaculture and Water Quality,* D. E. Brune and J. R. Tomasso, eds. The World Aquaculture Society, Baton Rouge, LA.

Tookwinas, S. 2000. Closed-circulating shrimp farming system. State-of-the Art Series, November 2000, Southeast Asian Fisheries Development Center (SEAFDEC) and Association of Southeast Asians Nation (ASEAN), Philippines.

Treybal, D. E. 1981. *Mass Transfer Operations*, 3rd ed. McGraw-Hill, Singapore.

van Rijn, J. 1996. The potential for integrated biological treatment systems in recirculating fish culture–a review. *Aquaculture* 139, 181-201.

Wijkstrom, U. N. and New, M. B. 1989. Fish for feed: A help or a hindrance to aquaculture in 2000? *Infofish International* 6189, 48-52.

Wheaton, F. W. 1977. *Aquacultural Engineering*. Wiley-Interscience, New York.

Wheaton, F. W, Hochheimer, J. N., Kaiser, G.E and Krones, M. J. 1991. Principles of biofiltration. *Engineering Aspects of Intensive Aquaculture, Proceedings from the Aquaculture Symposium*, Ithaca, New York, Cornell University, April 4-6, 1991.

Zlokarnik, M. 1979. Scale-up of surface aerators for wastewater treatment. *Adv. Biochem. Eng.* 11, 157-179.

APPENDICES

Appendix A

Table A-1: Atomic Weights of Major Elements in Water, their Usual Ionic Form and Equivalent Weights

Element	*Symbol/Ionic Form*	*Atomic Weight*	*Equivalent Weight*
Hydrogen	H^+	1	1
Carbon	C	12.0	
Nitrogen	N	14.0	
Oxygen	O	16.0	
Sodium	Na^+	23.0	23.0
Magnesium	Mg^{2+}	24.3	12.16
Aluminium	Al^{3+}	26.98	8.99
Silicon	Si	28.1	
Phosphorus	P	31.0	
Sulphur	S^{2-}	32.1	
Chlorine	Cl^-	35.5	35.5
Potassium	K^+	39.1	39.1
Calcium	Ca^{2+}	40.1	20.04
Manganese	Mn^{2+}	54.9	27.5
Iron	Fe^{2+}	55.9	27.9
Copper	Cu^{2+}	63.5	
Zinc	Zn^{2+}	65.4	

Table A-2: Oxygen Solubility in Water Saturated with Air at Atmospheric Pressure

Temperature, °C	*Oxygen Solubility, mg/l*					
	Chlorinity: 0	*5.0*	*10.0*	*15.0*	*20.0*	*25.0*
0.0	14.621	13.728	12.888	12.097	11.355	10.657
1.0	14.216	13.356	12.545	11.783	11.066	10.392
2.0	13.829	13.000	12.218	11.483	10.790	10.139
3.0	13.460	12.660	11.906	11.195	10.526	9.897
4.0	13.107	12.335	11.607	10.920	10.273	9.664
5.0	12.770	12.024	11.320	10.656	10.031	9.441
6.0	12.447	11.727	11.046	10.404	9.799	9.228
7.0	12.139	11.442	10.783	10.162	9.576	9.023
8.0	11.843	11.169	10.532	9.930	9.362	8.826
9.0	11.559	10.907	10.290	9.707	9.136	8.636
10.0	11.288	10.656	10.058	9.493	8.959	8.454
11.0	11.027	10.415	9.835	9.287	8.769	8.279
12.0	10.777	10.183	9.621	9.089	8.586	8.111
13.0	10.537	9.961	9.416	8.899	8.411	7.949
14.0	10.306	9.747	9.218	8.716	8.242	7.792
15.0	10.084	9.541	9.027	8.540	8.079	7.642
16.0	9.870	9.344	8.844	8.370	7.922	7.496
17.0	9.665	9.153	8.667	8.207	7.770	7.356
18.0	9.467	8.969	8.497	8.049	7.624	7.221
19.0	9.276	8.792	8.333	7.896	7.483	7.090
20.0	9.092	8.621	8.174	7.749	7.346	6.964
21.0	8.915	8.456	8.021	7.607	7.214	6.842
22.0	8.734	8.297	7.873	7.470	7.087	6.723
23.0	8.578	8.143	7.730	7.337	6.963	6.609
24.0	8.484	7.994	7.591	7.208	6.844	6.498
25.0	8.263	7.850	7.457	7.083	6.728	6.390
26.0	8.113	7.711	7.327	6.962	6.615	6.285
27.0	7.968	7.575	7.201	6.845	6.506	6.184
28.0	7.827	7.444	7.079	6.731	6.400	6.085
29.0	7.691	7.317	6.961	6.621	6.297	5.990
30.0	7.559	7.194	6.845	6.513	6.197	5.896
31.0	7.430	7.073	6.733	6.409	6.100	5.806
32.0	7.305	6.957	6.624	6.307	6.005	5.717
33.0	7.183	6.843	6.518	6.208	5.912	5.631
34.0	7.063	6.732	6.415	6.111	5.822	5.546

Contd...

Table A-2–Contd...

Temperature, °C	*Oxygen Solubility, mg/l*					
	Chlorinity: 0	*5.0*	*10.0*	*15.0*	*20.0*	*25.0*
35.0	6.950	6.624	6.314	6.017	5.734	5.464
36.0	6.837	6.519	6.215	5.925	5.648	5.384
37.0	6.727	6.416	6.119	5.835	5.564	5.305
38.0	6.620	6.316	6.025	5.747	5.481	5.228
39.0	6.515	6.217	5.932	5.660	5.400	5.152
40.0	6.412	6.121	5.842	5.576	5.321	5.078
41.0	6.312	6.026	5.753	5.493	5.243	5.005
42.0	6.213	5.934	5.667	5.411	5.167	4.933
43.0	6.116	5.843	5.581	5.331	5.091	4.862
44.0	6.021	5.753	5.487	5.252	5.017	4.793
45.0	5.927	5.665	5.414	5.174	4.944	4.724
46.0	5.835	5.578	5.333	5.097	4.872	4.656
47.0	5.744	5.493	5.252	5.021	4.801	4.589
48.0	5.654	5.408	5.172	4.947	4.730	4.523
49.0	5.565	5.324	5.094	4.872	4.660	4.457
50.0	5.477	5.242	5.016	4.799	5.591	4.392

Source: Data from *Standard Methods for the Examination of Water and Wastewater*, 1980. American Public Health Association

Table A-3. Some common physical properties of water

Property	Value	Condition
Boiling point, °C	100.0	101.325 kPa
Melting point, ° C	0.0	101.325 kPa
Density, kg/m³		
–water	997.05	25° C
–ice	916.8	0° C
–gas	0.5976	100° C, 101.325 kPa
Surface tension, N/m	0.07198	25° C
Vapor pressure,		
kPa–water	3.165	25° C
Pa–water	611.657	273.16 K
Pa–ice	611.657	273.16 K
Viscosity, dynamic, mPa s		
–water	0.8909	25° C
–water	1.0016	20° C
–gas	0.0123	100° C
Viscosity, kinematic, stokes	0.008935	25° C
m²/s	0.8935×10^{-6}	
Specific heat capacity (C_p), J/mol.K		
–water	75.327	25° C
–ice	38.09	0° C
–gas	37.47	100° C
Conductivity, electrolytic, µS/cm	0.05501	25° C
Conductivity, thermal, W/m.K,		
–water	0.610	25° C
–ice	2.4	- 20° C
–gas	0.025	100° C
Ionic dissociation constant, mol/l	1.821×10^{-16}	25° C

Source: Website www.lsbu.ac.uk/water/data.html.

Appendix B

Model Questions for the Course: *Water Quality Management Practices*

B–1: Objective Type Multiple Choice Questions

I. Multiple Selection Type (with one correct answer)

1. The water quality parameter that is altered by biological activity in ponds is:
   A. Salinity B. DO C. Alkalinity D. CO_2 concentration
2. The gas that is directly utilized by all the aquatic organisms is:
   A. CO_2 B. NH_3 C. O_2 D. N_2
3. Acidity of pond bottom and water results from:
   A. reaction of liming compounds B. nitrification reaction
   C. addition of phosphatic fertilizer D. microbial decay of organic matter
4. Dissolved oxygen (DO) concentration of 4 mg/l in water at 30°C (saturation value = 7.54 mg/l) will be in equilibrium with oxygen tension of:
   A. 80.5 mm Hg B. 84.5 mm Hg C. 90 mm Hg D. 85 mm Hg
5. If the loss of alkalinity in pond water by addition of diammonium hydrogen phosphate fertilizer is 1.14 mg calcium carbonate, the amount of calcium carbonate to be required for neutralizing this loss from 15 kg fertilizer would be about:
   A. 171 kg B. 17 kg C. 34.2 kg D. 1.71 kg

II. Fill in the Blanks with any of the Four Options

1. The primary source of DO in most ponds is____________
   A. O_2 diffusion from atmosphere B. Photosynthesis
   C. Artificial aeration D. Natural aeration

2. At about pH 8 nearly 100 % of the carbon dioxide is converted to____________
   A. HCO_3^- B. CO_3^2 C. H_2CO_3 D. OH^-

3. Fish production by fertilizer application does not increase in direct proportion to____________
   A. Increased fertilizer input B. Increased fertilizer and lime input
   C. Increased organic manure input D. Increased lime input

4. Nitrogen added to pond during waste production in intensive and semi-intensive aquaculture is____________that added in fertilizer
   A. Same as B. Less than C. Nearly twice D. Nearly half of

5. In aquaculture effluent treatment normally the finer particles are removed by____________
   A. Sedimentation B. Screening
   C. Screening and filtration D. Gravitational settling

III. Matching the Answers Type

1. (*i*) Alkalinity	A. HCO_3^- and CO_3^{2-} ions
(*ii*) Hardness	B. Free CO_2 content
(*iii*) Decrease in pH	C. Ca and Mg ions
(*iv*) Increase in pH	D. Ca and Mg carbonate
2. (*i*) Ionic strength	A. First order reaction
(*ii*) Kinetics of bacterial growth	B. Water analysis
(*iii*) Ionic balance	C. Bacterial decomposition
(*iv*) C/N ratio of organic matter in pond sediment	D. Activity coefficient
3. (*i*) Chlorinity	A. Pond sediment
(*ii*) Ammonia	B. Acidity of pond water
(*iii*) Orthophosphate concentration in water	C. Dissolved ions
(*iv*) Liming treatment	D. Toxic effect on aquatic life
4. (*i*) DO concentration in water	A. Nitrification
(*ii*) Bacterial growth	B. pH balance
(*iii*) Suspended solids	C. Photosynthesis
(iv) $CO_2 + H_2O \leftrightarrow H^+ + HCO_3^-$	D. Fish disease

5. (*i*) Standard oxygen transfer rate (SOTR) A. Break power
   (*ii*) Standard aeration efficiency (SAE) B. Log deficit of DO
   (*iii*) Volumetric oxygen transfer coefficient (k_La) C. Maximum DO deficit
   (*iv*) α and β factors D. Saturation DO concentration

B – 2 Questions with Short Answers

1. Why is phosphorus, in general, the most effective nutrient for pond fertilization, but nitrogen is more suitable than phosphorus in shrimp ponds?
2. How is the pond liming rate decided from neutralization of exchange acidity of the pond mud sample?
3. How is the supplemental diet feeding is related to the standing crop and availability of natural food organisms in a pond?
4. Explain that fertilization with nitrogenous fertilizers leads to loss of alkalinity, whereas lime treatment increases the total alkalinity of pond water.
5. What are the various processes that are involved in the dynamics of DO in pond water?
6. What are the major treatment processes followed in a water reuse culture system?
7. Derive a relation to show how the disinfection process is dependent on the disinfectant concentration.
8. Compare the traditional and polarographic measurement of DO in water.
9. During air contact aeration oxygen transfer takes place at the air-water interface through three physical steps or resistance paths. What are those steps, and which step is rate controlling and why?
10. Which type of wastes are produced in pond, if efficiency of the feed decreases? What is the other type of feed-derived waste?

Index

Aeration:
- efficiencies, standard, 226
- experiment, 223
- fundamentals, 220–228
- process, 237–241
- types of, 231–234

Aeration systems, 228–237
- rating of, 226

Aerator:
- comparative performance of, 234–237
- design, 237–245
- gravity, 231
- surface, 229–230
- types of, 228

Alkalinity and hardness, as water quality parameter, 115–116, 117

Aquaculture:
- definition, 54
- in enclosures, pens and cages, 75
- extensive, 60
- farms, operation of, 67
- global production, 62
- inland, 60
- intensive, 60
- land and water-based, 58–59
- marine, 60
- production system, 56
- production systems in Asia and Europe, 60–61
- protein food and, 10–11
- in tanks, 73
- types of, 57, 59

Aquaculture effluents:
- treatability of, 271–276
- treatment practices, 280–285
- standards and regulations, 276–280

Atomic weights of major elements, 343

Biofilter:
- comparison of, 318
- configuration, 313–317
- operating parameters of, 317–318

Biological parameters, 121

Capture fisheries:
- classification, 55
- inland, 55
- production of, 6–8

Chlorination:
- chemistry of, 325–329
- disadvantages of, 329

Chlorine:
- compounds used in practice, 330
- forms of, 325
- removal, 330

Coastal:
- agriculture, 78–80

aquaculture, 56, 57
water management in, 68
areas and zones, 19–20
ecological divisions of, 21–26
development activities, 52
environment, 19–26
fisheries, 54
forestry, 80 -84
pollution, 84–88
population, 52
regulation zone, 25–26
species cultured, 60–63
Coastal resources:
use, 53, 78
integrated approach of, 100–102
Coastal water:
environmental characteristics of, 27–30
eutrophication in, 91–92
Coastal zones,
for aquaculture farms, 67
Crabs, 64
Culture systems:
basic approach of closed, 199–200
comparison of effluents from, 274–276
types of, 197–199

Disinfection:
mechanisms and kinetics of, 331–333
methods of, 324–325
Dissolved gasses, as water quality parameter, 116–121
Dissolved oxygen:
concentration balance in pond water, 165–169
content of natural waters, 38–39
dynamics, 165–168
mass balance (in biofilter), 320–321
parameter, 110–112
Dynamics: 152
of fed, 178–192
of fertilized, 169–174
of limed, 174–178
pond, 152

Equilibrium relationships:
of acids and bases, 131–134
of ammonia, 133
of carbon dioxide in water, 131–133
ionic in water, 126–131
of phosphates, 133–134
thermodynamic concepts of, 122
Estuary:
Gautami-Godavari, 33
species of, 55
circulation of, 34
Hooghly-Matlah, 24, 33, 34
species of, 55
Mahanadi, 33, 35
species of, 55
Narmada, 33
species of, 56
Estuarine circulation, 34
Estuarine environment, 46–48

Feed:
conversion and utilization, 182
formulation, 251–256
ingredients, 247–248
preparation, 256–257
supplementary, 248–250
Feeding:
complete diet, 181–182, 250–257
devices, 262–266
methods, 257
rate and frequency, 257–259
rate, relationship of water quality with, 191–192
supplementary diet, 179–181
tables, 260–262, 264
types of and options, 178–179, 246
Fertilizer:
application, 206–207
methods of, 207–208
requirement of, 203–206
types, properties and sources of, 202–203
Fertilization:
effects on phosphorus cycle, 170–172, 175
effects on pond dynamics, 169–174
and supplemental feeding, 246–250
Filter:
diatomaceous earth, 293
gravity, 292
rapid, 292
types of, 290–295
Filtration:
biological, 306
process, 295–298
mechanical, 290–298
Fisheries:
capture, 54, 55
definition, 54

marine, 56
potential in India, 58
production, 6–8
types, 54–55

Impacts on coastal environment, 90–93
of agriculture, 4–6, 93, 95
of aquaculture, 93, 95–99
Impacts of capture fisheries on environment, 5, 93, 95–96

Kinetics:
of biological reaction systems, 142–151
of chemical reaction, 135–142
of enzyme catalyzed reaction, 144–146
of microbial growth, 146–151

Lakes:
brackish water, Kerala, 35–37
species, 56
temperature and salinity variation of, 36
Chilka, 35
fisheries, 56
species, 56
temperature and salinity variation of, 35
Pulicat, 35
species of, 56
temperature and salinity variation of, 36
Land resource use, 2–4
Lime requirement, 213–217
Liming:
effects on pond dynamics, 174–177
methods, 217–219
rate, 213–217
Lobster, 63
Macrobrachium rosenbergii, 63, 64
Maritime States of India, 57
Ministry of Environment and Forests, 54
Molluscs, 64–65

Nitrification:
organisms, 306
principle of ammonia removal by, 306–312
reactions, 307–308
Nitrification rate, environmental factors affecting, 308–312
Nutrients:
and BOD, 269
dynamics of, 153–158
and organic substances, 41–43
primary productivity and, (as water quality parameter), 113
and solids budget, 183

Organic manures, 209–212
Oxygen requirement by fish, 111–112
Oxygen solubility, 110–111
in water saturated with air, 344–345
Oxygen transfer:
coefficient, evaluation of, 223–226
rate, standard, 226
theory, 220–226

Penaeus:
monodon, *farming systems in India, 60–61*
species cultured, 63–64
pH:
and carbon dioxide content of natural waters, 39–41
concept, 131
as water quality parameter, 115
Plant and animal community, 43–49
Pollution:
types of, in coastal areas, 4
potential of effluents, 271–274
Pond culture, 70
examples of, 73
traditional, 71
modern, 72
Pond management methods, 201

Questions, model for course, 347–349

Raceways culture, 73
Raceways design, 75

Salinity, as water quality parameter, 113–114
Sedimentation:
tanks, or basins, types of, of solids, 298–305
Settling:
mathematical analysis of, 299–305
types of, 298–299

Screens:
- design characteristics of, 288–290
- types of, 286–288

Seaweed, 66

Shrimps:
- culture of marine, 73
- species cultured, 64

Solids:
- removal, 281–282
- removal mechanisms, 295–296
- sedimentation of, 298–305
- suspended, 114–115
- suspended, in aquaculture effluents, 267–269

Suspended solids:
- in aquaculture effluents, 207–209
- as water quality parameter, 114–115

Tank culture, 73

Temperature:
- and circulation, 31–38
- as water quality parameter, 113

Treatment:
- effluent, technologies in use, 281
- methods for water quality management, 201–201
- potassium permanganet, 331
- sludge, 284

Turtle, 66

Waste:
- agricultural, 87
- components, 185–186
- disposal in coastal areas, 84
- industrial, 86
- materials, in aquaculture effluents, 267–270
- production, 182, 187
- production from fertilization, 188–189
- solid, 88

Water:
- coastal (neritic), 43–45
- common physical properties, 346
- ground, 18–19
- ground, characteristics of, 49–51
- quality parameters, 109–110
- resources, characteristics of, 27
- source, general characteristics of, 30–51
- surface, 16–18
- temperature and circulation, 31–38